Single Variable CalcLabs with the TI-89/92

for Stewart's

FOURTH EDITION

Calculus
Single Variable Calculus
Calculus: Early Transcendentals
Single Variable Calculus: Early Transcendentals

Selwyn Hollis
Armstrong Atlantic State University

BROOKS/COLE PUBLISHING COMPANY
I(T)P® An International Thomson Publishing Company

Pacific Grove • Albany • Belmont • Bonn • Boston • Cincinnati • Detroit • Johannesburg • London
Madrid • Melbourne • Mexico City • New York • Paris • Singapore • Tokyo • Toronto • Washington

Assistant Editor: *Carol Ann Benedict*
Marketing Manager: *Caroline Croley*
Marketing Assistant: *Debra Johnston*
Production Coordinator: *Dorothy Bell*
Cover Illustration: *dan clegg*
Printing and Binding: *West Publishing*

For more information, contact:

BROOKS/COLE PUBLISHING COMPANY
511 Forest Lodge Road
Pacific Grove, CA 93950
USA

International Thomson Editores
Seneca 53
Col. Polanco
11560 México, D. F., México

International Thomson Publishing Europe
Berkshire House 168-173
High Holborn
London WC1V 7AA
England

International Thomson Publishing GmbH
Königswinterer Strasse 418
53227 Bonn
Germany

Thomas Nelson Australia
102 Dodds Street
South Melbourne, 3205
Victoria, Australia

International Thomson Publishing Asia
60 Albert Street
#15-01 Albert Complex
Singapore 189969

Nelson Canada
1120 Birchmount Road
Scarborough, Ontario
Canada M1K 5G4

International Thomson Publishing Japan
Palaceside Building, 5F
1-1-1 Hitotsubashi
Chiyoda-ku, Tokyo 100-0003
Japan

Printed in the United States of America

10 9 8 7 6 5 4 3 2 1

ISBN 0-534-36437-3

Contents

Introduction

THE TEXAS INSTRUMENTS **TI-89** and **TI-92** calculators bring the capabilities of a powerful software package for symbolic computation (DERIVE) to the world of the hand(s)-held calculator. Probably the most significant implication of this is that computing power previously available only on the metaphoric desktop of a computer can now very easily and inexpensively move to the literal desktop of the classroom.

Now with that bit of profundity out of the way, the **TI-89** and **TI-92** calculators are amazingly powerful calculators that are just plain fun to use!

This is a manual written to accompany the fourth edition of *Calculus* by James Stewart, one of the most respected and successful calculus texts of recent years—a time during which the wide availability of powerful computational tools such as the graphing calculator has had a significant effect upon the way calculus is taught and learned.

The primary goal of this manual is to show you how the **TI-89/92** can help you learn and use calculus. The approach (we hope) is not to use the **TI-89/92** as a black box, but rather as a tool for exploring the way calculus works and the way calculus can be used to solve problems—without having to rely entirely upon "cooked-up" problems with nice, clean solutions.

Two secondary goals of this manual are: 1) to present in a very concise manner many of the central ideas of calculus, and 2) to introduce some of the capabilities of this computer called the **TI-89/92**. Programming can play an important role on both of these fronts, and there are numerous programs here that hopefully illustrate some of the fundamentals of calculus. Some of the programs are actually quite useful in other ways as well. Rather than a single chapter devoted to programming, tucked away at the end of the manual, sections on programming are included in each of the first six chapters. It is not essential that you understand all the nuances of programming; much will be gained by simply getting the programs to work and using them.

The **TI-89/92** will not "do calculus" for you. It cannot decide the proper approach to a problem nor can it interpret results for you. In short, *you* still have to do the thinking, and *you* need to know the fundamental concepts of calculus. You must learn calculus from lectures and your textbook—and most importantly by working problems. That's where the **TI-89/92** comes in, allowing you to work interesting problems without getting bogged down on algebraic and computational details that can sometimes distract from the calculus concepts.

The last chapter of this manual contains twenty-three "projects," which cover a wide range of the topics and are arranged roughly in the same order as the corresponding material in Stewart's *Calculus*. The projects vary considerably in length, level of difficulty, and the amount of guidance provided. We hope that these projects will be interesting—and occasionally fun—while reinforcing important calculus concepts.

Until you've had a lot of experience using your **TI-89/92**, you will probably need to consult your **TI-89/92 Guidebook** very often. In fact, the *first* thing you should do is work very carefully through the first two chapters of the **TI-89/92 Guidebook** and make at least a cursory pass through Chapters 3 and 6. Also familiarize yourself with the table of contents and the information in the appendices. Even after you've been using your **TI-89/92** for some time, you will still occasionally need to refer to your **TI-89/92 Guidebook**.

A few comments about notation and typographic conventions: It will become obvious right away that words and characters in **bold sans-serif type** indicate **TI-89/92** commands, functions, variables, keystrokes, modes, etc. Also, we have used the symbols **[⇒]**, **[⇑]**, **[⇐]**, and **[⇓]** to indicate pressing the cursor keys/pad in one of the four horizontal and vertical directions. Also, brackets (**[ ]**) typically indicate reference to one of the keys on the **TI-89/92** keyboard—for example, **[+]**, **[−]**, **[×]**, etc.

It might help you avoid some frustrations early on, if we point out some typical causes of **trouble**.

- The letters ***d*** and ***e*** are the names of the derivative operator ***d*()** and the base of the exponential function ***e*^()**. These have special locations on the keyboard. *The letters* **d** *and* **e** *do not give the same thing.*
- Be aware of the difference between the subtraction key **[−]** and the minus (i.e., negation) key **[(−)]**.
- Finally, be aware that strange or unexpected results—and sometimes error messages—are often caused by a variable having been previously assigned a value. This kind of problem is easily fixed by clearing the guilty variable(s). (For more on this see page 4.)

Information related to the **TI-89/92**, including many programs, can be found at `www.ti.com/calc/docs/92.htm`. Material related to this manual, including programs and answers to exercises, can be found at `www.math.armstrong.edu/ti92`.

This manual, including all graphics, was created and typeset—with nary a computer-related headache—by a complete amateur (me) on an Apple Macintosh computer with the help of the **TI-Graph Link** software (which you can download from the aforementioned Texas Instruments web site) and two great shareware programs, OzTEX by Andrew Trevorrow and Graphic-Converter by Thorsten Lemke.

Finally, my thanks go to Jeff Morgan, who threw this project my way, and whose many helpful suggestions made this manual better than it otherwise would have been.

S.L.H.

1 TI-89/92 Basics

This initial chapter is a brief introduction to many of the capabilities and uses of the **TI-89/92** that will be important throughout the remainder of this manual. This is not intended to be a substitute for the **TI-89/92 Guidebook**. Always keep your **TI-89/92 Guidebook** handy for reference. In particular, you should work carefully through Chapters 1 and 2 and make at least a cursory pass through Chapters 3–6 of the **TI-92 Guidebook** or Chapters 3, 5, 6, and 13 of the **TI-89 Guidebook** before proceeding any further in this manual. In conjunction with Section 1.2 of this chapter, you should read Section 1.3, "Graphing Calculators and Computers," in Stewart's *Calculus*.

1.1 The Home screen

The Home screen is the starting point for both symbolic and numerical computation on the **TI-89/92**. Let's begin by clearing the Home screen (press **F1-8**) and working with a simple example. If we enter an expression such as $(1+\sqrt{3})/2$, the **TI-89/92** does nothing to it, because it is already in its simplest exact form. (An expression such as $(4+\sqrt{12})/\sqrt{2}$ would be simplified. Try it yourself.)

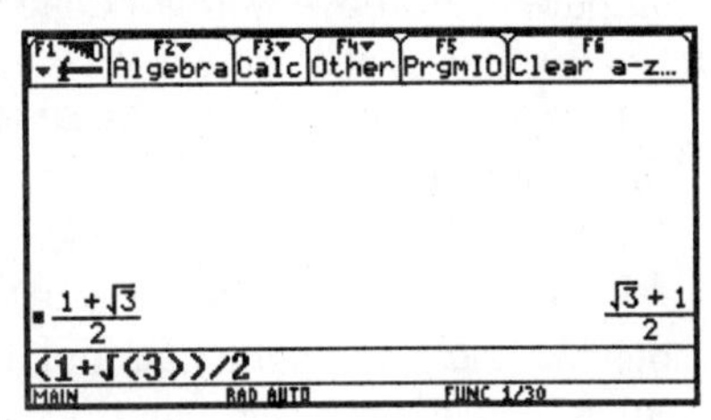

There are three ways to get a numerical value for this expression. The first is via the **approx()** function in the **Algebra** menu (**F2-5**). (Notice the use of **ans(1)** (press **[2nd]-[(−)]**) to avoid re-entering the expression.)

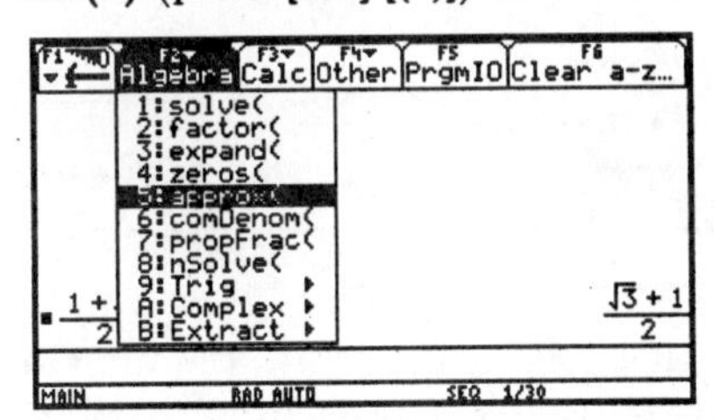

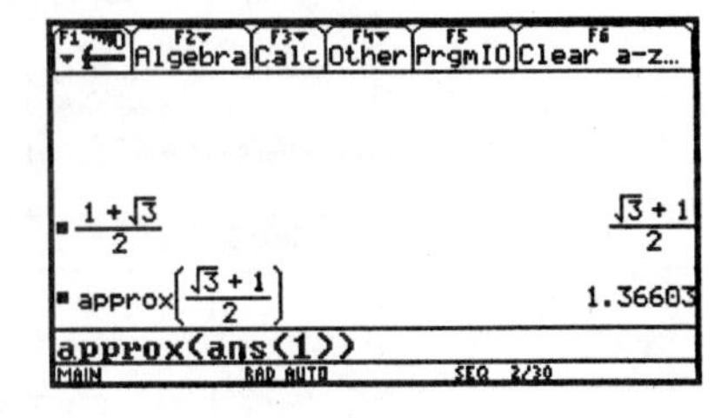

A numerical value for this expression can also be computed by placing a decimal point after any of the whole numbers in the expression. Finally, a numerical value can be found by pressing the ◇ key prior to pressing **ENTER**.

(Notice the green "≈" above the **ENTER** key to the right of the QWERTY keyboard.)

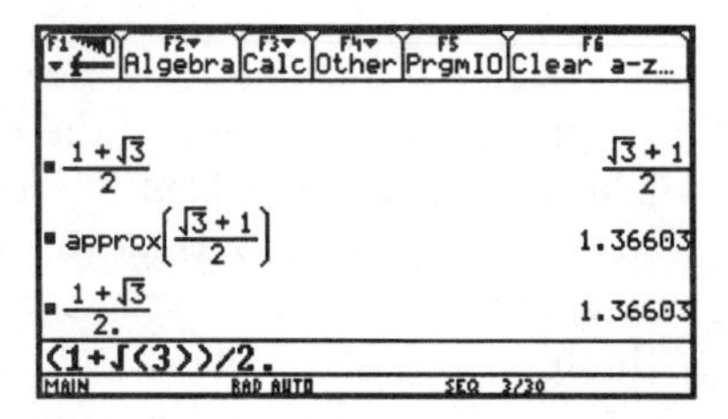

The manner in which the **TI-89/92** handles such computations can also be controlled by setting the **Exact/Approx MODE**. On **Page 2** of the **MODE** dialog box (which appears upon pressing the **MODE** key and then **F2**) one can set **Exact/Approx** to **1:AUTO**, **2:EXACT**, or **3:APPROXIMATE**.

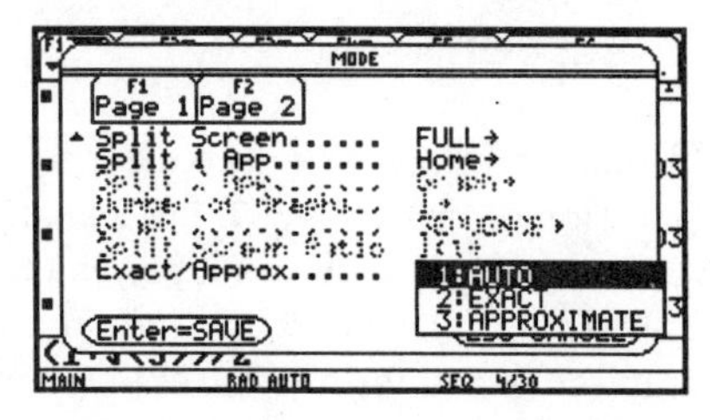

The computations above were all done with **Exact/Approx** set to **AUTO**. You should experiment with the same computations after setting **Exact/Approx** first to **EXACT** and then to **APPROXIMATE**. We recommend using **AUTO** in most circumstances.

The Algebra menu. This menu (**F2**) allows easy access to the **TI-89/92**'s functions for doing symbolic manipulation of algebraic (and trigonometric) expressions and for solving equations. The screens shown below illustrate the algebra functions **factor()**, **expand()**, **comDenom()**, and **propFrac()**.

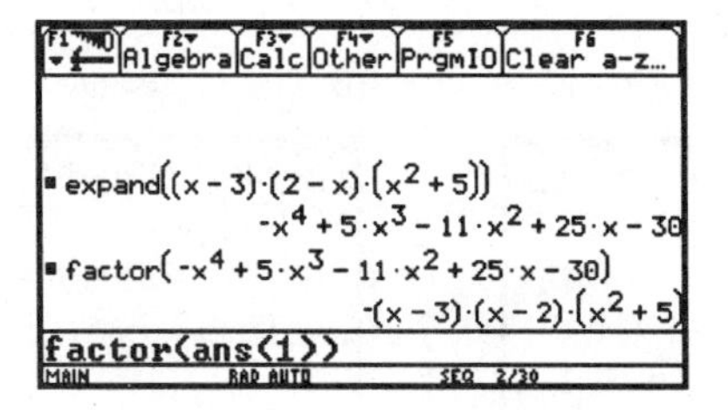

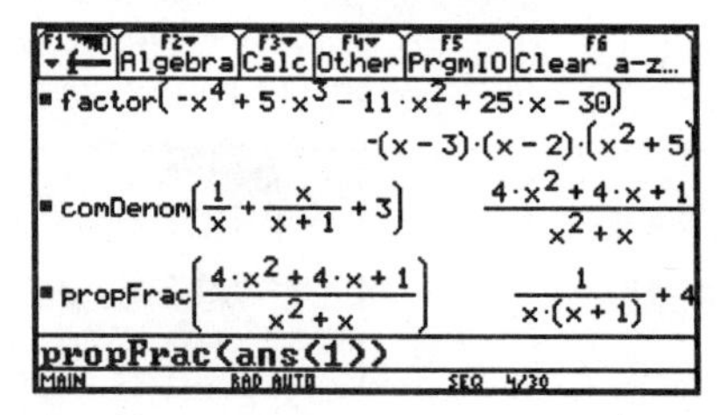

The left-hand screen that follows illustrates the functions **tExpand()** and **tCollect()**, which manipulate trigonometric expressions. (These are found in the **Trig** submenu.) The right-hand screen illustrates the **solve()**, **zeros()**, and **nSolve()** functions for solving equations.

We will have more to say about **solve()**, **zeros()**, and **nSolve()**—and solving equations in general—in Section 2.3.

Defining functions. There are two ways to define a function on the Home screen. One is with the **Define** command. You can type this in or access it in the **Other** menu (**F4-1**). The second way is by using the **[STO▷]** key. Once a function is defined, evaluations can be performed on the Home screen.

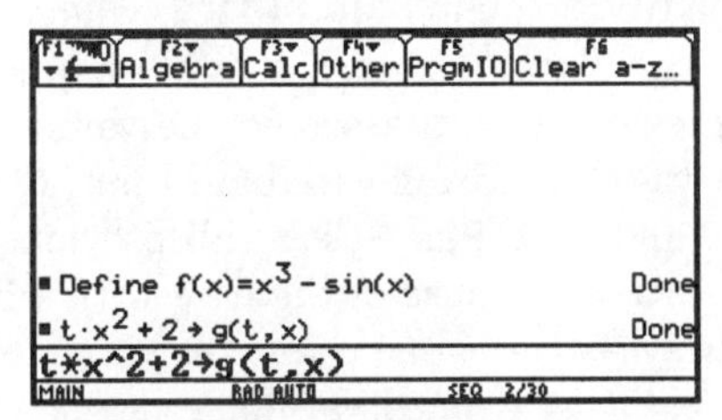

The "With" operator. Use of the "with" operator is essential to taking advantage of the **TI-89/92**'s advanced capabilities. The "with" operator is the vertical bar (|), accessed by pressing **[2nd]-K**. One use of the "with" operator is to make substitutions in expressions—including function evaluations.

Warning: Improper use of the "with" operator can throw the **TI-89/92** into an infinite recursive loop. An infinite recursion can occur when a substituted expression involves the variable that is substituted for—for example, when entering a command such as **f(x) | x=x+1**. For more on this issue and the "with" operator in general, see page 55 of the **TI-89 Guidebook** or page 93 of the **TI-92 Guidebook**.

Another important use of the "with" operator is to restrict variable values in certain computations. For example, **nSolve()** returns a numerical approximation to one solution of an equation. If there is more than one solution,

nSolve() needs help in order to locate them all. Also, **solve()** attempts to find *all* solutions of an equation, while many equations have more solutions than we are really interested in finding.

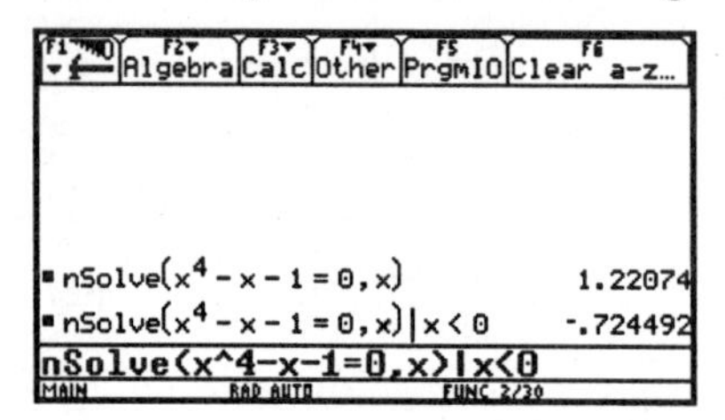


```
F1 F2 Algebra F3 Calc F4 Other F5 PrgmIO F6 Clear a-z...
sin(x)·cos(x^2) = 1/3 → eqn
                              sin(x)·cos(x^2) = 1/3
solve(eqn, x)
 x = 8.74144 or x = 7.01546 or x = 4.8187▸
solve(eqn, x) | x > 0 and x < π/2
                   x = 1.08861 or x = .342278
solve(eqn,x)|x>0 and x<π/2
MAIN            RAD AUTO            FUNC 3/30
```

Clearing variables. Unexpected or erroneous results sometimes occur when one or more variables have been assigned values previously. Consider, for example, the following situation. We define the function $f(x) = x^2$ by entering enter **x^2 →f(x)**. We then enter **(f(x)+1)^2**, which should result in $(x^2+1)^2$, but instead the calculator returns a 4! What happened? We forgot that the variable **x** had previously been assigned the value 1.

There are two simple ways of clearing variables. First, **Clear a–z... (F6)**—or **NewProb** on the **TI-89** and **TI-92 Plus**—clears all previously defined one-character Roman variables **a–z**. Because of the ease with which this can be done, it is a good idea always to use a one-character variable **a–z** for naming any "throwaway" variable. Other defined variables can be cleared by using the **DelVar** command from the **Other** menu (**F4-4**). The second screen below illustrates the use of **DelVar**.

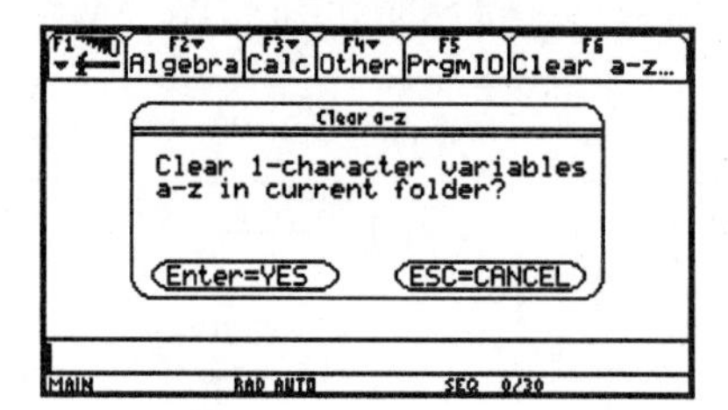

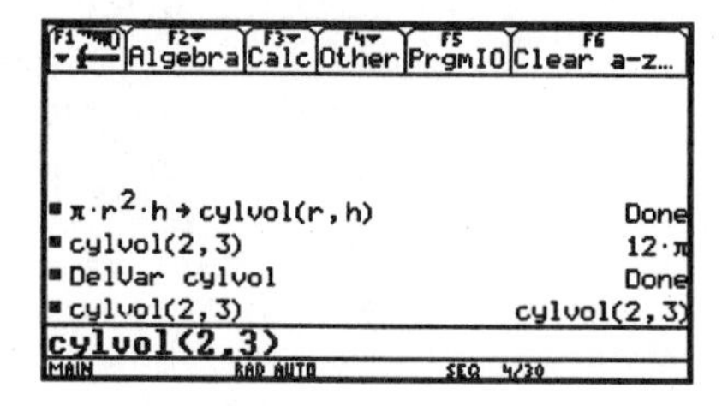

Very important: You will find that clearing variables will cure many of the problems you encounter while using the **TI-89/92**. *Remember this!*

Exercises

Do each of the following with the **Exact/Approx MODE** set to **AUTO**. Before beginning, clear all variables **a–z**.

1. Enter **sin(**π**/k)** for each of **k** $= 2, 3, 4, \ldots, 12$. Which of these is the **TI-89/92** able to evaluate?

2. Compute $\ln(e^2)$, $\ln(2e^3)$, $e^{2\ln(3)}$, and $e^{x\ln(2)}$.

3. a) Enter **expand(ln(a*b))** and then **expand(ln(a*b))|a>0 and b>0)**.

b) Enter **ln(a^x)** and then **ln(a^x)|a>0**.

4. Expand each of the polynomials:

   a) $(2x-1)^2(3-x)^3$ b) $(x-1.23)(3.73-x)(x+7.77)$

5. Factor each of the polynomials:

   a) $6x^3+47x^2+71x-70$ b) $3x^3+4x^2+5x-6$

6. Enter:

   a) **factor(180047)**; b) **factor(1234567)**; c) **factor(1235711)**;

   d) **factor(*ssn*)**, where *ssn* is your social security number.

6. a) Define a function **cylsurf(r)** that gives the surface area of a closed, right circular cylinder with radius r inches and volume 100 cubic inches. How might the "with" operator be used in doing this?

   b) Which of the radii $r = 1, 2, 3, 4$ gives the least surface area?

7. The function $f(x) = 5\sin(4x) + x$ has several zeros between $x = 0$ and $x = 2\pi$. Use **solve()** and the "with" operator to find them all.

1.2 Graphs and tables

The most straightforward way to define a function for graphing is to use the **Y=** Editor. This can be accessed by pressing ◇**Y=** or by pressing **APPS-2**. However, let's first press the **MODE** key to bring up the **MODE** dialog box. The Graph mode should be set to **FUNCTION**. Now we'll press ◇**Y=** to bring up the **Y=** Editor and define the function $f(x) = \sin(x)\cos(x)$ as **y1(x)**. Note that the variable in the function *must* be x.

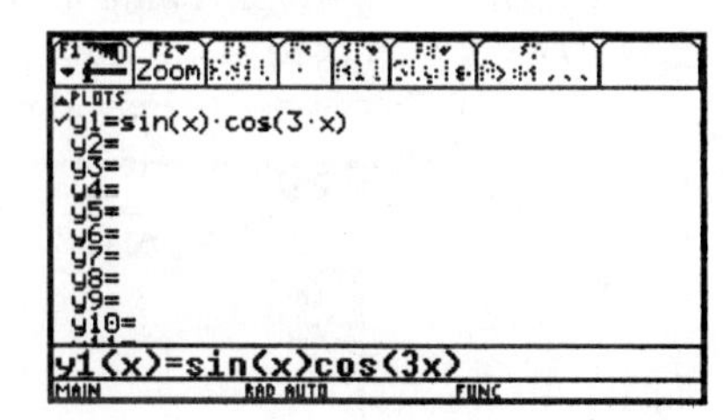

We now need to set appropriate **window variables** for the graph. This is done in the Window Editor, which we access by pressing ◇**WINDOW**. Because of the nature of the particular function we're graphing, we'll set **xmin, xmax, ymin** and **ymax** to give us a $[0, 2\pi] \times [-1.5, 1.5]$ window. The variables **xscl** and **yscl** determine the spacing between axis tick-marks. **xres** determines the horizontal spacing between the pixels where function evaluations are done.

With window variables set, we then press ◇**GRAPH** to plot the graph.

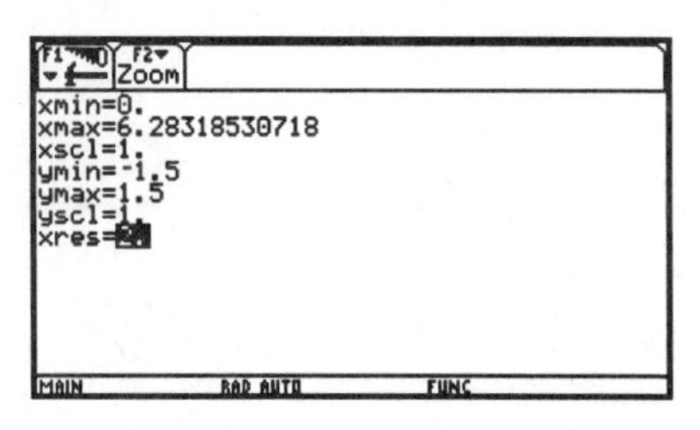

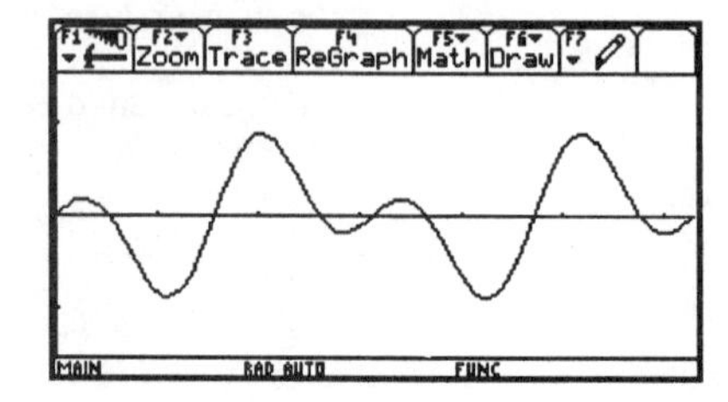

We can also plot several functions at once.

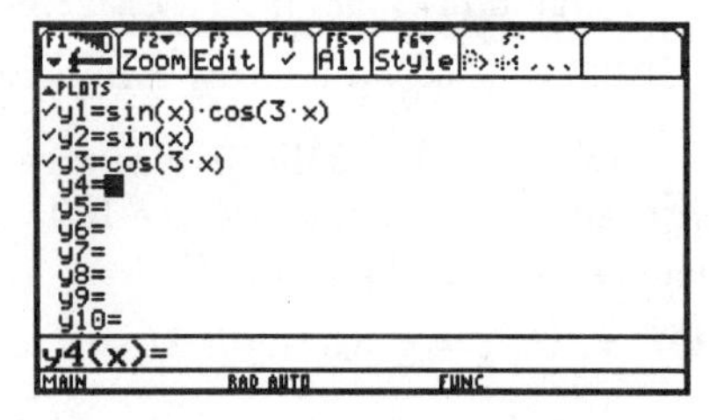

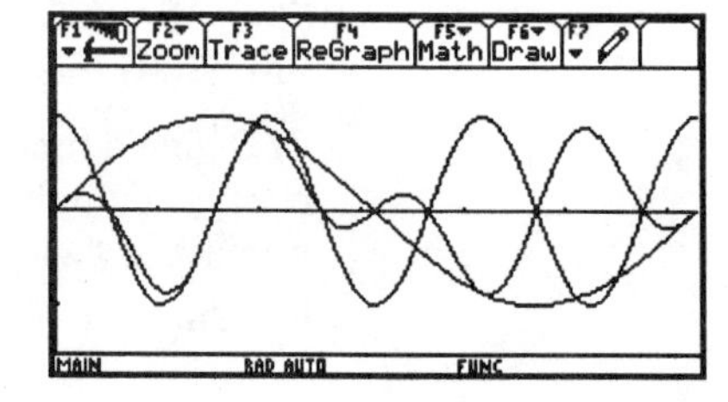

It is often convenient to define functions on the Home screen for graphing, particularly if the functions we wish to plot are each one of a family of functions. For example, let's plot $y = \sin(t - k\pi/4)$, for each of $k = 0, 1, 2,$ and 3, on the interval $0 \le t \le 2\pi$.

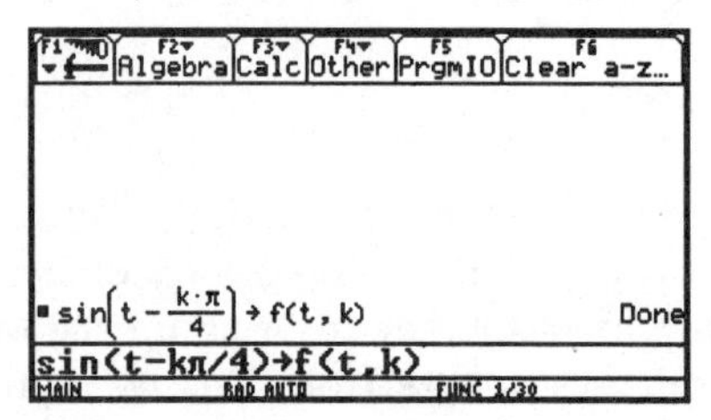

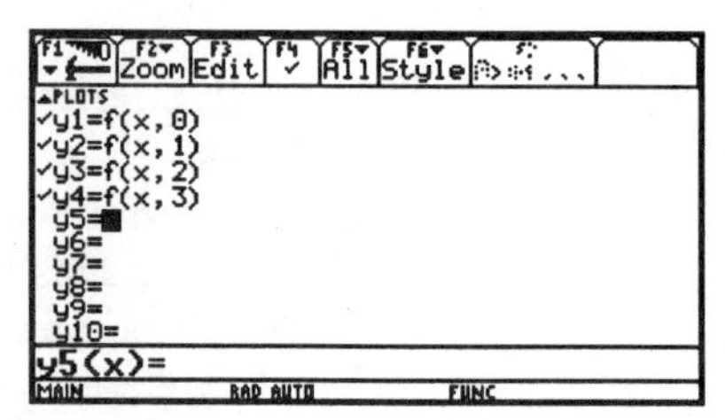

Notice that we are able to define a function on the Home screen using any variable name that we choose. Then we simply use **x** as the variable when entering the functions in the **Y=** Editor. Finally, we'll press ◇**WINDOW**, set the desired window variables, and then press ◇**GRAPH** to plot the graph.

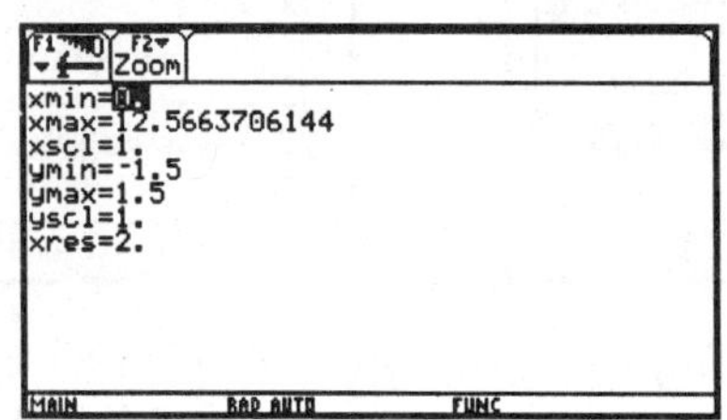

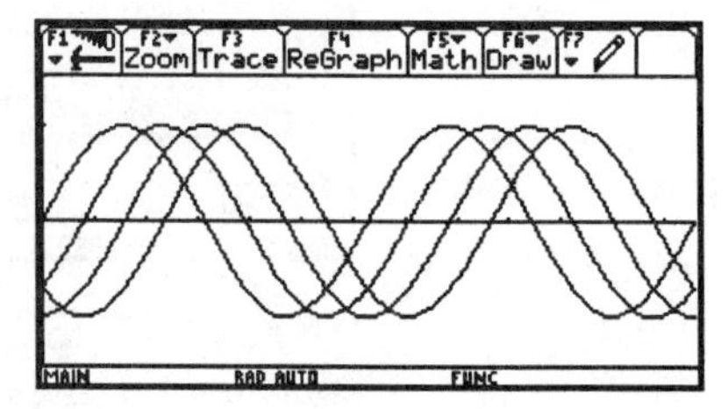

We should mention here that there is a much more convenient way of plotting several members of such a family of functions. It involves storing parameter values in a *list*. (The next section discusses lists in more detail.) The following screens illustrate this, as well as the use of the **Graph** command from the Home screen's **Other** menu (**F4-2**).

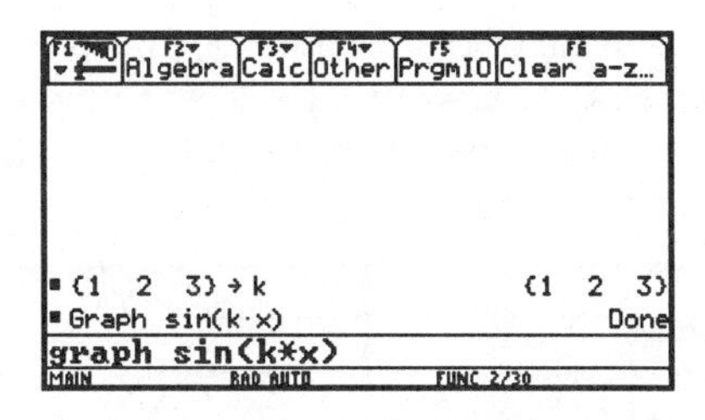

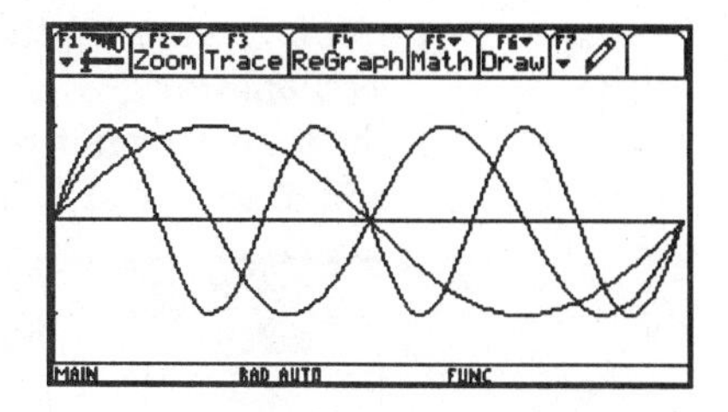

Zoom Tools. The **Zoom** menu (**F2**) is accessible from each of the **Y=** and **WINDOW** Editors as well as from the Graph screen. A variety of Zoom tools are available in this menu—see page 107 of the **TI-89 Guidebook** or page 59 of the **TI-92 Guidebook** for a complete description. Here we will only illustrate the use of the **ZoomBox** tool.

The graph of the cubic polynomial $f(x) = 10x^3 - 121x^2 + 221x - 110$ seems to indicate that f has just two real zeros. The window shown here is $[-2, 11] \times [-1200, 500]$. Let's use the **ZoomBox** tool to go to a graph that shows the behavior of the function near the first of the two apparent zeros.

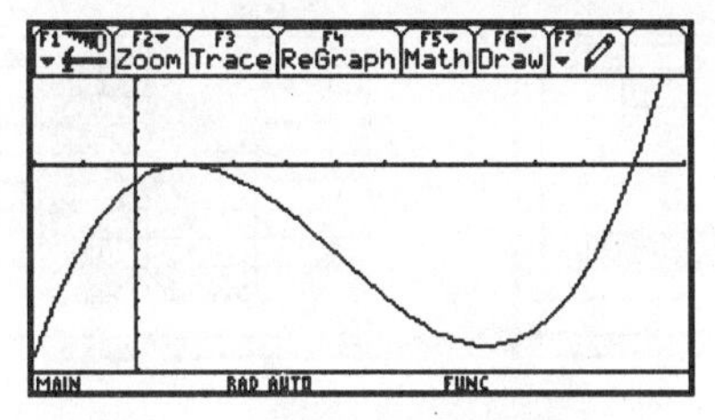

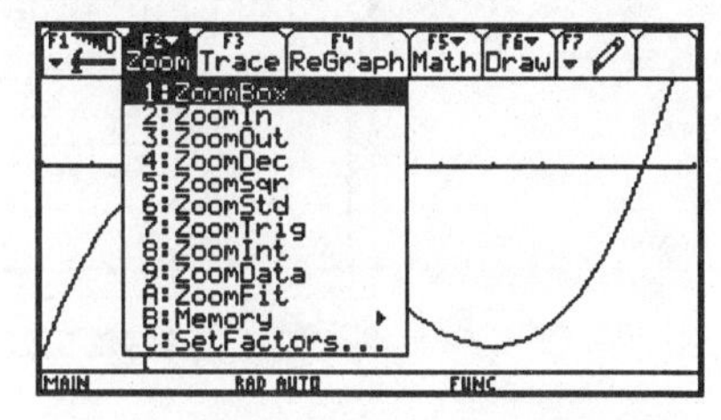

When we use the cursor pad to draw a box around the area of interest and **ENTER** the second corner point, a new graph is drawn.

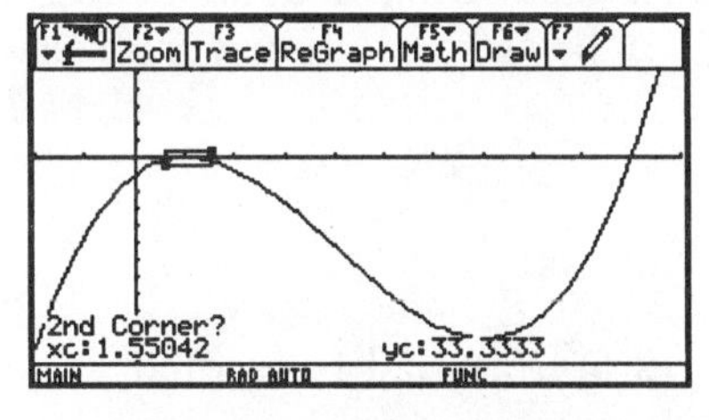

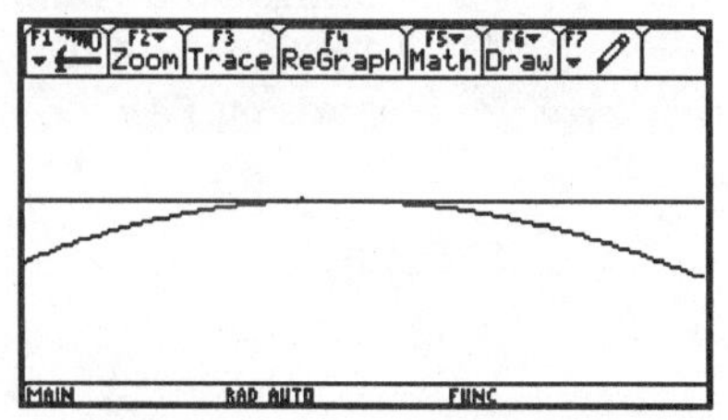

Repeating the process once more finally reveals two zeros of the function that are quite close together.

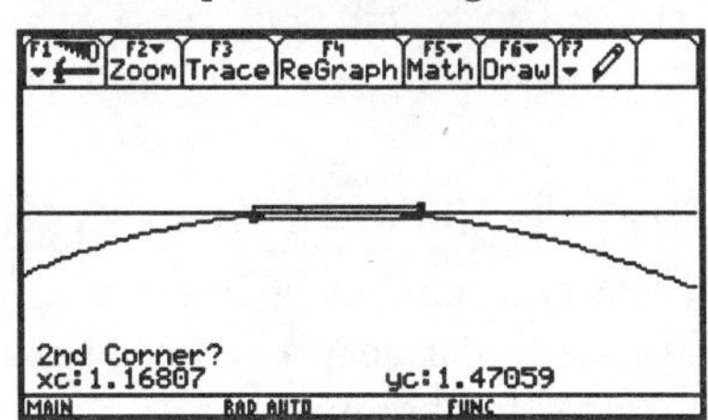

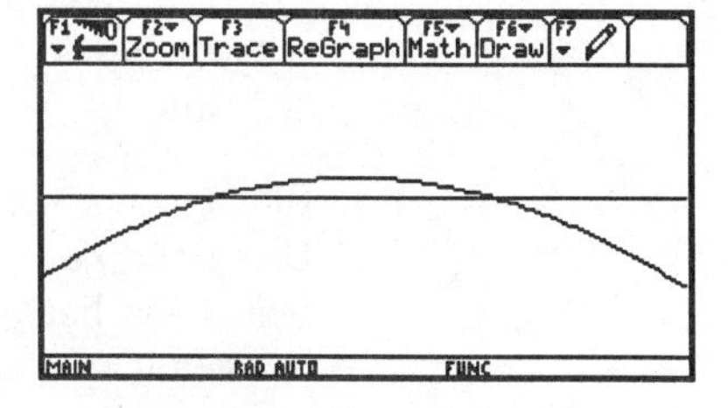

Plot Styles. In the **Style** menu of the **Y=** Editor, four plot basic styles are available for graphs: **Line, Dot, Square,** and **Thick**. For illustration, let's plot the same four functions as above after choosing plot styles **Line, Dot, Square,** and **Thick** for **y1, y2,** and **y3**, respectively.

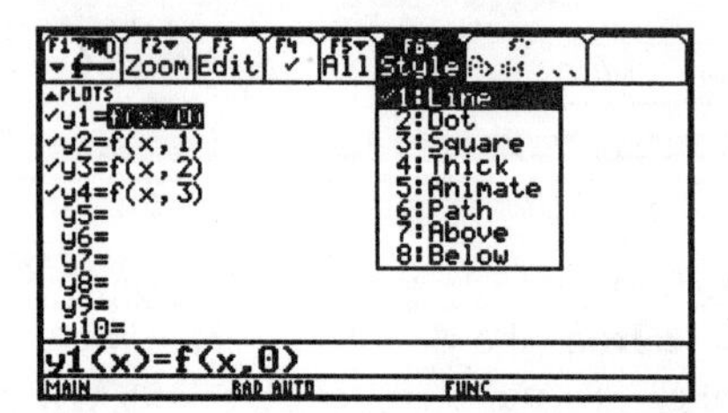

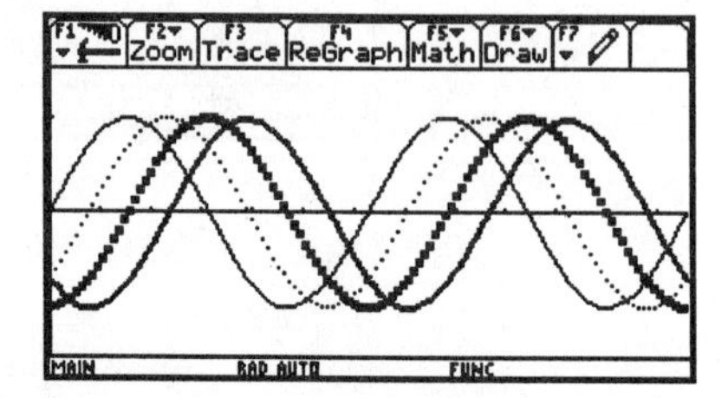

Tables. A table of function values is as easy to create as a graph. First enter the functions in the **Y=** Editor. Then press ⋄**TblSet** and enter the starting value **tblStart** of **x** and the stepsize **Δtbl** between variable values. With this done, press ⋄**TABLE** to create the table.

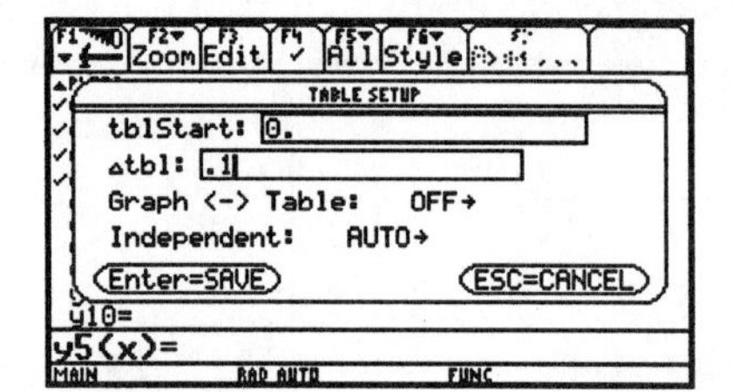

x	y1	y2	y3	y4
0.	0.	-.7071	-1.	-.7071
.1	.09983	-.633	-.995	-.7742
.2	.19867	-.5525	-.9801	-.8335
.3	.29552	-.4666	-.9553	-.8845
.4	.38942	-.3759	-.9211	-.9266
.5	.47943	-.2815	-.8776	-.9595
.6	.56464	-.1843	-.8253	-.9829
.7	.64422	-.0853	-.7648	-.9964

x=0.

A shortcut for scrolling down the table is to press **[2nd]-[⇓]**. This, in effect, is a "page down" command. Notice too that, from the Table screen, **F1-9** brings up a **FORMATS** dialog that allows us to set the cell width. Also, the **TABLE SETUP** dialog box can be accessed directly from the Table screen by pressing **F2**.

Exercises

1. Plot $y = x^n$, for $n = 1, 2, 3, 4, 5, 8,$ and 11 in a $[-1, 1] \times [-1, 1]$ window.

2. Plot $y = \sin k\pi x$, for $k = 1, 2, 3,$ and 4 in a $[0, 2] \times [-1, 1]$ window.

3. Plot the function $y = \sin 2\pi x$ with **Dot Style (F6-2** in the **Y=** Editor) in the window $[0, 1] \times [-1, 1]$, first with **xres** $= 10$ and then with **xres** $= 5$, 3, 2, and 1.

4. Plot $y = x \sin(x/(x^2 + .001))$ in a $[-1, 1] \times [-1, 1]$ window with **xres** $= 1$. Using the **ZoomIn** tool **(F2-2)** a number of times, magnify the graph to reveal the behavior of the function near $x = 0$. What are the values of the window variables for your final graph?

5. Create a table of values for the function $f(x) = (1+x)^{1/x}$, using **tblStart** $= -.02$ and **Δtbl** $= .005$. Notice the strange value of 1 shown at $x = 0$ and the warning message at the bottom of the screen. Change **tblStart** to $-.001$ and **Δtbl** to $.00025$ and recreate the table. If you had to assign $f(0)$ a value, approximately what should it be (rather than 1)?

6. Plot $y = \sqrt{1-x^2}$ and $y = -\sqrt{1-x^2}$ in each of the windows

$$[-2,2]\times[-1,1]; \qquad [-2,2.011]\times[-1,1]; \qquad [-1.95,1.95]\times[-1,1].$$

Can you think of a good reason for what you've observed?

1.3 Lists and matrices

We have already encountered two of the **TI-89/92**'s data types: *expressions* and *functions.* Two other important data types are *lists* and *matrices.*

Lists. A list is literally a list of objects. The objects—called *elements*—in a list may be numbers, expressions, functions, or even other lists, and they need not be all of the same type. A list can be defined on the Home screen by entering the elements of the list separated by commas and enclosed by braces ({}). Entering *listname*[*i*] accesses the *i*th element of a list.

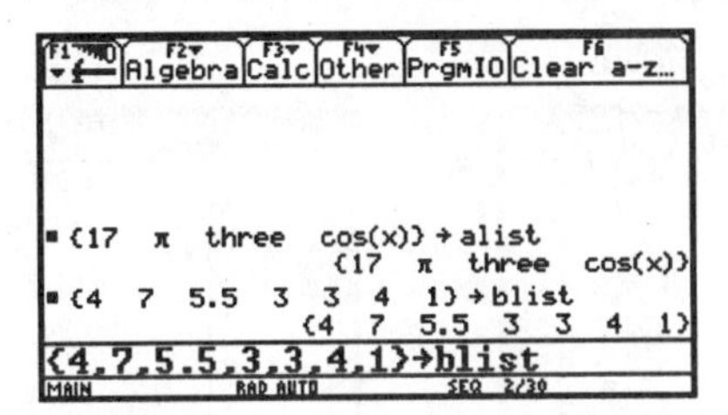


```
{17 π three cos(x)} → alist      {17 π three cos(x)}
{4 7 5.5 3 3 4 1} → blist        {4 7 5.5 3 3 4 1}
alist[3]                          three
alist[4]                          cos(x)
blist[3]                          5.5
blist[1]                          4
blist[1]
```

A list can be multiplied by an expression. It is also possible to apply various functions to a list. All such operations are done "component-wise." Also, two lists with the same number of element can be added, subtracted, multiplied, or divided.

```
k·alist      {17·k  k·π  k·three  k·cos(x)}
alist^2      {289  π^2  three^2  (cos(x))^2}
sin(alist)   {sin(17)  0  sin(three)  sin(cos(x))}
alist^-1     {1/17  1/π  1/three  1/cos(x)}
alist^-1
```

```
{1 2 3 4 5} → a     {1 2 3 4 5}
{3 2 0 2 1} → b     {3 2 0 2 1}
a+b                 {4 4 3 6 6}
a-b                 {-2 0 3 2 4}
a·b                 {3 4 0 8 5}
a/b                 {1/3 1 undef 2 5}
a/b
```

The length of a list (i.e., the number of elements in it) can be found by entering **dim(*listname*)**. Also, the $\Sigma()$ operator (**[2nd]-[4]**, or **F3-4**) is handy for summing the elements in a list. This makes it easy, for example, to compute the average (or mean) of a list of numbers.

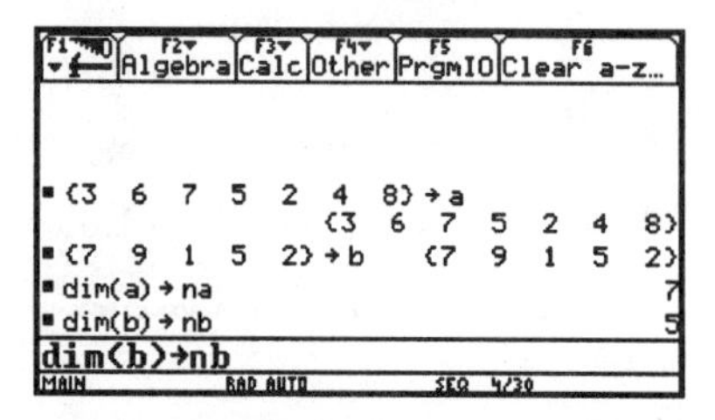

In addition to **dim()**, there are numerous other built-in functions for performing operations on lists. Among these are **augment(), max(), min(), product(), sum(), mean(), median(),** and **stdDev()**. Also, there are operators for sorting lists—**SortA** sorts a list in ascending order, and **SortD** sorts a list in descending order. (All of these are found in either the **List** or **Statistics** submenus of the **MATH** menu.) For descriptions of these, see Appendix A of the **TI-89/92 Guidebook.**

Matrices. A matrix is a rectangular array of elements. Just as braces ({ }) were used as delimiters for lists, brackets (**[ ]**) are the delimiters for matrices. When defining a matrix on the Home screen, elements of each row are separated by commas and rows are separated from each other either by semicolons or sets of brackets.

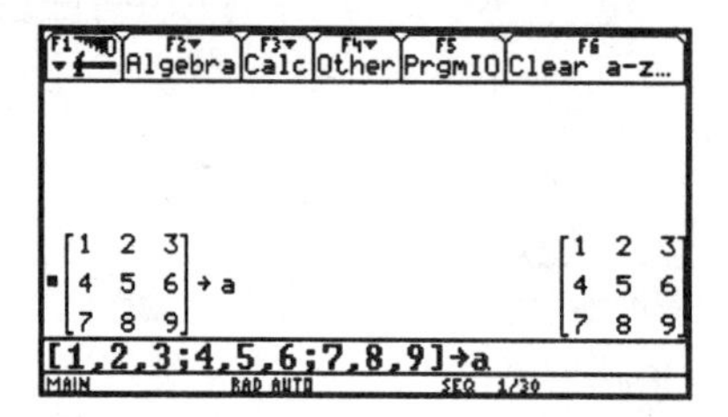

The ith row of a matrix is referenced by *matrixname*$[i]$. The element in the ith row and jth column is referenced by *matrixname*$[i, j]$. Also, the **TI-89/92** handles many algebraic operations with matrices automatically.

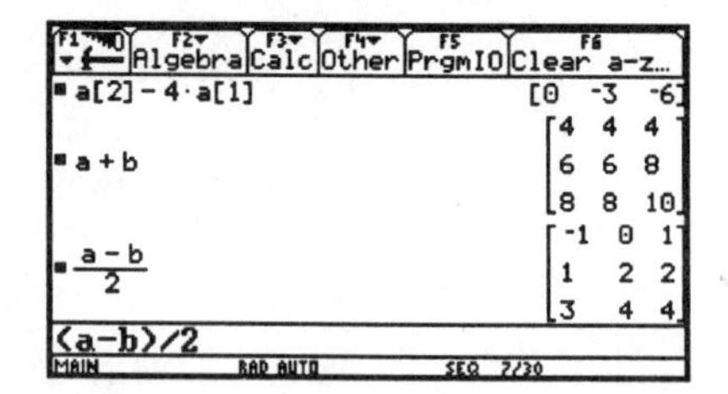

The Data/Matrix Editor. The **TI-89/92** provides a handy editor for creating and editing lists and matrices. To create a list in the Data/Matrix Editor, we first bring up the editor by pressing the **APPS** key, and then selecting **Data/Matrix Editor** and **New**... . In the resulting dialog box, we'll choose **Type: List** and **Folder: main** and then type in the name of our list.

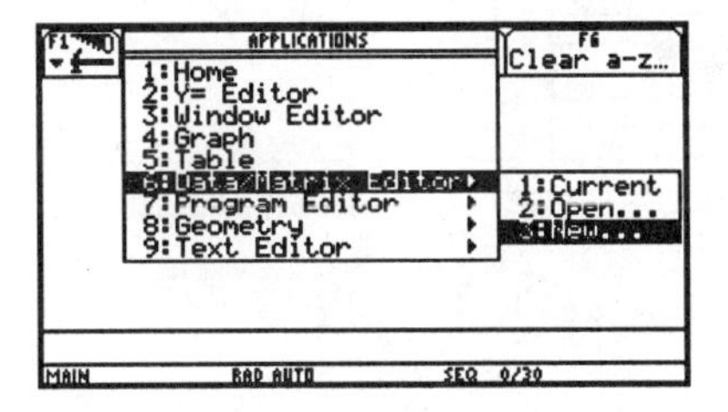

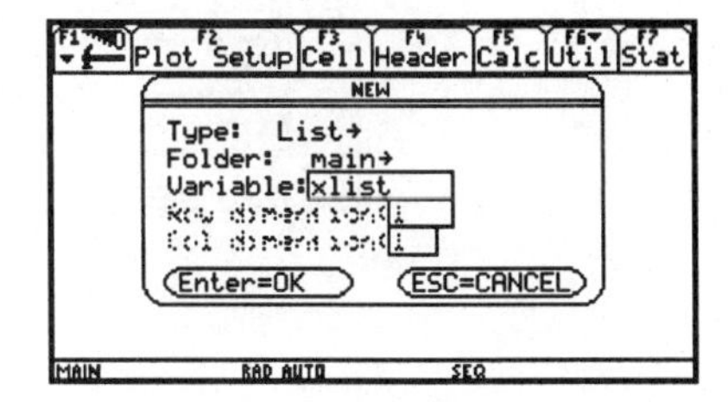

This takes us to the Data/Matrix Editor, ready to begin entering elements into the list. This is done by filling in the first column of the grid. (If items are entered into another column, the list variable is automatically converted to a data variable, which is essentially a collection of lists. Data variables are useful for doing Statistics.) After we've entered the elements of the list, we can go back to the Home screen and refer to the list we just created.

One advantage of using the Data/Matrix Editor is that it is easy to go back and add elements or otherwise modify the list. Simply press **APPS-6** and select **Current**.

Let's now use the Data/Matrix Editor to create a matrix. As before, we'll press **APPS-6-3** to bring up the **NEW** dialog box. There we choose **Type: Matrix** and **Folder: main**, name the matrix, and enter the number of rows and number of columns. This takes us to the Data/Matrix Editor, ready to begin entering elements into the matrix. Notice that the matrix is initially filled with zeros.

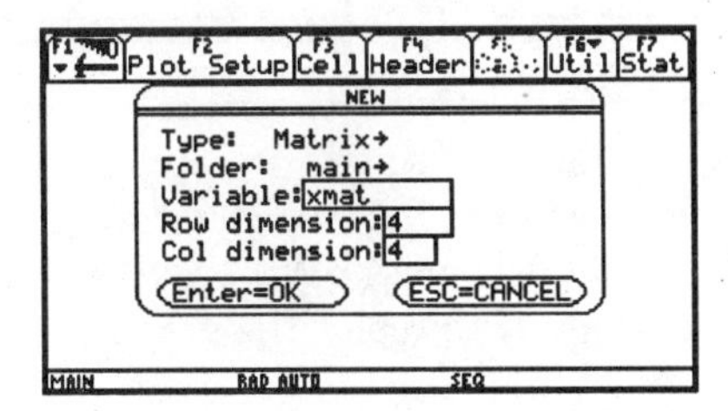

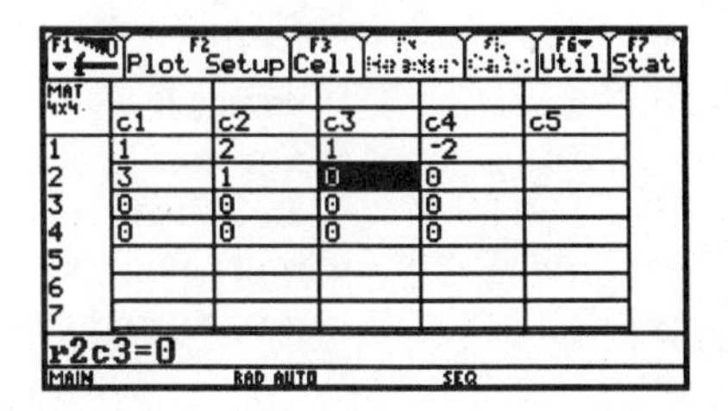

Once we've finished entering the elements of the matrix, we return to the Home screen ready to work with the matrix we've created.

1.4 Split screens

One of the most attractive and useful features of the **TI-89/92** is its Split Screen capability. To set the Split Screen Mode, press **MODE** and then **F2** to get to page 2 of the **MODE** dialog box. There we can select either **FULL**, **TOP-BOTTOM**, or **LEFT-RIGHT** as the **Split Screen MODE**. After selecting **LEFT-RIGHT**, let's then select the **Y= Editor** as the **Split 1 App**, **Graph** as the **Split 2 App**, and **1:1** as the **Split Screen Ratio**.

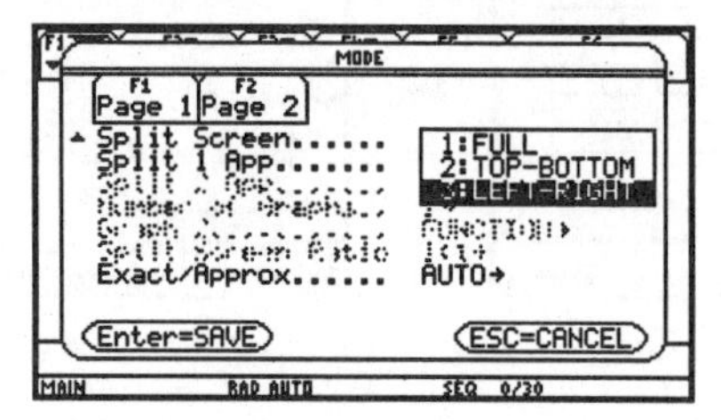

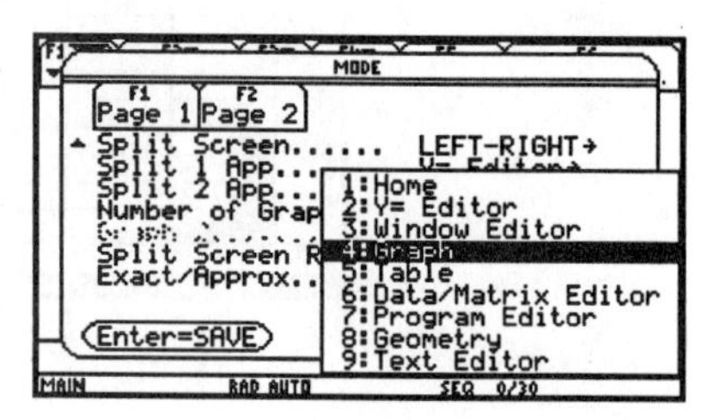

This takes us to a split screen with the **Y=** Editor active. We can then enter functions and plot them by pressing ◇**GRAPH**.

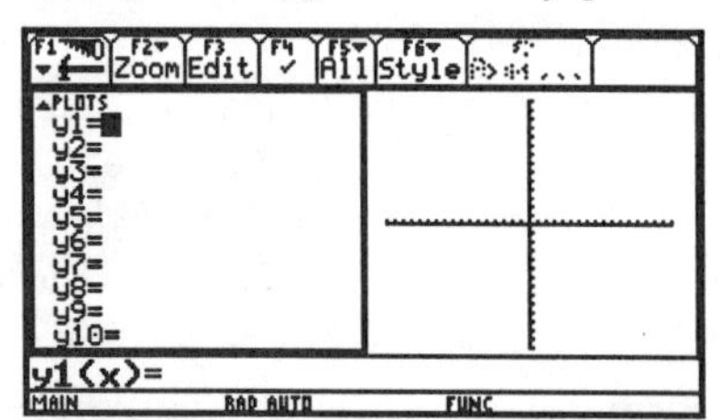

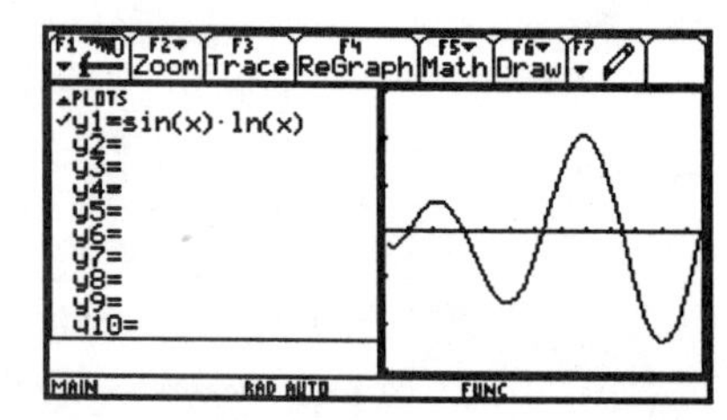

Pressing **[2nd]-APPS** makes the inactive split screen active. The active split screen can be switched to a different screen or application in the same way as in Full Screen mode. For instance, here we might want to change window variables for the graph. So we just press ◇**WINDOW**, enter the new window variables, and then press ◇**GRAPH** to replot the graph.

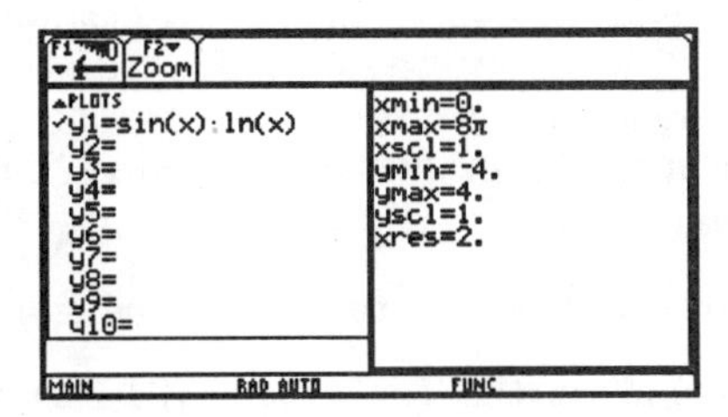

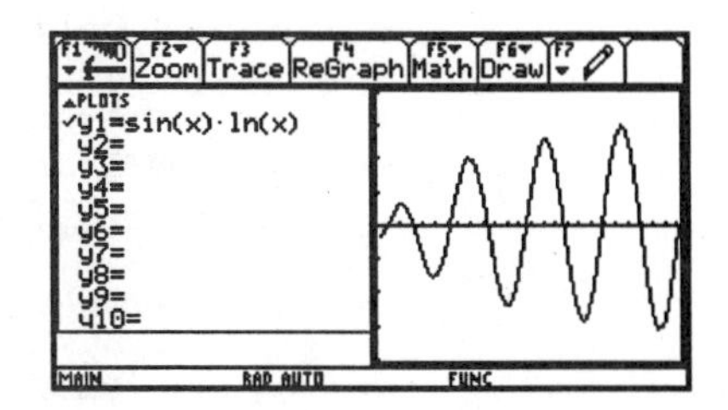

An interesting kind of Split Screen for examining function behavior is one that shows both a table of values and a graph. Let's look closely at the function $f(x) = \sin(x)\ln(x)$ on the interval $0 \leq x \leq 1$. We'll first change the window variables for the graph. Then we press ◇**GRAPH** to replot the graph, **[2nd]-APPS** to switch to the other split screen application, and then ◇**Table** to create the table.

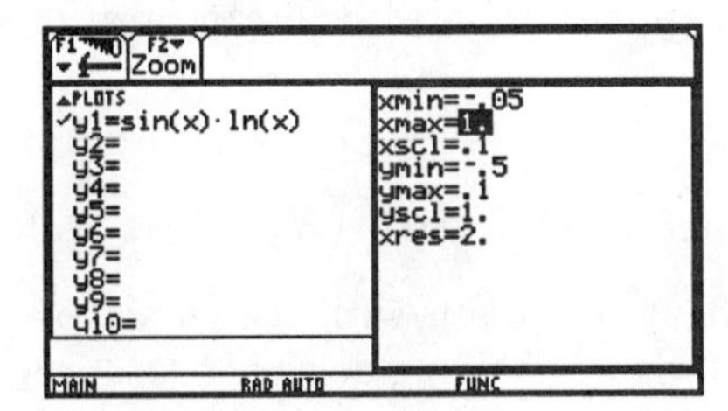

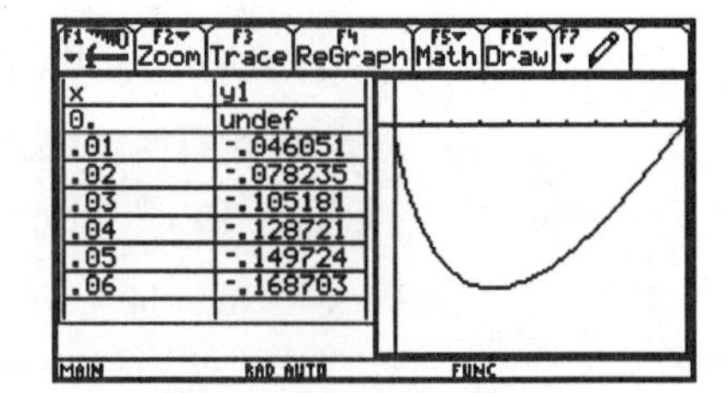

You should experiment with **TOP-BOTTOM** Split Screen **MODE** and with different **Split Screen Ratios**.

Throughout the rest of this manual, we will use Split Screens at will, often simply to convey more information in less space on the page.

1.5 Shortcuts and special characters

We have already been using a number of the **TI-89/92**'s shortcuts. For example, each of ◇**HOME**, ◇**Y=**, ◇**WINDOW**, ◇**GRAPH**, ◇**TblSet**, and ◇**TABLE** is a shortcut to accessing an item in the **APPS** menu. The following is a list of several other helpful shortcuts and tips.

On the Home screen:

- **ans**(i) references the "answer" from the ith most recent entry in the Home screen history area. Pressing **ANS** (**[2nd]-[(−)]**) puts **ans(1)** on the entry line. Notice also that pressing any of **[+]**, **[−]**, **[×]**, **[÷]**, **[ˆ]**, **[x^{-1}]**, or **[STO▷]** with a blank entry line automatically inserts **ans(1)**.
- **entry**(i) references the ith most recent entry in the history area. Pressing **ENTRY** (**[2nd]-ENTER**) puts the last entry on the entry line.
- **F1-8** clears the history area of the Home screen (or the contents of the **Y=** Editor).

- Press **[2nd]-[⇒]** or **[2nd]-[⇐]** to move to the end or beginning of an entry on the entry line.
- Press **◇ ENTER** to force numerical evaluation of an entry. This is equivalent to the **approx()** function (**F2-5**).
- Multiple entries, separated by colons, can be entered simultaneously.
- **Copy** and **Paste** commands in the Tools menu (**F1-5,6**) often help you avoid a lot of typing. Shortcuts for these commands are **◇C** and **◇V**, respectively.
- Press **[↑]-[⇒]** or **[↑]-[⇐]** to highlight characters for copying and pasting. (**[↑]** is the "shift" key.)
- After using **[⇑]** and **[⇓]** (up and down on the cursor pad) to highlight an entry or an answer in the history area, you can press **ENTER** to place it on the entry line.

On the Graph screen:

- The **ON** key stops a plot, and the **ENTER** key pauses a plot.
- Moving the cursor about the screen with the cursor pad (in tracing or zooming) is much faster while holding down the **[2nd]** key.
- **◇F** brings up the Graph Formats dialog box.

Special Characters. Pressing **CHAR** (**[2nd]-[+]**) reveals a large collection of symbols and characters that are not seen on the standard keyboard.

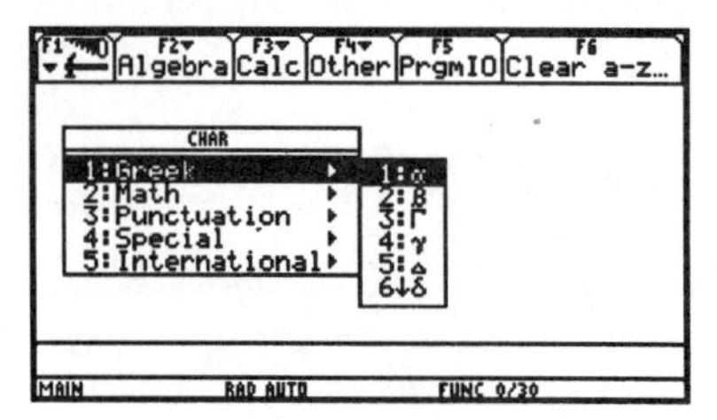

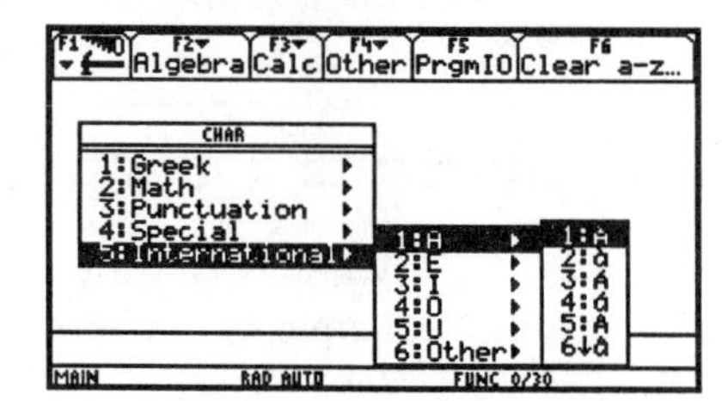

Greek letters are especially useful since many Greek letters are commonly used in mathematics. Many of these letters can be accessed on the **TI-92** QWERTY keyboard by pressing an appropriate prefix key combination first. The prefix for all Greek letters is **[2nd]-G**. After pressing **[2nd]-G**, simply press the Roman "equivalent" of the desired letter. For example, **[2nd]-G-G** produces γ (gamma), and **[2nd]-G-A** produces α (alpha). The analogous prefix key combination for the **TI-89** is **[2nd]-[( ]**.

A collection of "International" characters, many of which are accented members of the standard Roman alphabet, provides an extra source of single-character variables on the **TI-92**. Prefixes for these accented letters are as follows.

[2nd]-A produces the grave accent for à, è, ì, ò, ù;

[2nd]-C produces the cedilla accent for ç;

[2nd]-E produces the acute accent for á, é, í, ó, ú, ý;

[2nd]-N produces the tilde accent for ñ, õ;

[2nd]-O produces the caret, or circumflex, accent for â, ê, î, ô, û;

[2nd]-U produces the umlaut accent for ä, ë, ï, ö, ü, ÿ.

1.6 The Program Editor

This section is an introduction to creating and running a program on the **TI-89/92**. However, if you've never had any programming experience, you should read Chapter 17 in your **TI-89/92 Guidebook** before going any further here.

The first thing we should do, before entering any programs, is to create a new folder to hold them. To do this, press **VAR-LINK** (**[2nd][–]**). In the resulting **VAR-LINK** dialog box, press **F1** and select **Create Folder**. Then enter **myprogs** (or whatever you like) as the name of the new folder and exit **VAR-LINK**.

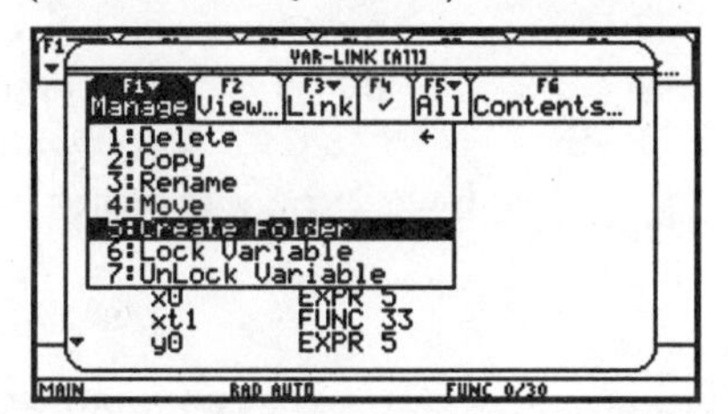

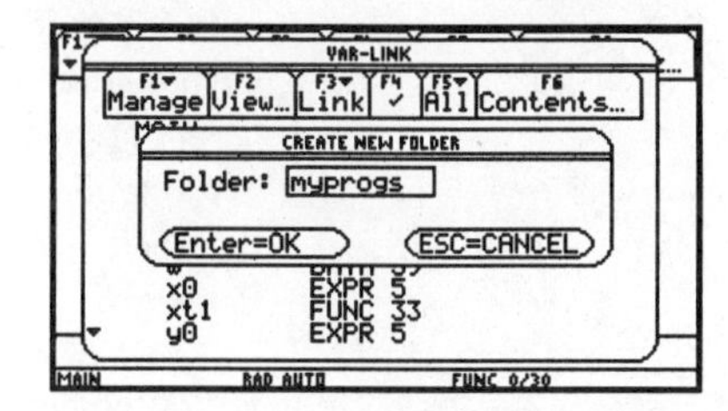

The **VAR-LINK** dialog box is where most memory management takes place. There we can see a listing of all defined variables (including functions and programs). We can also delete and rename variables, move a variable to a different folder, and send and receive files to and from another **TI-89/92** or a computer.

Now, before getting to a somewhat more useful program, let's first create a very simple program that does nothing more than draw random circles on the screen. To enter a new program, first press the **APPS** key and then select **Program Editor** and **New...** . In the subsequent dialog box select **Type: Program**, **Folder: myprogs**, and enter the name of the program, **circles**, in the box beside Variable:.

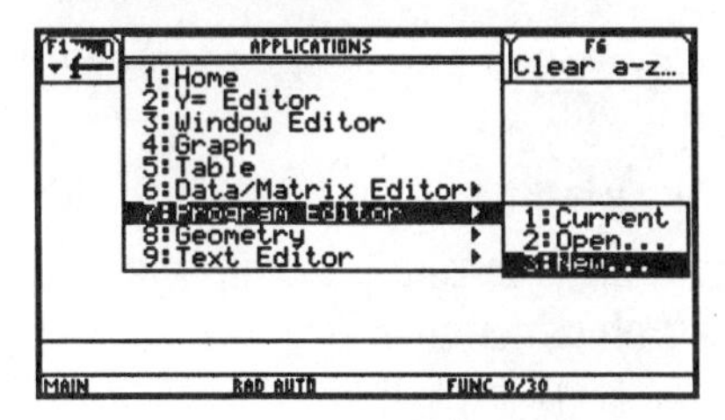

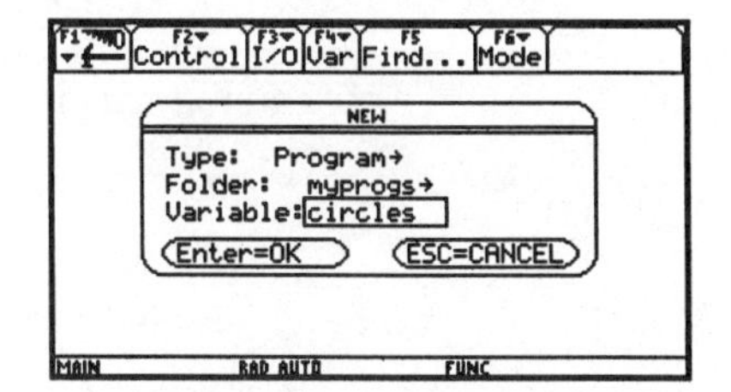

This takes us to the Program Editor, ready to begin entering the program.

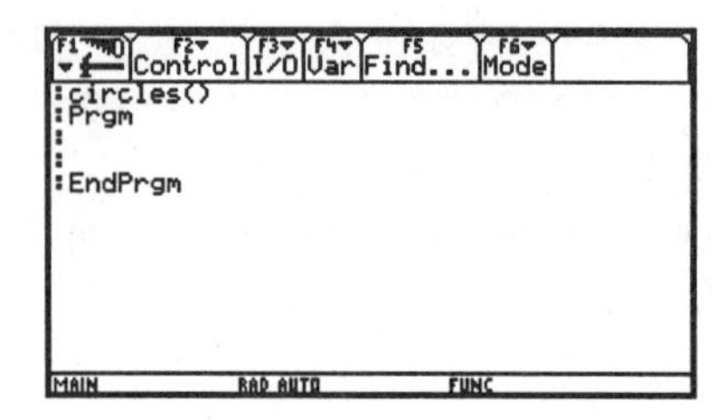

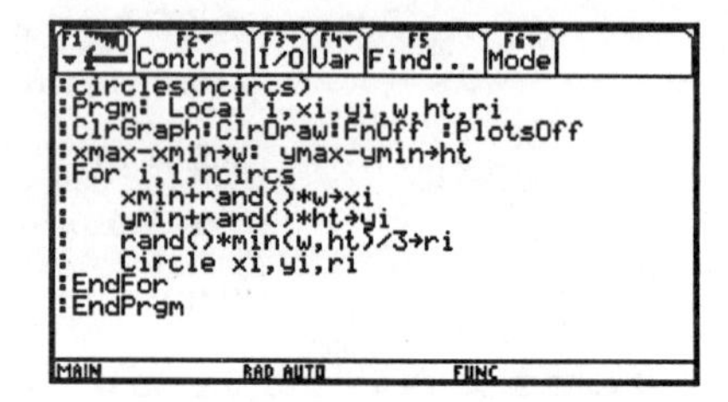

Enter the program as listed below. The argument *ncircs* is the number of circles to be drawn by the program.

```
: circles(ncircs)                          © programname(arguments)
: Prgm                                     © beginning of program block
: Local i, xi, yi, ri, w, ht               © declares local variables
: ClrGraph: ClrDraw: FnOff: PlotsOff       © prepares Graph screen
: xmax−xmin →w: ymax−ymin →ht              © computes screen width and height
: For i, 1, ncircs                         © beginning of For...EndFor loop
:    xmin+rand()*w →xi                     © computes random x coordinate
:    ymin+rand()*ht →yi                    © computes random y coordinate
:    rand()*min(w,ht)/3 →ri                © computes random radius
:    Circle xi, yi, ri                     © draws circle with center (xi,yi) and radius ri
: EndFor                                   © end of For...EndFor loop
: EndPrgm                                  © end of program block
```

Once you've entered the program, press **◇HOME** or **QUIT** to return to the Home screen. The program is run by entering **myprogs\circles(***ncircs***)** from the Home screen. Entering **myprogs\circles(30)** will produce (something like) the following picture in a $[-20, 20] \times [-10, 10]$ window.

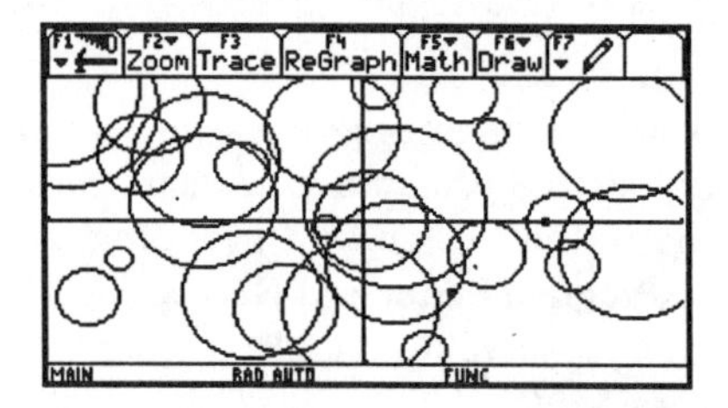

This program has a few features that will be common in the remainder of the programs in this manual:

- declaration of **Local** variables;
- the commands **ClrGraph: ClrDraw: FnOff: PlotsOff** to prepare the Graph screen for graphics commands;
- a **For** . . . **EndFor** loop.

Read the descriptions of each of these in Appendix A of your **TI-89/92 Guidebook**. Also read the descriptions there for the statements **Prgm**, **rand()**, **min()**, and **Circle**.

In Exercise 1 at the end of this section, you will be asked to make a few straightforward modifications to this program.

Now that we've had a touch of experience programming the **TI-89/92**, let's create a program that will connect an ordered collection of points with line segments. The program might be used, for example, to draw a triangle or a quadrilateral. The collection of points will be stored as an n by 2 matrix, where n is the number of points. Each row of the matrix will contain the coordinates of a point.

As before, to enter a new program first press the **APPS** key and then select **Program Editor** and **New....** In the subsequent dialog box, select **Type: Program**, **Folder: myprogs**, and enter the name of the program—which here will be **polyline**—in the box beside **Variable:**. This takes us to the program editor, ready to begin entering the program.

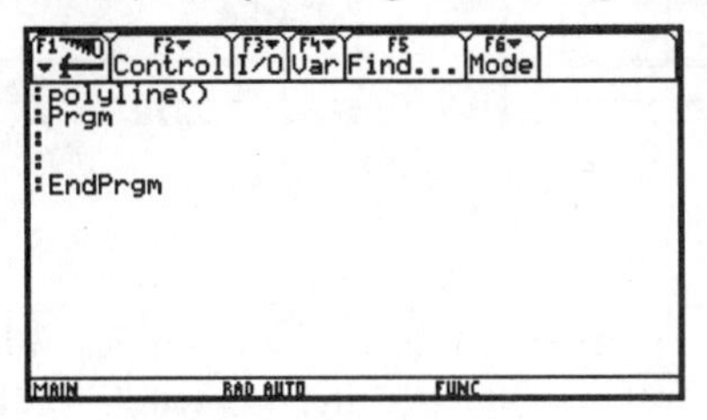

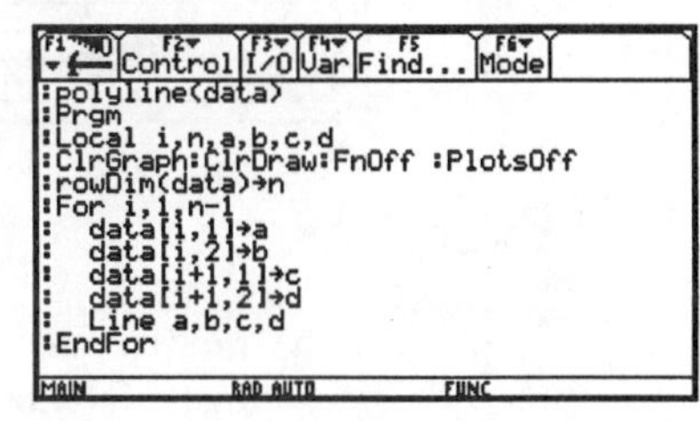

Carefully enter the program, which is listed with comments below.

```
: polyline(data)                          © programname(arguments)
: Prgm                                    © beginning of program block
: Local i,n,a,b,c,d                       © declares local variables
: ClrGraph:ClrDraw:FnOff :PlotsOff        © prepares Graph screen
: rowDim(data) →n                         © finds number of rows in the data matrix
: For i,1,n–1                             © beginning of For loop
:    data[i,1] →a                         © extracts x coordinate of initial point
:    data[i,2] →b                         © extracts y coordinate of initial point
:    data[i+1,1] →c                       © extracts x coordinate of terminal point
:    data[i+1,2] →d                       © extracts y coordinate of terminal point
:    Line a,b,c,d                         © draws the segment
: EndFor                                  © end of For loop
: EndPrgm                                 © end of program block
```

Before running the program, we need to enter coordinates into a matrix. Suppose we want to draw a diamond with vertices $(2,1)$, $(3,3)$, $(2,5)$, and $(1,3)$. To set up the matrix, enter

[2,1; 3,3; 2,5; 1,3; 2,1] →a .

Note that the first point is also entered as the last point. (Why?) Now we're ready to draw the picture.

```
F1 F2 Algebra F3 Calc F4 Other F5 PrgmIO F6 Clear a-z...

  [2 1]                    [2 1]
  [3 3]                    [3 3]
■ [2 5] → a                [2 5]
  [1 3]                    [1 3]
  [2 1]                    [2 1]
■ myprogs\polyline(a)       Done
myprogs\polyline(a)
MAIN        RAD AUTO        FUNC 2/30
```

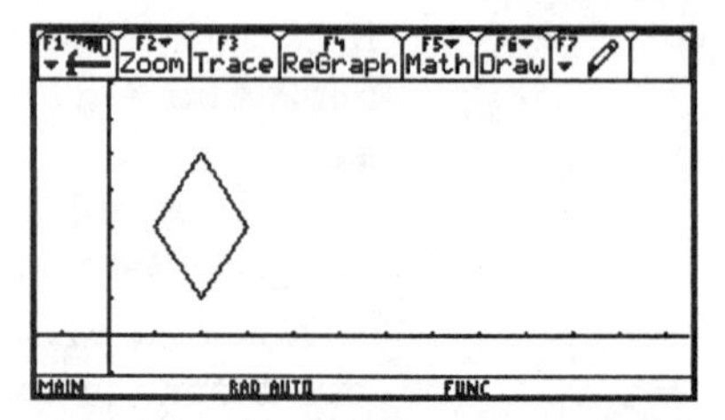

The **polyline** program can be used to plot a *piecewise-linear function.* This is very useful for plotting a linear interpolation of a function whose values are known only at discrete points, as shown below, where we use the Data/Matrix Editor to set up the matrix.

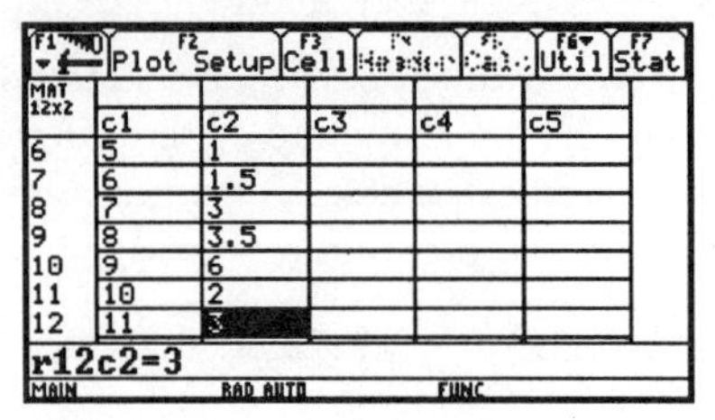

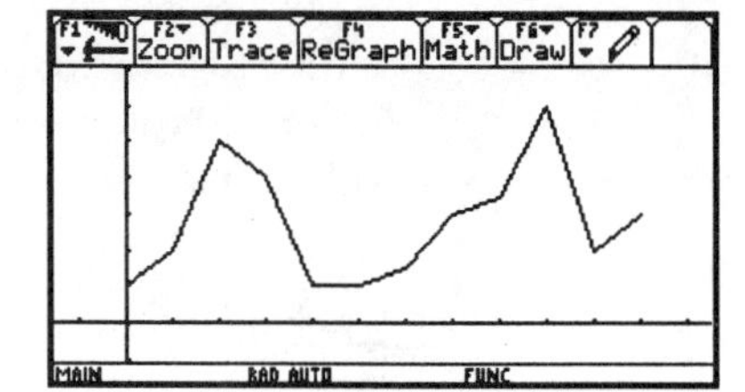

Exercises

1. a) Modify the **circles()** program so that the circles all have the same radius, **min(w,ht)/5**. Test by entering **myprogs\circles(30)**.

   b) Modify **circles()** so that the circles are centered at $(0,0)$ (with the same random radii as in the original program). Test as in part a.

   c) Modify **circles()** so that the circles are centered at $(0,0)$ and have radii given by **(i/ncircs)∗min(w,ht)**. Test as in part a.

2. Use **polyline()** to draw each of the following figures:

   a) The triangle with vertices $(0,1)$, $(4,2)$, and $(3,3)$.

   b) The regular pentagon with vertices $\big(\cos(2k\pi/5), \sin(2k\pi/5)\big)$ for $k = 0, 1, \ldots, 5$.

3. A factory discharges effluent into a river. The rate of effluent discharge (in cubic meters per minute) is recorded hourly over a 24-hour period. The measurements are shown in the following table.

t	0	1	2	3	4	5	6	7	8	9	10	11	12	13	14	15	16	17	18	19	20	21	22	23	24
$r(t)$	3.2	2.4	4.3	5.5	5.8	6.3	4.7	4.0	5.9	7.2	8.3	8.0	7.3	6.6	5.1	3.8	3.5	3.1	2.2	4.1	5.5	5.1	4.8	4.1	3.7

   Use **polyline()** to plot the effluent discharge versus time.

4. Read the descriptions in Appendix A of your **TI-89/92 Guidebook** for **Circle, ClrGraph, ClrDraw, DelVar, FnOff, For, Line, Local, min(), PlotsOff, Prgm, rowDim, colDim, rand()**, and the "copyright" symbol ©.

2 Functions and Equations

This chapter introduces a number of techniques for defining and plotting graphs of functions and solving equations on the **TI-89/92**. There is a great deal of interesting related material in Chapters 6, 7, and 12 of the **TI-89 Guidebook**, Chapters 3, 11, and 15 of the **TI-92 Guidebook**, and in Sections 1.3 and 1.4 of Stewart's *Calculus*.

2.1 Functions

Functions may be defined on the **TI-89/92** either in the **Y=** Editor or on the Home screen. A function of a variable x is defined on the Home screen either by storing (**STO ▷**) an expression in *fnctname*(**x**) or by entering **Define** *fnctname*(**x**) = expression. By defining a function on the Home screen, we can name the function as we like (using up to eight characters). If we define a function on the Home screen, we can graph that function by specifying it in the **Y=** Editor and then pressing **◇GRAPH**. (Some adjustment of the window variables is usually necessary to get a nice plot.)

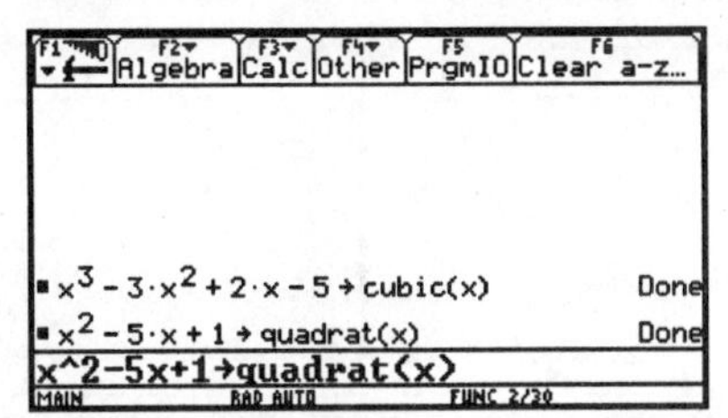

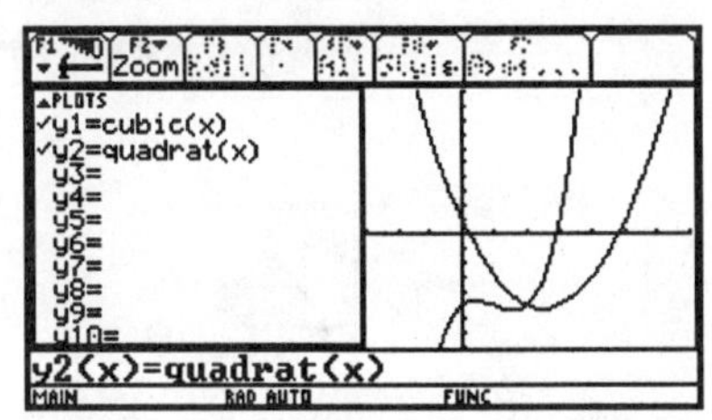

If we define a function in the **Y=** Editor, we can not only plot the graph readily but also perform evaluations on the Home screen. (Any function defined in the **Y=** Editor must be defined in terms of the variable **x**.)

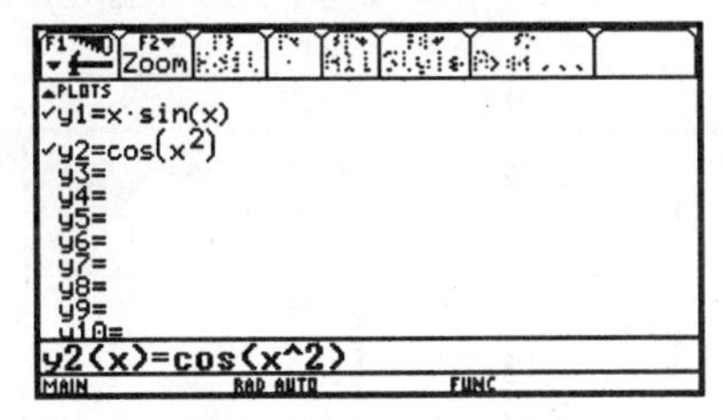

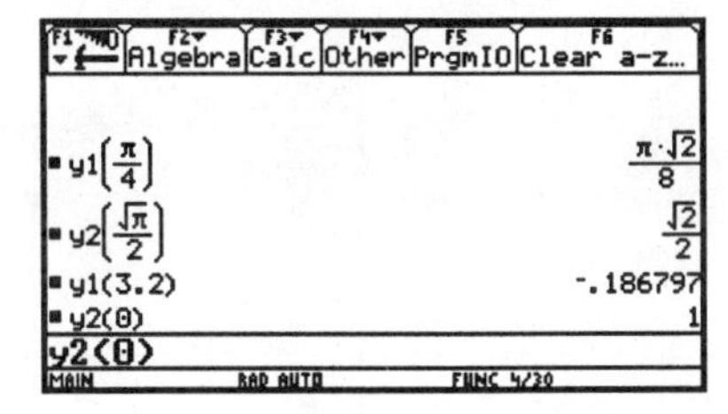

Piecewise-defined functions. Piecewise-defined functions are easy to define by means of the logical **when()** statement. The structure of **when()** is:

when(*condition*, *trueResult*, *falseResult*).

• EXAMPLE 1. The function

$$f(x) = \begin{cases} x, & \text{if } x < 1 \\ -x, & \text{if } x \geq 1 \end{cases}$$

can be defined by means of **when(x<1, x, −x)**, which returns **x** if **x<1** and returns **−x** if **x≥1**.

```
y1(-2)     -2
y1(-.5)    -.5
y1(.5)     .5
y1(1)      -1
y1(2)      -2
y1(7)      -7
y1(π)      -π
y1(x)=when(x<1,x,-x)
```

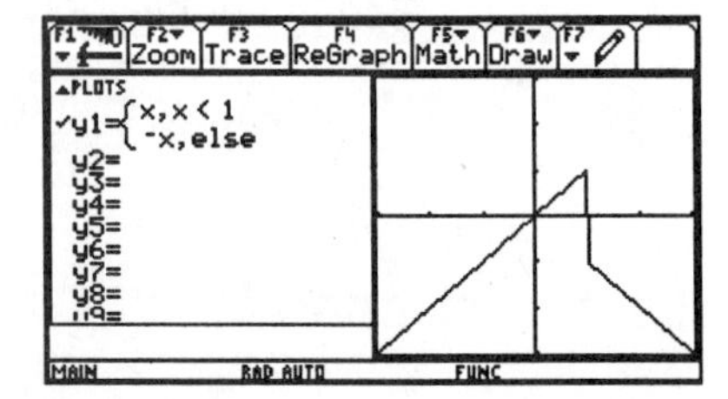

Nested **when()** statements can be used to define piecewise-defined functions with more than two "pieces."

• EXAMPLE 2. The function

$$f(x) = \begin{cases} x, & \text{if } x < 0 \\ 0, & \text{if } 0 \leq x < 1 \\ 1, & \text{if } x \geq 1 \end{cases}$$

can be defined by means of **when(x<0, x, when(x<1, 0, 1))**.

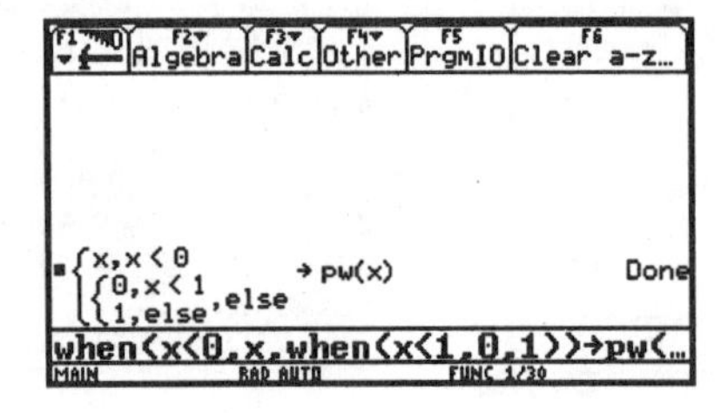

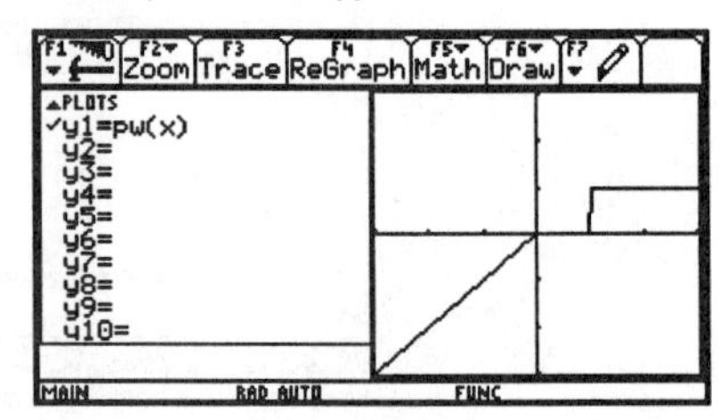

Another way of defining certain piecewise-defined functions is by means of the **floor()** (or greatest integer) function:

$$\text{floor}(x) = n, \quad \text{where } n \leq x < n + 1 \text{ and } n \text{ is an integer.}$$

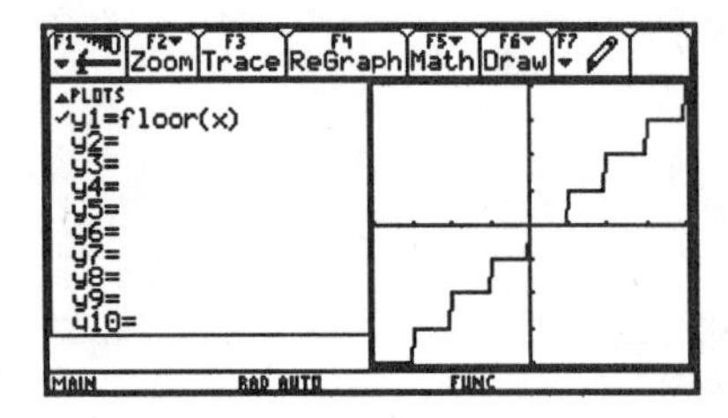

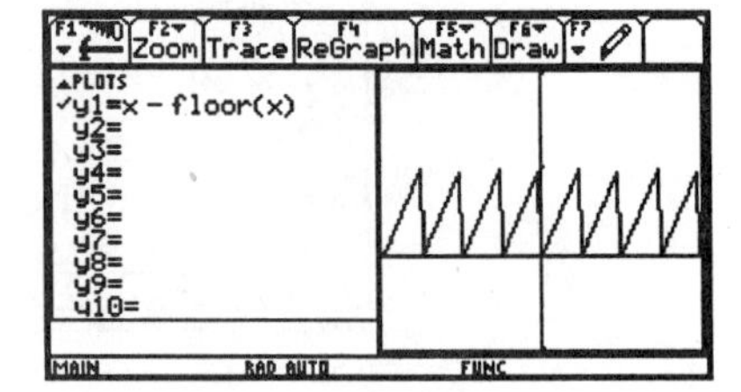

Notice the way that the **TI-89/92** draws vertical (or near-vertical) segments wherever these piecewise-defined functions undergo abrupt changes in value. Strictly speaking, such vertical segments can never be part of the graph of a function, but they are an unavoidable consequence of the way in which graphs are drawn on a (rather course) grid of "pixels." However, once you get use to seeing these vertical segments, they simply become indications of "jumps" in the graph. In order to make them as vertical as possible, thereby avoiding as much chance for misinterpretation as possible, you should always set the window variable **xres** to 1 when plotting such a function.

Later, in the context of approximating the area under the graph of a function, we will be interested in plotting a "piecewise constant," or "step-function" approximation to a given curve. This can be done quite easily with the **floor()** function.

• EXAMPLE 3. Consider the function $f(x) = \sin \pi x$ for $0 \leq x \leq 1$. To obtain a step-function approximation of f with n "steps," plot the function $f\big(\text{floor}(nx)/n\big)$. With $n = 10$ and $n = 20$ steps, we see the following.

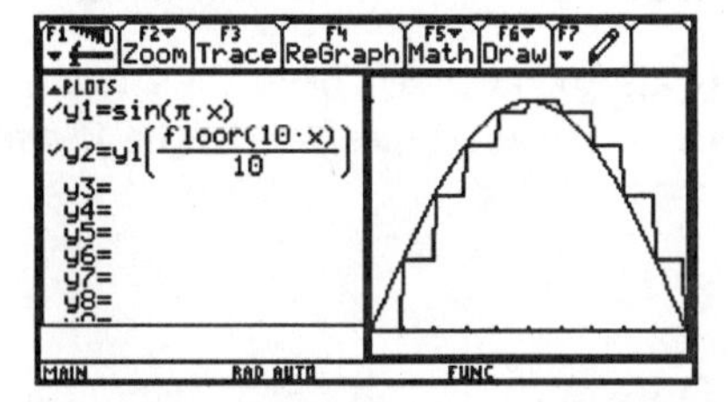

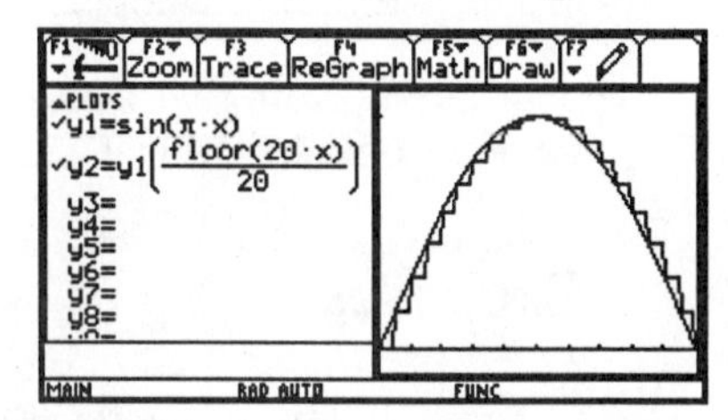

Notice that the "steps" all lie to the right of the curve. To have steps that lie to the left of the curve, plot instead $f\big((\text{floor}(nx) + 1)/n\big)$. The graph of $f\big((\text{floor}(nx) + .5)/n\big)$ will intersect the curve at the midpoint of each step.

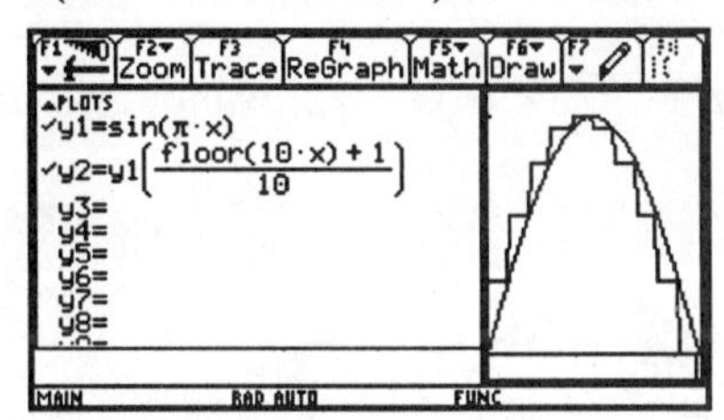

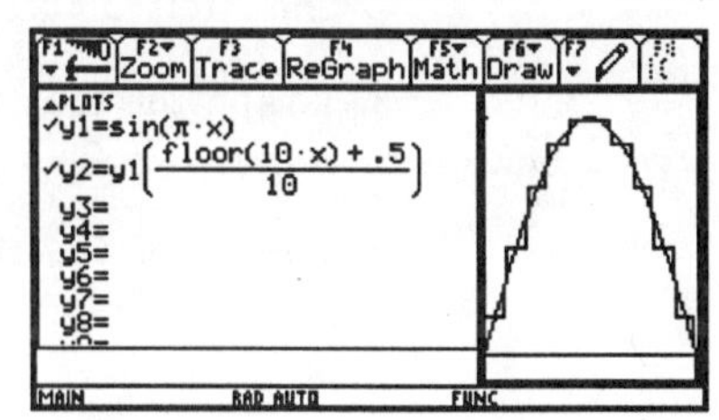

Inverse functions. With the **DrawInv** command, we can plot the graph of an inverse function.

• EXAMPLE 4. Let's plot $f(x) = x^3$ along with $f^{-1}(x) = \sqrt[3]{x}$. Enter **y1(x)= x³** in the **Y=** Editor and then **DrawInv y1(x)** on the Home screen.

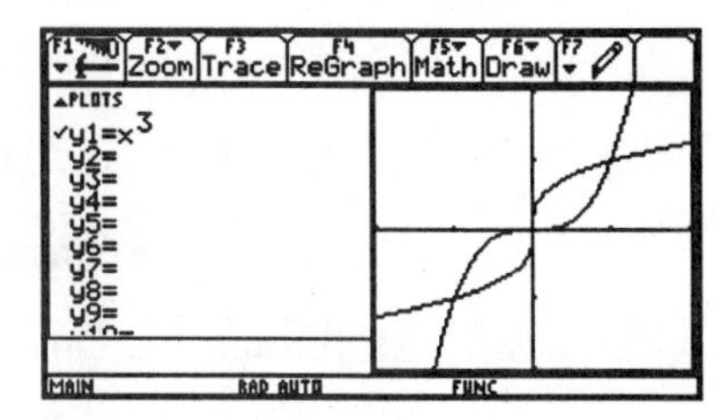

DrawInv is especially useful for plotting equations in which x is a function of y.

• EXAMPLE 5. Suppose we are interested in the region in the plane that is bounded by the graphs of $y = x - 1$ and $x = y^2 - 1$. We can get the graph of $x = y^2 - 1$ by entering and unchecking **y2(x)= x²– 1** in the **Y=** Editor (**F-4**) and then entering **DrawInv y2(x)** on the Home screen.

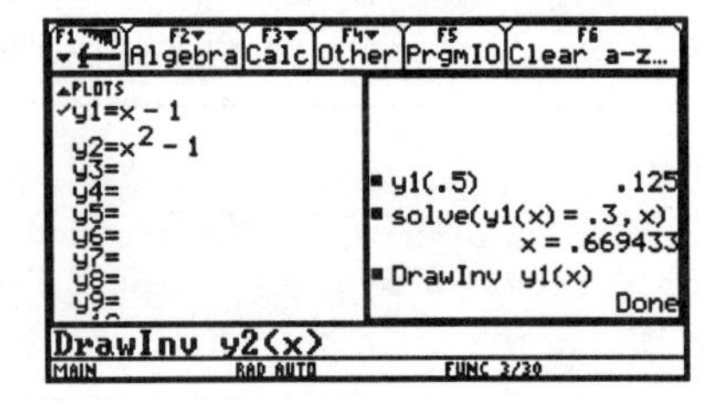

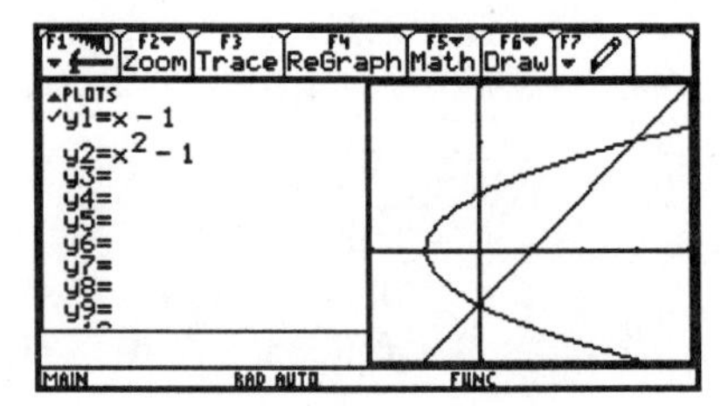

Note that **DrawInv** may also be accessed from the **Draw** menu (**F6**) of the Graph screen. Doing so automatically takes you to the Home screen.

Exercises

1. Graph the functions $\sin \pi x$, $\sin 2\pi x$, and $\sin 3\pi x$ simultaneously over the interval $[-2, 2]$.

2. Graph the functions $\sin\left(2\pi\left(x - \frac{k}{4}\right)\right)$, $k = 0, 1, 2, 3$, simultaneously over the interval $[0, 1]$.

3. Graph the functions $x^3 - kx$, $k = 0, 1, 2, 3$, simultaneously over the interval $[-2, 2]$.

4. Graph the function

$$f(x) = \begin{cases} -x, & \text{if } x < 0 \\ x^2 - 1, & \text{if } x \geq 0 \end{cases}$$

5. Graph the function

$$f(x) = \begin{cases} 1, & \text{if } x < \pi/2 \\ \sin x, & \text{if } x \geq \pi/2 \end{cases}$$

6. Graph the function $f(x) = (x - \text{floor}(x))^2$ on the interval $[-2, 2]$.

7. Graph the function $f(x) = (-1)^{\text{floor}(x)}$ on the interval $[-4, 4]$.

8. Graph the function $f(x) = x(-1)^{\text{floor}(x)}$ on the interval $[-4, 4]$.

9. Graph the function $f(x) = x/\text{floor}(x)$ on the interval $[-10, 10]$.

10. a) Let $g(x) = x^2$. Graph the function $f(x) = g(\text{floor}(nx)/n)$, together with g, on the interval $[0, 1]$ for each of $n = 2, 4$, and 8.

    b) Repeat part a with $f(x) = g(\text{floor}(nx + 1)/n)$.

    c) Repeat part a with $f(x) = g(\text{floor}(nx + .5)/n)$.

11. Repeat Exercise 10 with $g(x) = \cos \pi x$.

12. Graph the equations $y = x^2 - 1$ and $x = y^2 - 1$ simultaneously. How many distinct regions in the plane are bounded by these two graphs?

13. Graph the equations $y = \sin x$ and $x = \sin y$ simultaneously in the window $[-2\pi, 2\pi] \times [-2\pi, 2\pi]$. Repeat with $y = \cos x$ and $x = \cos y$.

14. Graph the equations $y = \tan x$ and $x = \tan y$ simultaneously in the window $[-\pi, \pi] \times [-\pi, \pi]$. Repeat with $y = \cot x$ and $x = \cot y$.

2.2 Parametric curves

Often it is convenient to express the x- and y-coordinates of points on a curve in terms of a third variable (or *parameter*) t. This means that x and y are given in terms of t by a pair of *parametric equations*

$$x = f(t), \quad y = g(t).$$

This is particularly useful when we want to think of each point on the curve as the position of a moving particle at time t. The **TI-89/92** has a **PARAMETRIC** Graph mode for defining and plotting parametric curves.

• EXAMPLE 1. Suppose that x and y are given by

$$x = \cos t, \quad y = \sin 2t.$$

First we'll press **MODE** and switch the Graph mode to **PARAMETRIC**. Then in the **Y=** Editor we define **xt1**= $\cos t$ and **yt1**= $\sin 2t$ prior to setting appropriate window variables.

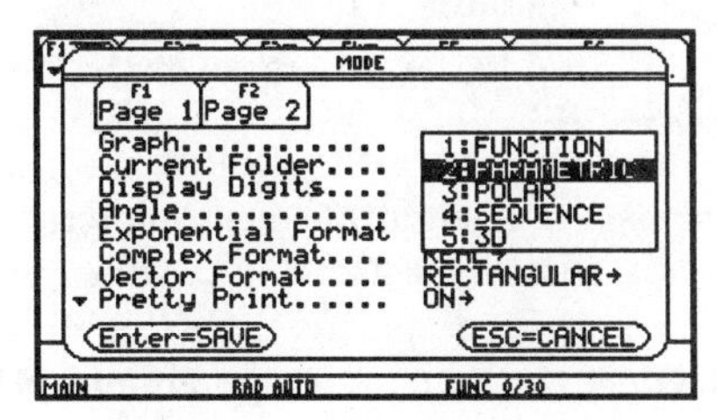

Now we're ready to graph the curve. It is particularly interesting to watch the curve being drawn if either **Path Style** is selected in the **Y=** Editor (**F6-6**) or **Leading Cursor** is set to **ON** in Graph **Format** (**F1-9**). Each of these does essentially the same thing. Also, note that the speed at which the curve is plotted can be adjusted by changing the value of the window variable **tstep**.

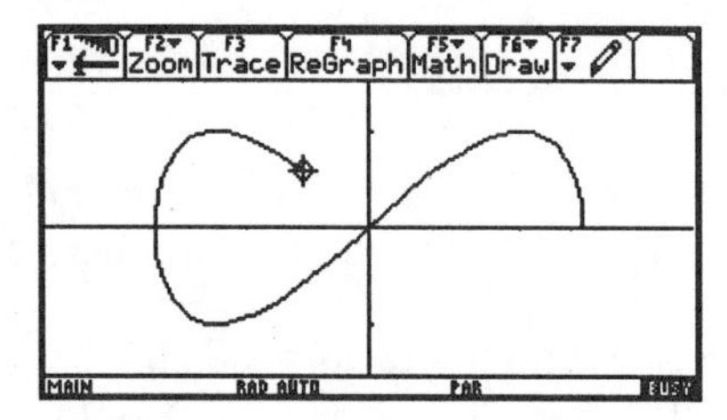

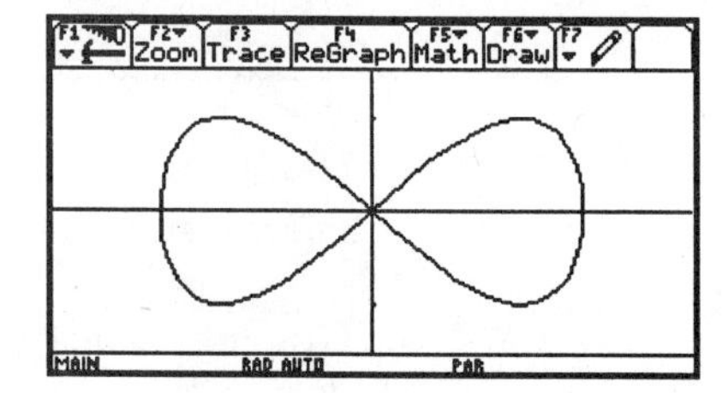

Projectile paths. Parametric equations for the path of a projectile with initial position $(0, 0)$ are

$$x = (v_0 \cos\theta_0)\,t, \quad y = -16\,t^2 + (v_0 \sin\theta_0)\,t,$$

where v_0 is the velocity of the projectile at time $t = 0$ and θ_0 is the angle between the trajectory and a horizontal line at time $t = 0$. (These equations arise only if we ignore air resistance.)

• EXAMPLE 2. Let's plot the trajectory of the projectile if $v_0 = 100$ feet per second and $\theta_0 = \pi/4$.

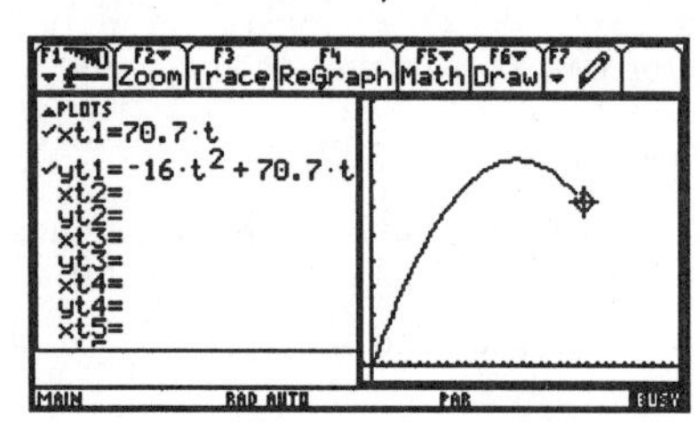

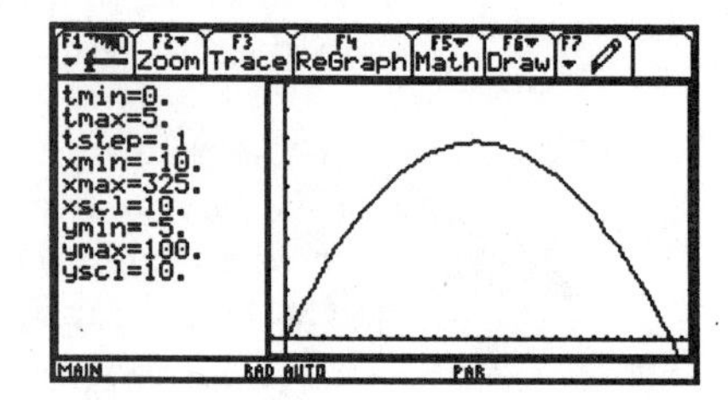

If air resistance is taken into account, it is possible to derive the following parametric equations for the path of the projectile:

$$x = \frac{v_0 \cos\theta_0}{k}\left(1 - e^{-kt}\right), \quad y = \frac{1}{k}\left[(v_0 \sin\theta_0 + 32/k)\left(1 - e^{-kt}\right) - 32t\right]$$

where k is a "drag coefficient" divided by the mass of the projectile.

• EXAMPLE 3. Let's plot the trajectory of the projectile if $v_0 = 100$ feet per second, $\theta_0 = \pi/4$, and $k = .1$, while simultaneously plotting the trajectory that results from ignoring air resistance. Note that we enter the functions in terms of the parameter k and then specify the value of k on the Home screen. Also, setting **Graph Order** to **SIMUL** in Graph **Format** (**F1-9**) causes the graphs to be plotted simultaneously rather than sequentially, which makes for a fun plot to watch.

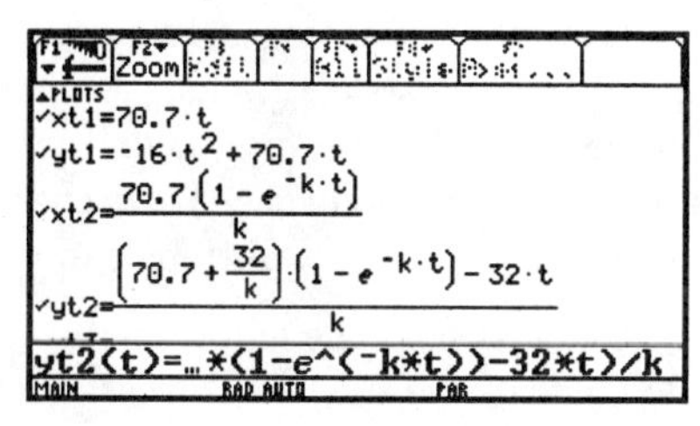

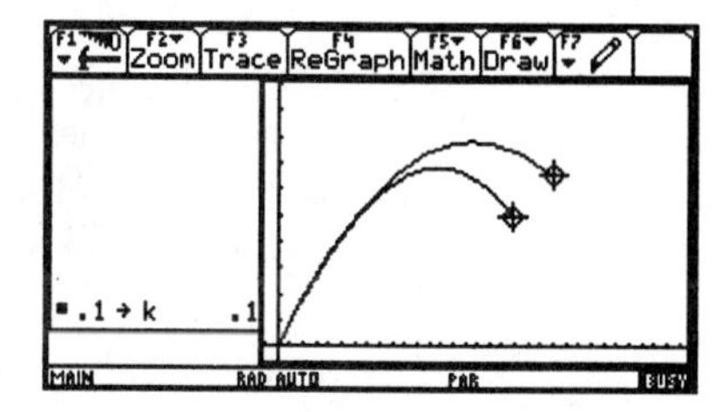

Exercises

In Exercises 1–5, plot the curve given by the parametric equations on the specified interval. Adjust window variables to get a nice graph. In particular, adjust the value of **tstep** to affect the speed at which the curve is plotted.

1. $x = \cos t,\ y = \sin t$ for $0 \le t \le 2\pi$
2. $x = \cos t,\ y = \sin 3t$ for $0 \le t \le 2\pi$
3. $x = \sin 2t,\ y = \cos 5t$ for $0 \le t \le 2\pi$
4. $x = e^{t-2},\ y = e^{2t-4}$ for $0 \le t \le 4$
5. $x = \cos(e^{t-2}),\ y = \sin(e^{t-2})$ for $0 \le t \le 4$

6. Simultaneously plot the paths of a projectile, ignoring air resistance, for initial velocity $v_0 = 100$ feet per second with $\theta_0 = 30°,\ 40°,\ 45°,\ 50°$, and $60°$.

7. For a projectile propelled initially from ground level, the y coordinate of the projectile's position represents height above the ground. At some positive time t the projectile's height becomes zero, indicating that the projectile has hit the ground (after which the parametric equations of the path are no longer in effect). Using the parametric equations that account for air resistance, and using $v_0 = 100$ feet per second and $\theta_0 = \pi/4$, graphically estimate the distance from $(0, 0)$ to where the projectile hits the ground, for each of the drag coefficient values $k = 0.10$ and 0.11.

8. A cannonball is fired from the edge of a cliff with initial velocity 100 feet per second. Using the parametric equations that account for air resistance and a drag coefficient of $k = .1$, graphically and experimentally estimate the angle θ_0 at which the cannonball should be fired in order to hit a target that is on the ground 100 feet below and 200 feet from the base of the cliff.

2.3 Solving equations

The **TI-89/92** has three built-in functions for solving equations: **solve()**, **zeros()**, and **nSolve()**. Each can be accessed in the **Algebra** menu (**F3**) of the Home screen. The first of these, **solve()**, attempts to find exact solutions. If the **Exact/Approx MODE** is set to **AUTO**, then **solve()** switches to a numerical scheme for approximating solutions, if exact solutions can't be found.

• EXAMPLE 1. Suppose we want to find the zeros of the function $f(x) = 9x^4 - 6x^3 - 15x + 10$. A quick plot of the function reveals that f has exactly two real zeros. The **solve()** function is able to find the exact value of each of them.

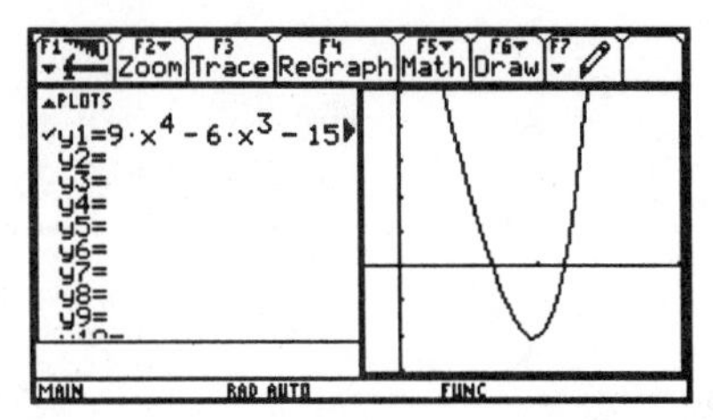

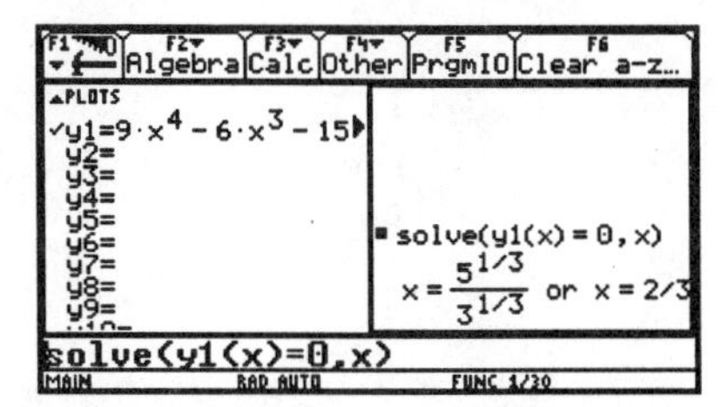

However if we change the function slightly to, say, $f(x) = 9x^4 - 6x^3 - 15x + 11$, then **Solve()** can find no exact solutions and so returns approximate solutions.

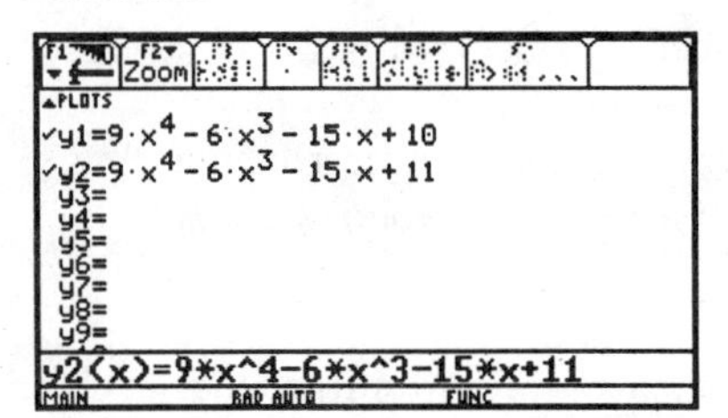

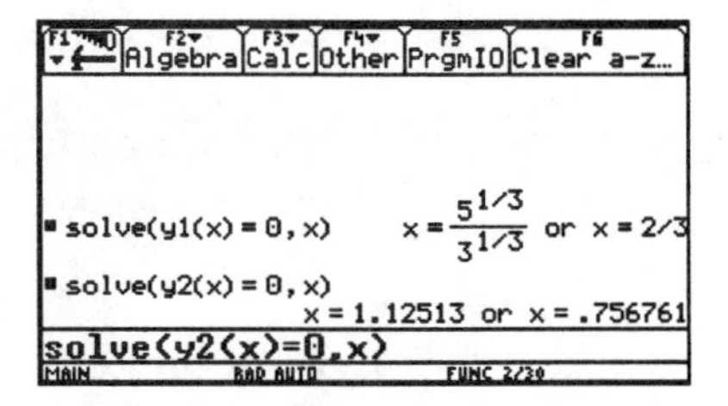

The **zeros()** function does nothing more than execute **solve()** and return the results in a list [**TI-89/92 Guidebook**, Appendix A]. Items in the list are easily extracted and saved. For example, entering **ans(1)[2]→root2** stores the second element of the last result in the variable **root2**. Note that the first argument in **zeros()** is not an equation, but rather an expression.

• EXAMPLE 2. Let's apply **zeros()** to the functions in Example 1. Note that the same solutions are found, but they are returned in the form of a list.

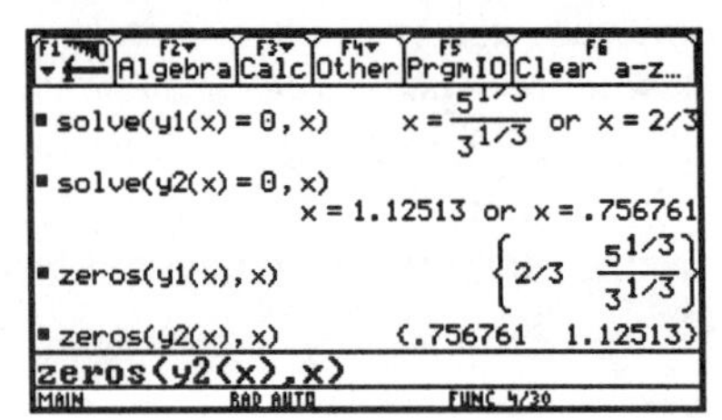

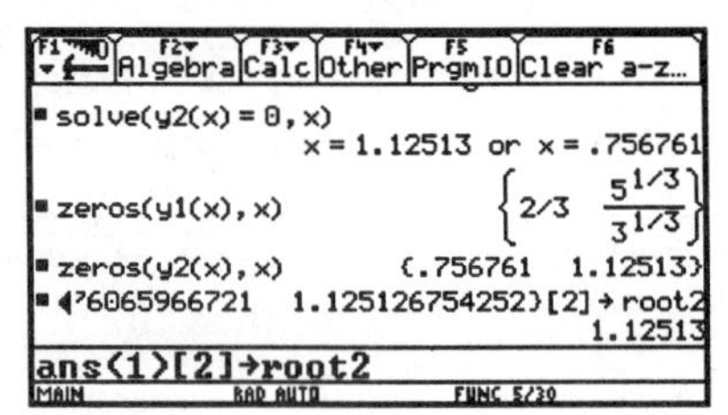

In situations where exact solutions are not expected—or not desired—quicker results can typically be obtained by using **nSolve()** instead of **solve()**. Notice that **nSolve()** returns only one solution at a time. By restricting **nSolve()**'s search to a certain range of the variable where a desired solution is known to live, we can force **nSolve()** to return the desired solution. Restriction of the search range also causes **nSolve()** to do its work faster, in general.

• EXAMPLE 3. Let's apply **nSolve()** to locate the zeros of the functions in Example 1, restricting the search as necessary to find all the zeros.

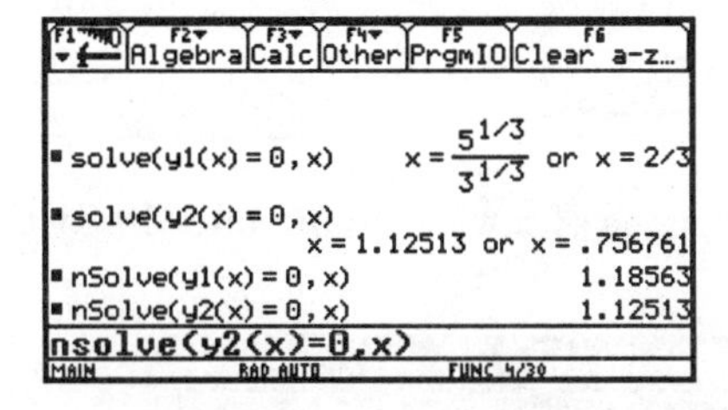


```
solve(y2(x) = 0, x)
          x = 1.12513 or x = .756761
nSolve(y1(x) = 0, x)           1.18563
nSolve(y2(x) = 0, x)           1.12513
nSolve(y1(x) = 0, x)|x < 1     .666667
nSolve(y2(x) = 0, x)|x < 1     .756761
nsolve(y2(x)=0,x)|x<1
MAIN      RAD AUTO      FUNC 6/30
```

There are also situations in which one would like to find solutions of an equation on some specified interval. A typical example involves solving a trigonometric equation that has infinitely many solutions.

• EXAMPLE 4. Suppose we wish to find all solutions of the equation

$$2 \sin x \cos^2 x = 2 \sin 2x + 1$$

in the interval $[0, 2\pi]$. Graphing each side of the equation reveals that there are four solutions in this interval, which **zeros()** is able to locate numerically. Notice that the first argument in **zeros()** is the expression **y1(x)−y2(x)**, not the equation **y1(x)=y2(x)**.

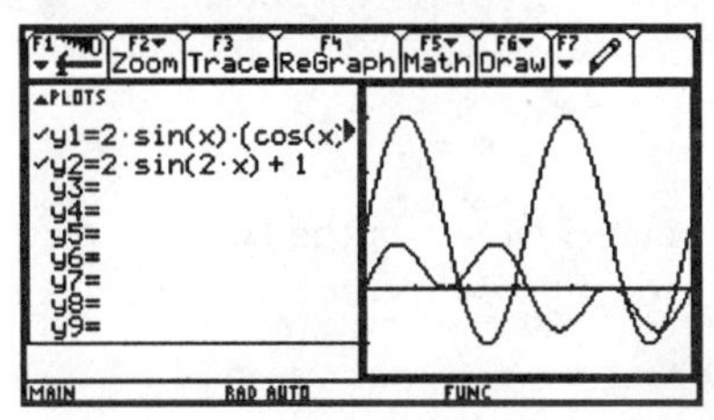

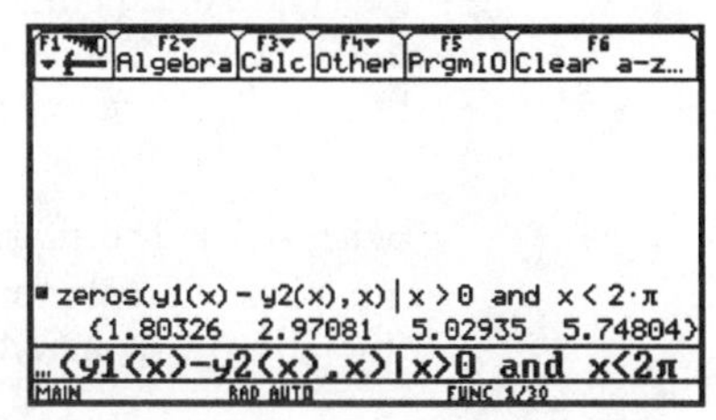

Systems of equations. A system of two (possibly nonlinear) equations in two unknowns can sometimes be solved by using **solve()** in conjunction with the "with" operator (displayed as |). This substitution technique is fairly effective for solving systems that allow us to seek out one solution at a time.

• EXAMPLE 4. Suppose we want to solve simultaneously the pair of equations

$$x^2 + y = 1$$
$$x^2 + y^2 = 4$$

Graphing the two equations reveals symmetry that allows us to simply focus our attention on the solution pair in which $x > 0$. So first we'll solve the first equation for y in terms of x and then solve the second equation for x, using the "with" operator to substitute the first result into the second and restrict x to be positive.

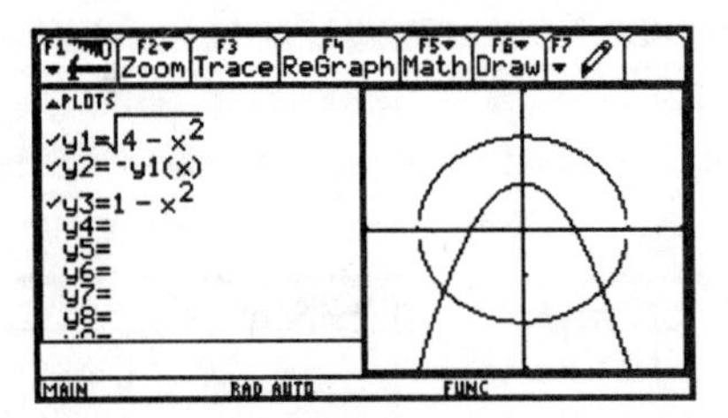

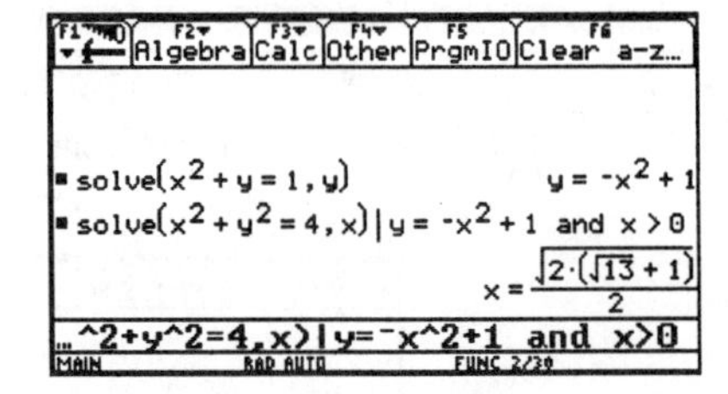

Finally, we substitute the value of x we've found into the formula for y from the previous step.

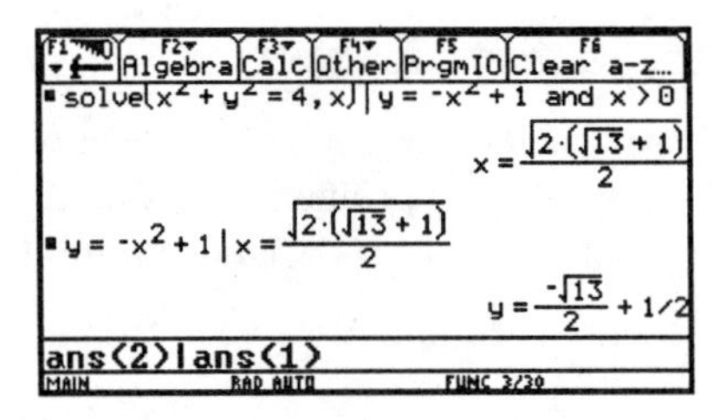

This substitution technique works very well on pairs of linear equations. This is outlined nicely on page 84 of the **TI-92 Guidebook** and on page 46 of the **TI-89 Guidebook**.

The **simult()** function provides a simple way of solving **systems of linear equations**. The system should first be written in matrix form

$$Ax = b,$$

where A is the matrix of coefficients, x is the column vector of unknowns, and b is the column vector of right-side constants. Such a system is solved by entering **simult(A,b)**.

• EXAMPLE 5. Let's solve the system

$$\begin{aligned} 2x - 3y + z &= 5 \\ 2x + y - 2z &= 1 \\ 3x - 7y + 3z &= 2 \end{aligned} \; .$$

First we'll define A and b. Note that arrays are delimited by square brackets and that rows of the matrix and elements of the column vector are separated by semicolons. (We could also use the Data/Matrix Editor to enter the matrix.)

```
F1 F2 Algebra F3 Calc F4 Other F5 PrgmIO F6 Clear a-z...

■ [2 -3 1; 2 1 -2; 3 -7 3] → a        [2 -3 1; 2 1 -2; 3 -7 3]
[2,-3,1;2,1,-2;3,-7,3]→A
MAIN    RAD AUTO    FUNC 1/30
```

```
F1 F2 Algebra F3 Calc F4 Other F5 PrgmIO F6 Clear a-z...
■ [2 -3 1; 2 1 -2; 3 -7 3] → a        [2 -3 1; 2 1 -2; 3 -7 3]
■ [5; 1; 2] → b                        [5; 1; 2]
[5;1;2]→b
MAIN    RAD AUTO    FUNC 2/30
```

Now we need only enter **simult(A,b)** to obtain the solution. (Notice that the upper-case "**A**" is automatically converted to a lower-case "**a.**") Also, we can check the answer by multiplying the result by **A**.

```
F1 F2 Algebra F3 Calc F4 Other F5 PrgmIO F6 Clear a-z...
 3 -7 3]                               3 -7 3]
■ [5; 1; 2] → b                        [5; 1; 2]
■ simult(a, b)                         [43/3; 15; 64/3]
simult(A,b)
MAIN    RAD AUTO    FUNC 3/30
```

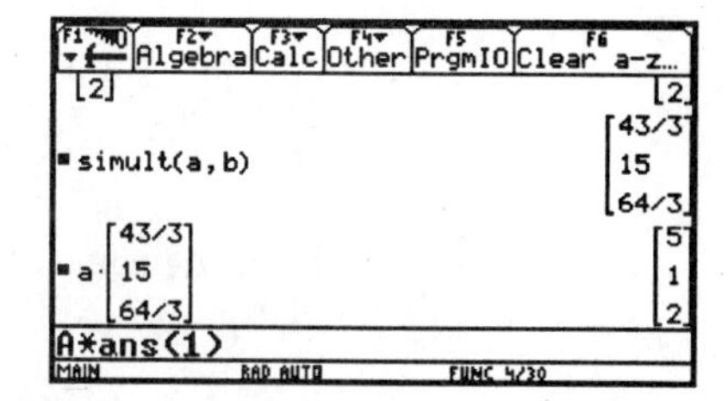

Exercises

In Exercises 1–5, graph the given function to determine the approximate location of each of its zeros; then find all the zeros of the function.

1. $f(x) = x^3 - x^2 - 2x + 1$
2. $f(x) = 2x^5 - 5x^3 + 2x^2 + 3x - 3$
3. $f(x) = x^2 e^{-x/2} - 1$
4. $f(x) = 9\cos x - x$
5. $f(x) = \tan^{-1} x - x^2$

In Exercises 6–8, graph both sides of the given equations to determine the approximate location of each of its solutions on the specified interval; then find all the solutions on that interval.

6. $\sin x \cos 2x = \cos x \sin 3x$, and $0 \le x \le 2\pi$
7. $\sin x^2 = \sin^2 x$, and $0 \le x \le \pi$
8. $\tan x = x$, and $0 \le x \le 3\pi$

9. Use the substitution method to solve the system $2x+3y = 1,\ x^2+y^2 = 1.$

In Exercises 10–12, use **simult()** to solve the given system of linear equations.

10. $3x - 2y = 5,\ 7x + 3y = 2$

11. $3x - 2y + z = 5,\ 7x + 3y - 2z = 2,\ x + y + z = 0$

12. $3x - 2y + z - 2w = 5,\ 7x + 3y - 2z + w = 2,\ x + y + z + w = 0$

13. A closed cylindrical can has a volume of 100 cubic inches and a surface area of 100 square inches. Find the radius and the height of the can.

14. Two spheres have a combined volume of 148 cubic inches and a combined surface area of 160 square inches. Find the radii of the two spheres.

15. An open-topped aquarium holds 40 cubic feet of water and is made of 60 square feet of glass. The length of the aquarium's base is twice its width. Find the dimensions of the aquarium.

16. Find the equation of the parabola that passes through the points $(-1, 1)$, $(1, 2)$, and $(2, 3)$.

17. Find the cubic polynomial $f(x)$ such that $f(1) = f(2) = f(3) = 1$ and $f(4) = 7$.

2.4 Exploring graphs

Once we plot a graph, or collection of graphs, a great deal of detailed information can be obtained easily without leaving the Graph screen. The **Math** menu on the Graph screen has several commands that are useful for exploring graphs. For example, suppose that we wish to study the function $f(x) = \sin x \cos 3x$ on the interval $0 \leq x \leq \pi$. First we enter the function in the **Y=** Editor and define appropriate window variables. Then we plot the graph and select the **Math** menu (**F5**).

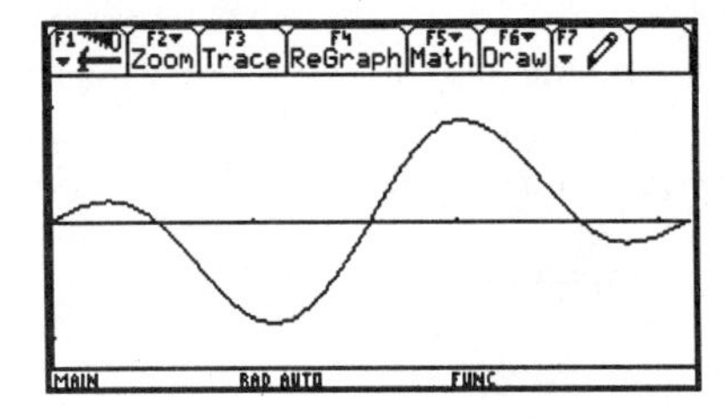

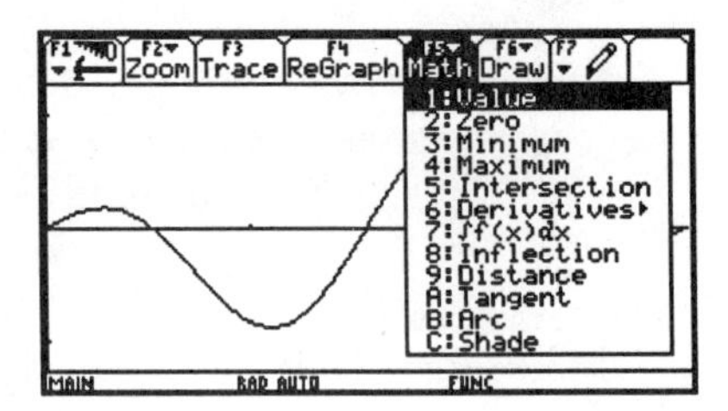

The first item in the **Math** menu is **Value** (**F5-1**). This command lets us compute function values and view the corresponding point on the graph. **Value** prompts us to enter an x. After doing so, we press **ENTER** and see the value of y and the point on the graph.

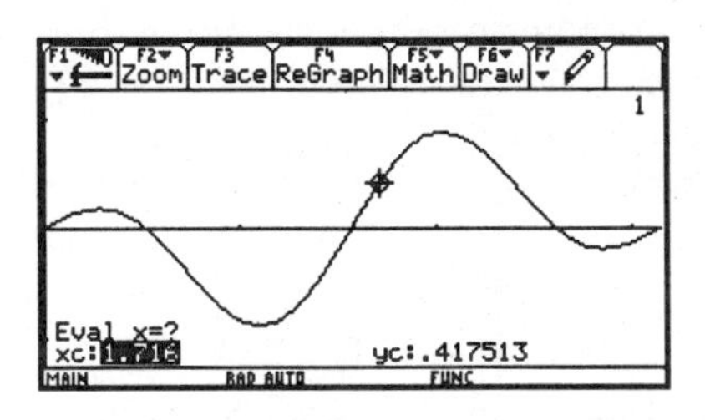

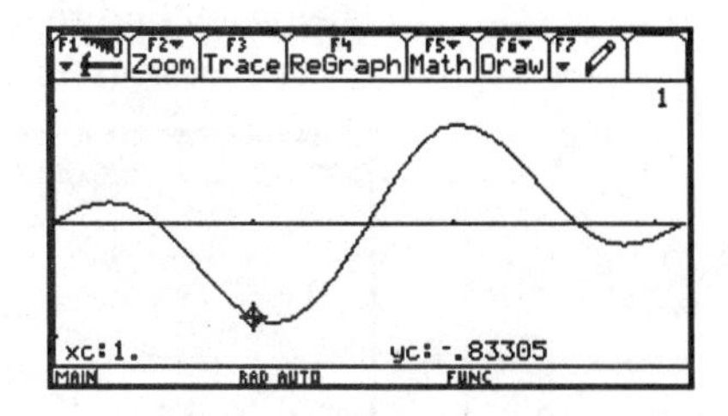

The next item in the **Math** menu is **Zero** (**F5-2**), which will find the zeros of the function. Selecting **Zero** brings a prompt to enter a lower bound for the zero we want to find. After entering that, we're prompted for an upper bound. When lower and upper bounds have been entered, a zero between them is found.

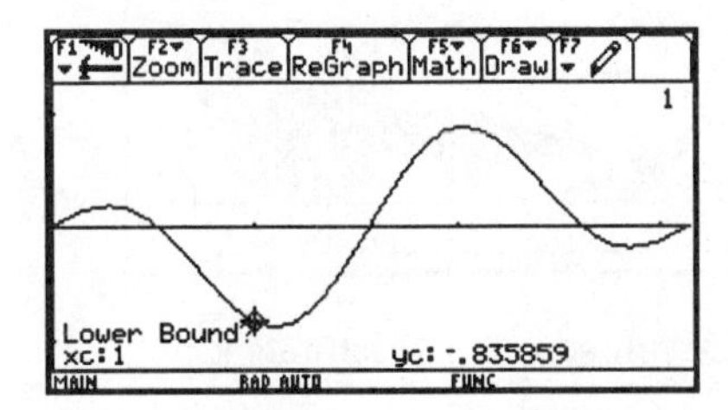

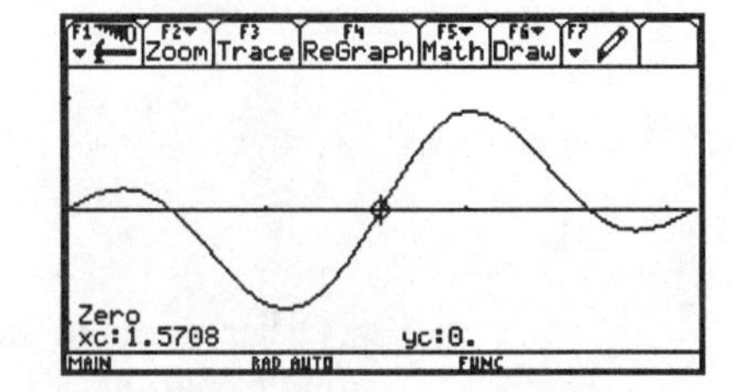

The next two items in the **Math** menu are **Minimum** (**F5-3**) and **Maximum** (**F5-4**). These are similar to **Zero** in that we are prompted to enter both lower and upper bounds for the point we're looking for. Entering lower and upper bounds of 0 and 2, respectively, lets us find the bottom of the first valley on the curve and the top of the first peak.

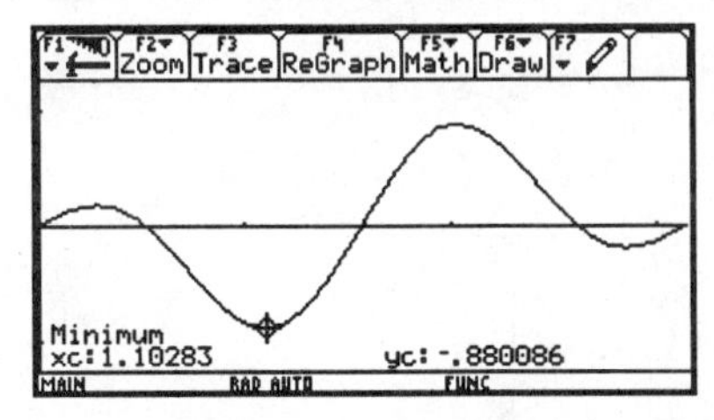

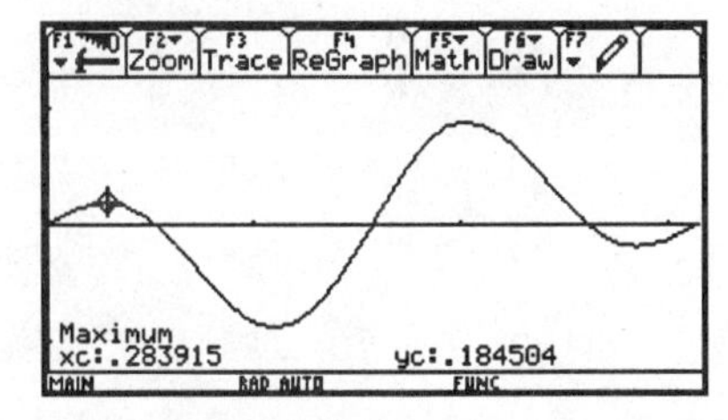

Now let's plot two graphs at once. Say we're interested in the region in the plane that is bounded by the graphs of $y = (x+1)^2(x-1)^2$ and $y = 2x$. First we'll plot the two graphs together.

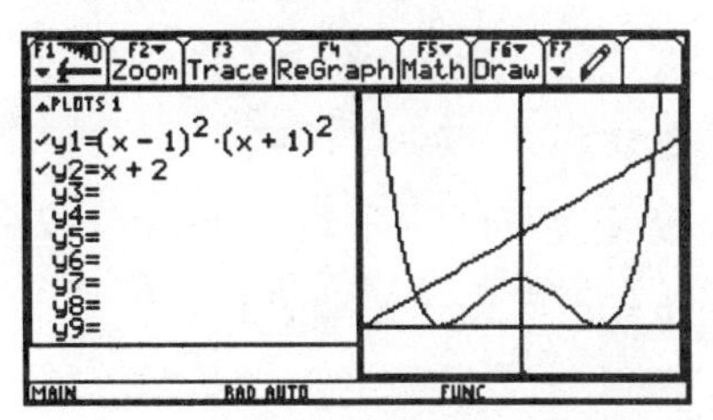

Then we'll find the points of intersection of the two graphs. For this we can use the **Intersection** command from the **Math** menu (**F5-5**).

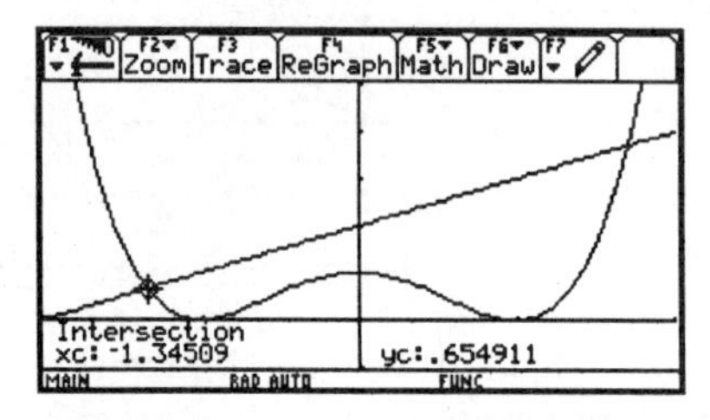

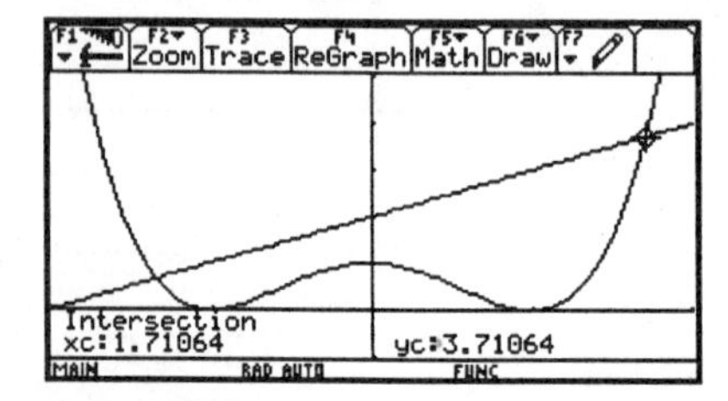

Finally, we'll use the **Shade** command (**F5-C**) to shade the region.

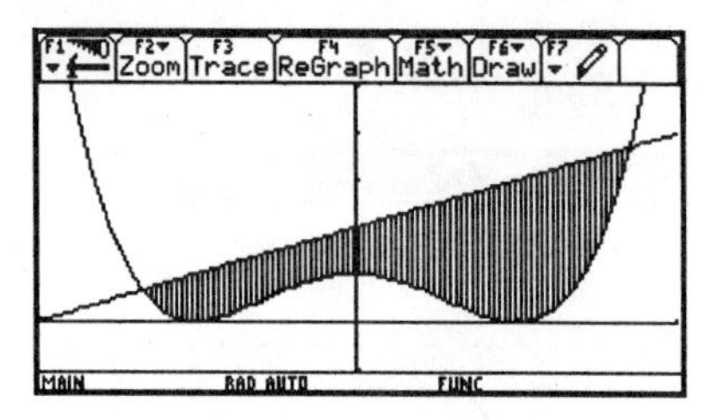

Other items in the Graph screen's **Math** menu will be discussed in subsequent chapters.

Exercises

In Exercises 1–4, graph the function on the indicated interval. Then find all zeros, locally minimum values, and locally maximum values.

1. $f(x) = x^3 - 3x^2 + x + 2$ on $[-1, 3]$

2. $f(x) = \sin x^2 - \sin^2 x$ on $[0, 3]$

3. $f(x) = e^{-x}\cos(\pi x)$ on $[-.5, 2.5]$

4. $f(x) = x^6 - 2x^3$ on $[-1, 1.5]$

In Exercises 5–8, graph both equations, showing the region bounded by the two graphs. Find the points of intersection, and then shade the region bounded by the two graphs.

5. $y = x^3 - 3x^2 + x + 2$ and $y = x + 2$

6. $y = \sin x$ and $y = x^2$

7. $y = (x+1)^2(x-1)^2$ and $y = 2 - (x+1)^2(x-1)^2$

8. $y = e^{-x}$ and $y = x(2 - x)$

2.5 Programming notes

In this section, we will illustrate a few programming ideas by creating a function, **sgnchng()**, that will search for a sign change in a given function. Such a function can be used to find an interval that contains a zero of the given function. Once such an interval is known, the endpoints can be used to help the **nSolve()** function find the zero much more efficiently. The arguments of our function **sgnchng()** will be as in

sgnchng(*f*, *xvar*, *bnds*, *step*).

The first and second arguments, f and *xvar*, will be expressions representing the given function and its independent variable, respectively. The third argument, *bnds*, will be a list containing the left and right endpoints of the interval in which we're searching for a sign change. The last argument, *step*, will be the distance between test-values of *xvar*. The function will return either a list containing the test-values of *xvar* just before and just after (or at) the found sign change or else a message that no sign change was found.

Press **APPS** and select **Program Editor** and **New...** . In the **NEW** dialog box select **Type: Function** and **Folder: myprogs**, and enter **Variable: sgnchng**.

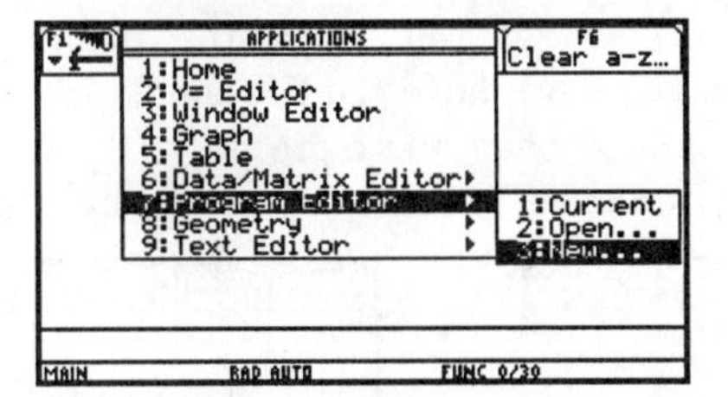

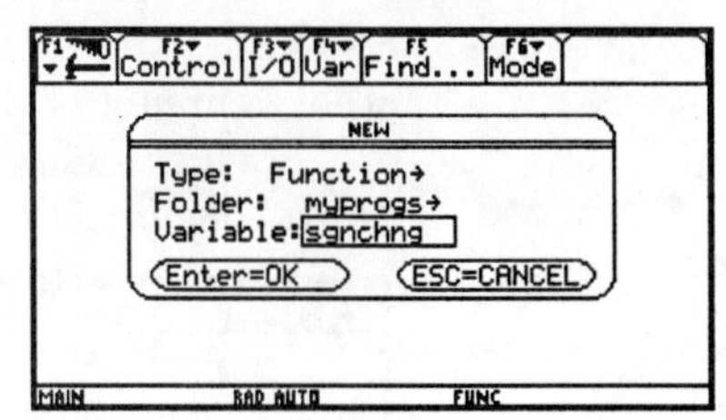

In the Program editor, carefully enter the following function.

```
: sgnchng(f,xvar,bnds,step)
: Func: Local i,n,xi,yi,ynxt,end
: bnds[1] →xi: bnds[2] →end
: f | xvar=xi →yi
: f | xvar=xi+step →ynxt
: If sign(yi)≠sign(ynxt)
:    Return {xi,xi+step}
: While sign(yi)=sign(ynxt) and xi<end
:    xi+step →xi: ynxt →yi
:    f | xvar=xi →ynxt
: EndWhile
: If xi≥end and sign(yi)=sign(ynxt) Then
:    Return "no sign change found"
: Else
:    Return {xi–step,xi}
: EndIf
: EndFunc
```

(Find $\neq$ and $\leq$ by pressing **MATH** and selecting **Test**.) Let's try out this program on the following problem.

• EXAMPLE 1. *The graphs of $y = x^3$ and $y = e^{x/10}$ intersect twice. For each of the points of intersection, find a pair of consecutive integers that bounds the x-coordinate. Then use* **nSolve()** *to find each of the zeros.*

Notice that we simply want to "bracket" the zeros of the function

$$f(x) = x^3 - e^{x/10}$$

with a pair of consecutive integers. So let's first try to bracket the lesser of the two zeros, and then the greater.

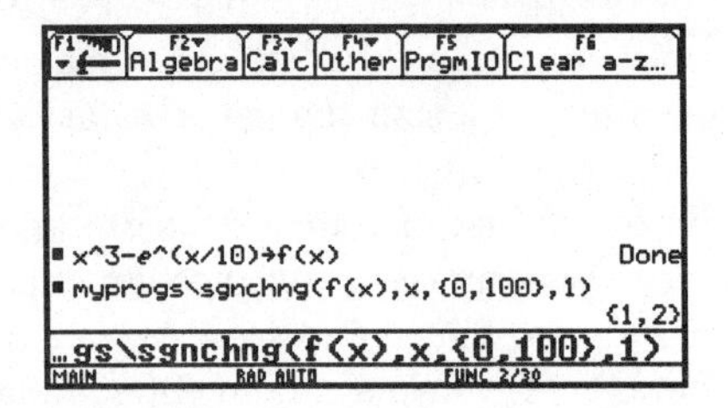

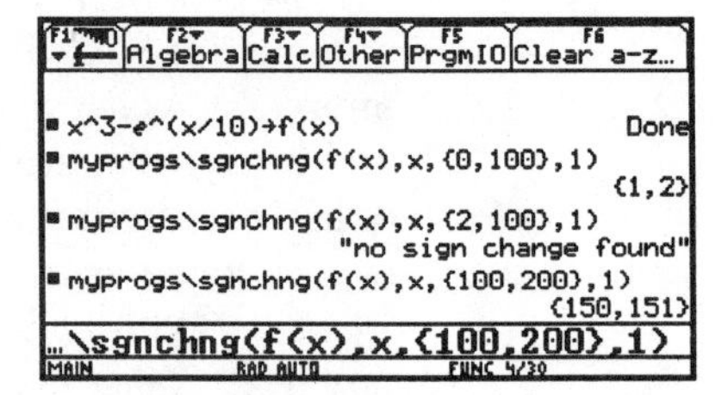

So we find that one of the two zeros of f is in the interval $(1, 2]$, and the other is in $(150, 151]$. Now that we have crude bounds on each of the zeros of f, we'll use **nSolve()** to locate them more precisely. To do this, we use the "with" operator to restrict **nSolve()**'s search to the desired interval.

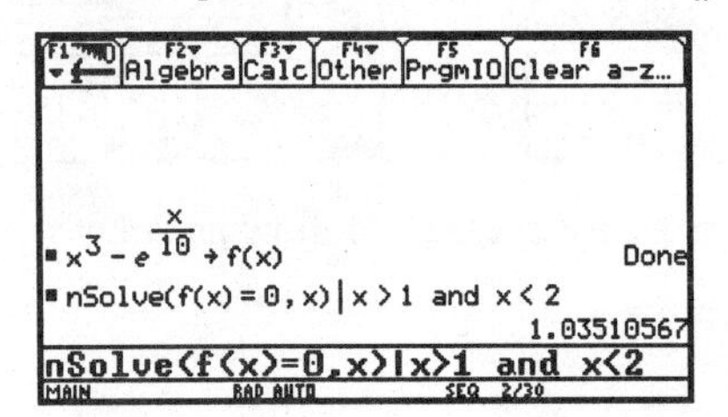

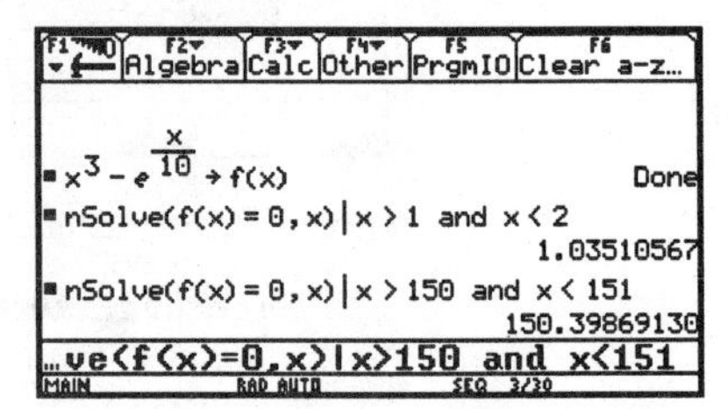

Theoretical stuff. In the last two examples we estimated a zero of a function by finding an interval at whose endpoints the values of the function had opposite signs. What is it that guarantees that such an interval must contain a zero of the function? It turns out that the answer is the continuity of the function. This is a special case of an important theorem known as the *Intermediate Value Theorem.* For more on this, see Section 2.4 of Stewart's *Calculus*.

Exercises

1. Use **sgnchng()** to bracket between consecutive integers the two zeros of $f(x) = \ln x - .005x - 4$ to three decimal places. Then use **nSolve()** and the "with" operator to find each of these zeros of f.

2. Use **sgnchng()** to bracket between consecutive integers the one and only real zero of $f(x) = x^3 - 231x - 83$. Then use **nSolve()** and the "with" operator to find it.

3. Suppose that the height above the ground of a projectile after t seconds is described by

$$y = 320\left[13\left(1 - e^{-t/5}\right) - t\right].$$

   a) Use **sgnchng()** to find an interval $[t_1, t_1 + 1]$ that contains the time t at which the projectile hits the ground.

   b) Use **sgnchng()** to find an interval $[t_2, t_2 + .1]$ that contains the time t at which the projectile hits the ground.

   c) Use **sgnchng()** to find an interval $[t_3, t_3 + .01]$ that contains the time t at which the projectile hits the ground.

4. Read the descriptions of **sign()**, **If**, **Return**, and **While** in Appendix A of your **TI-89/92 Guidebook.**

3 Limits and the Derivative

The graphical, algebraic, and numerical capabilities of the **TI-89/92** allow us to explore the basic concepts behind limits and derivatives. This chapter describes these capabilities and concepts and parallels much of the material in Chapter 2 of Stewart's *Calculus*.

3.1 Limits

The notion of limit is the basis of all of calculus. From one practical point of view, we can think of a limit as a means of describing a function's behavior at a point where the function cannot be evaluated.

• EXAMPLE 1. The function

$$f(x) = \frac{\sin x}{x}$$

is undefined at $x = 0$, but defined at all other x. The question arises, then, as to how the function behaves for x nearby, but not equal to, 0. More precisely, what do the values of the function do as x *approaches* 0? Let's investigate this by making a table of values and then by plotting the graph.

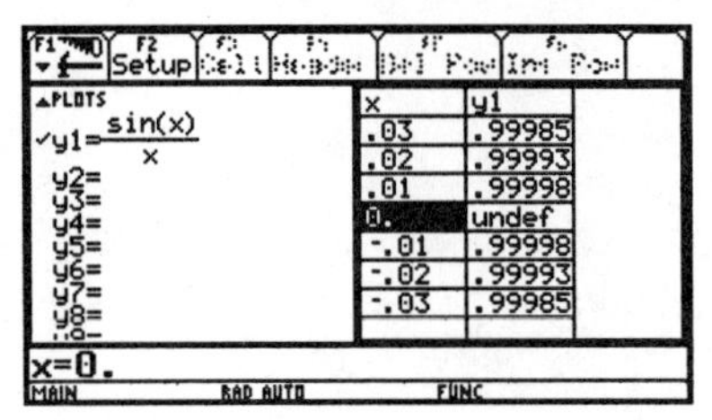

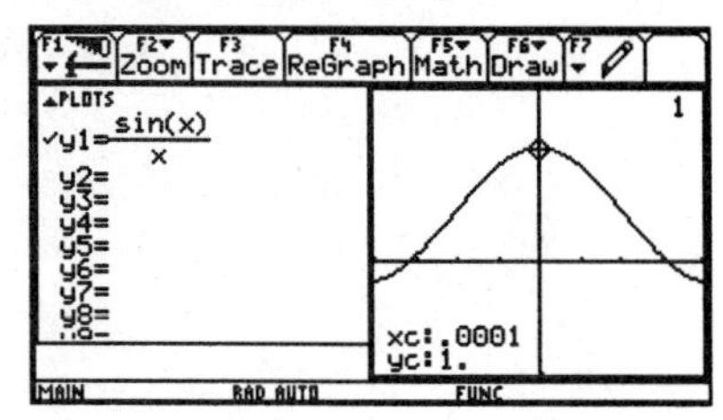

So we see that even though $f(0)$ is undefined, values of x near 0 produce values of $f(x)$ near 1. Moreover, it appears that the closer x is to 0, the closer $f(x)$ is to 1. We describe this by saying that 1 is the *limit of* $f(x)$ *as* x *approaches* 0 and express it mathematically by writing

$$\lim_{x \to 0} \frac{\sin x}{x} = 1.$$

We can also use the **TI-89/92**'s **limit()** function to calculate this limit.

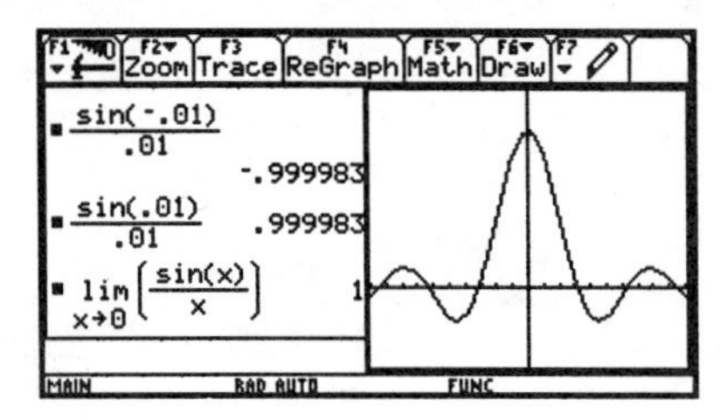

• Example 2. Consider the function

$$f(x) = \frac{(x^3-1)^{4/3}}{x-1}.$$

This function is undefined at $x = 1$ but defined elsewhere. Let's investigate the behavior of this function near $x = 1$.

```
F1 F2 Algebra F3 Calc F4 Other F5 PrgmIO F6 Clear a-z...
■ (x^3 - 1)^(4/3) / (x - 1) → f(x)      Done
■ f(.99)                               -.919803
■ f(.999)                              -.432098
■ f(1)                                    undef
■ f(1.001)                              .433252
■ f(1.01)                               .944661
f(1.01)
MAIN        RAD AUTO        FUNC 6/30
```

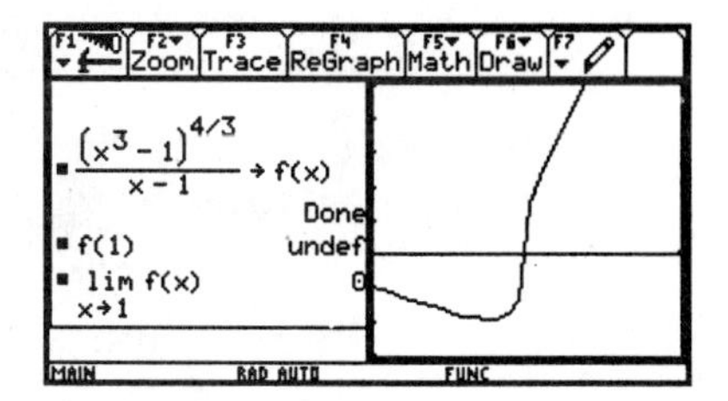

Thus we see that

$$\lim_{x \to 1} f(x) = 0.$$

Moreover, the graph indicates that x must be extremely close to 1 in order for $f(x)$ to be at all close to 0.

It is important to note here that we have have not *proved* that either of our stated limits in the previous examples is correct. Nor have we even given a meaningful definition of we mean by the limit of a function f as x approaches a number a—we have merely relied on an intuitive notion. Section 2.2 of Stewart's *Calculus* gives a correct, though informal, definition (as well as similar definitions for one-sided limits). For the precise technical definition, see Section 2.4 of Stewart's *Calculus*.

One-sided limits. It is not uncommon for a limit to fail to exist. Sometimes when a limit does not exist, one-sided (i.e., left- or right-sided) limits do exist. The **TI-89/92**'s **limit()** function finds left- or right-sided limits, when -1 or 1, respectively, is inserted as a fourth argument.

• Example 3. Consider the function

$$f(x) = \frac{2x + |x|}{|x|} \cos x,$$

which is undefined at $x = 0$. To compute the left-sided limit as x approaches zero, we'll enter

limit((2x+abs(x))/abs(x)*cos(x),x,-1) .

For the right-sided limit we'll enter the same except with 1 as the last argument.

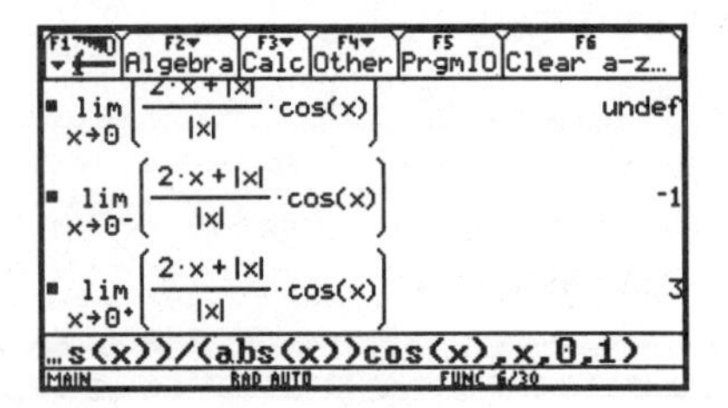

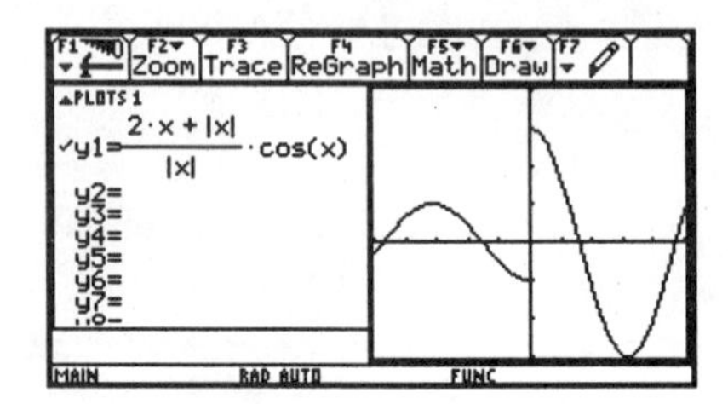

This function has what is sometimes called a "jump discontinuity" at $x = 0$ by virtue of the fact that both one-sided limits exist and have different values.

Infinite limits. Sometimes one-sided limits fail to exist because the function in question has a vertical asymptote. In such cases, one-sided limits can be assigned a value of either ∞ or $-\infty$ to described the manner in which the graph of the function approaches the vertical asymptote.

• EXAMPLE 4. Each of the functions

$$f(x) = \frac{1}{x-1} \quad \text{and} \quad g(x) = \frac{1}{(x-1)^2},$$

has a vertical asymptote at $x = 1$. One-sided limits decribe the behavior of each function at the vertical asymptote.

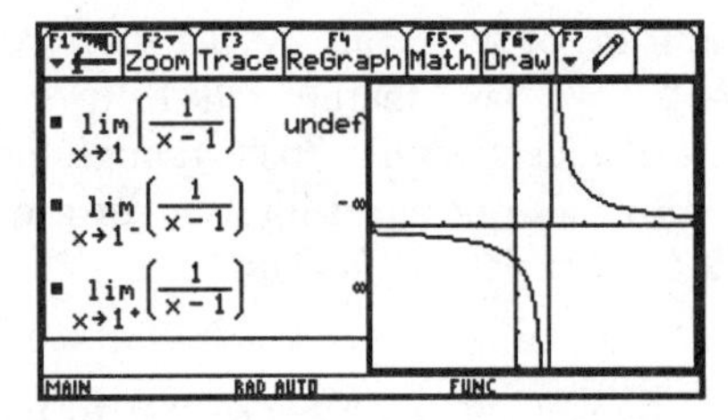

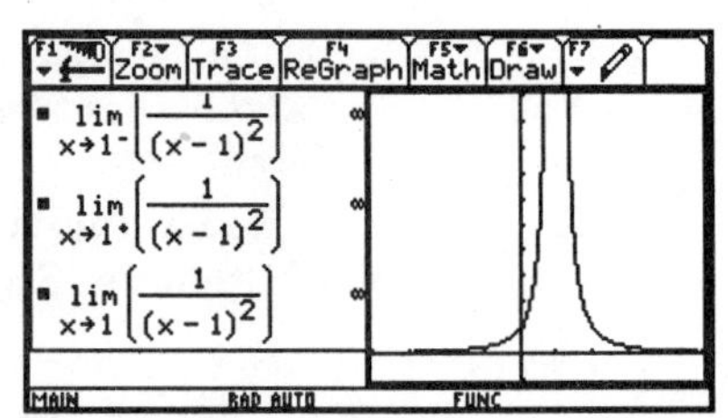

Limits at $\pm\infty$**.** Just as infinite limits occur at vertical asymptotes of a function, limits at $\pm\infty$ determine any horizontal asymptote that a function might have. The limit of $f(x)$ as $x \to \infty$ is a number to which $f(x)$ approaches as x increases without bound. The limit of $f(x)$ as $x \to -\infty$ is a number to which $f(x)$ approaches as x decreases without bound.

• EXAMPLE 5. The function $f(x) = \tan^{-1} x$ has distinct limits at $\pm\infty$.

x	y1
400.	1.5683
500.	1.5688
600.	1.56913
700.	1.56937
800.	1.56955
900.	1.56969
1000.	1.5698

limit(tan⁻¹(x),x,∞)

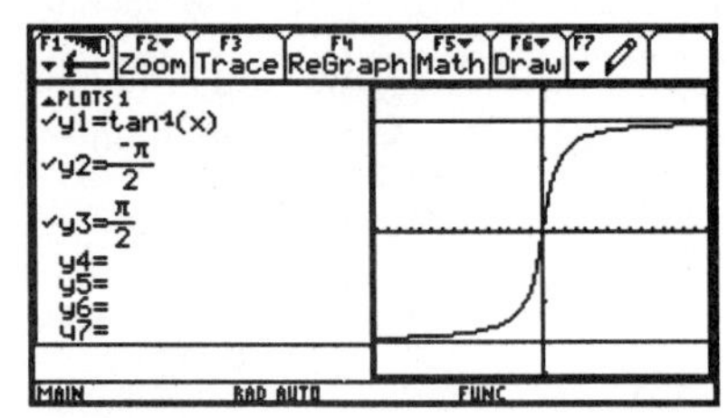

• EXAMPLE 6. The function $f(x) = e^{-x/10}\cos x^2$ has a limit at ∞ but no limit at $-\infty$.

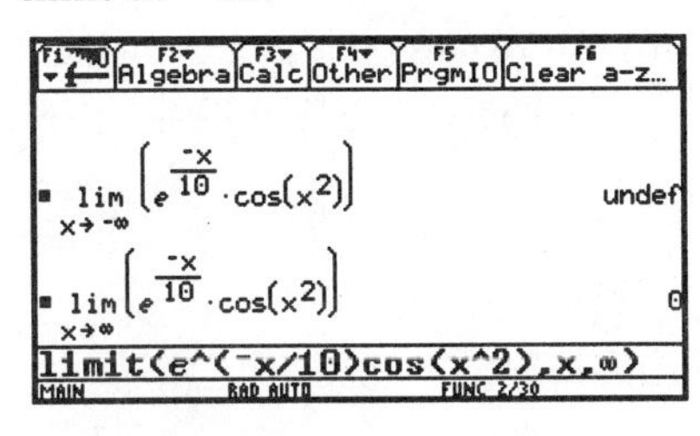

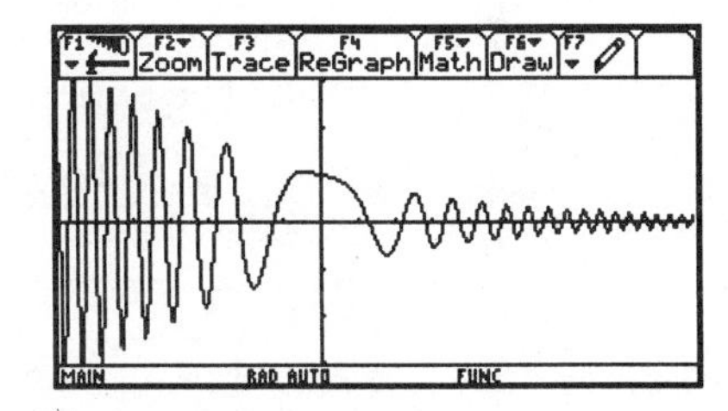

• EXAMPLE 7. The rational function $f(x) = \frac{2x^2+5x-1}{x^2+1}$ has the same limit at both $\pm\infty$.

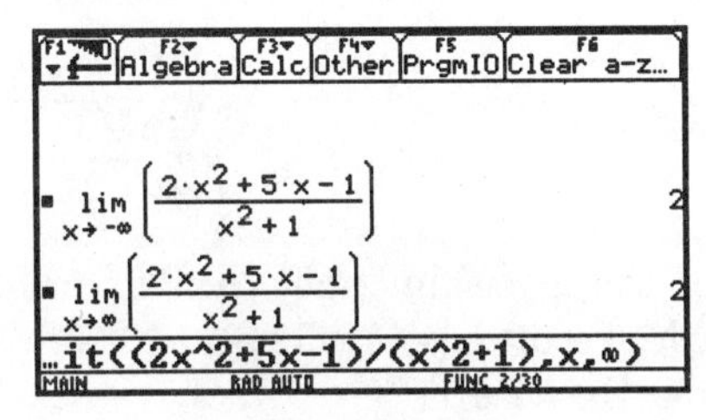

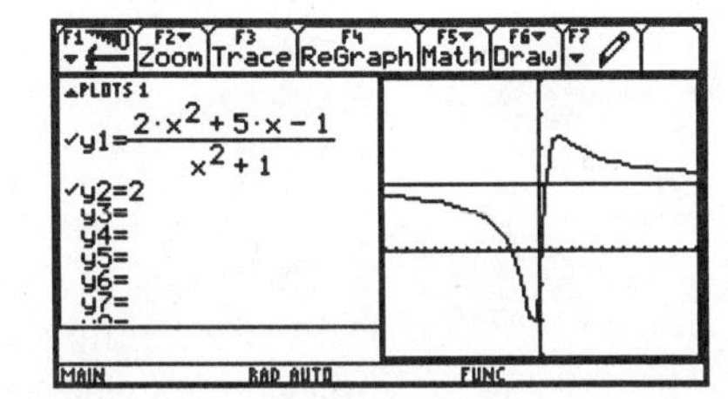

Limits of expressions with symbolic parameters. As we will see in the next section, limits such as

$$\lim_{h\to 0}\frac{\sqrt{4+h}-2}{h}$$

are very important in calculus. This particular limit is easily found to be $\frac{1}{4}$. Notice, however, that this limit is just the value at $x = 4$ of the function f defined by

$$f(x) = \lim_{h\to 0}\frac{\sqrt{x+h}-\sqrt{x}}{h}.$$

Let's see what happens if we try to evaluate f symbolically at x. First we define the quotient $\left(\sqrt{x+h}-\sqrt{x}\right)/h$ as a function $q(h)$ and then find the limit of $q(h)$ as $h \to 0$.

```
F1  F2▾ Algebra  F3▾ Calc  F4▾ Other  F5 PrgmIO  F6 Clear a-z...

■ (√(x+h) - √x)/h → q(h)                Done
■ lim q(h)                              1/(2·√x)
  h→0
limit(q(h),h,0)
MAIN        RAD AUTO        FUNC 2/30
```

```
F1  F2▾ Algebra  F3▾ Calc  F4▾ Other  F5 PrgmIO  F6 Clear a-z...
■ (√(x+h) - √x)/h → q(h)                Done
■ lim q(h)                              1/(2·√x)
  h→0
■ 1/(2·√x) → f(x)                       Done
■ f(4)                                  1/4
f(4)
MAIN        RAD AUTO        FUNC 4/30
```

Note that once the limit is computed in terms of x, we can evaluate $f(x)$ from the resulting algebraic expression.

Exercises

In Exercises 1–10, estimate the requested limit by using **Trace (F3)** on the graph. Then use the **TI-89/92**'s **limit()** function to find the limit.

1. $\lim_{x\to 1} \dfrac{x^3+2x-3}{x-1}$

2. $\lim_{x\to 1} \dfrac{\sqrt{2x^3-x^2-4x+3}}{x-1}$

3. $\lim_{x\to 0} x\sin(1/x)$

4. $\lim_{x\to 0} \dfrac{\sin(e^{2x}-1)}{x}$

5. $\lim_{x\to 0} \dfrac{\sin 5x}{x^2}$

6. $\lim_{x\to 0^-} \dfrac{|\sin x|}{x}$

7. $\lim_{x\to 0^+} \dfrac{|x|-x^2}{x}$

8. $\lim_{x\to 1^-} \dfrac{x^3+2x-4}{x-1}$

9. $\lim_{x\to 0^+} \dfrac{x^3+2x-3}{x}$

10. $\lim_{x\to \infty} \dfrac{\sqrt{3x^2+1}}{2x-5}$

In Exercises 11–13, plot the given function; then find both the left- and right-sided limits at each specified point. The **TI-89/92** definition for the function is given to you in Exercise 11.

11. $f(x) = \begin{cases} -x-2, & \text{if } x < -1 \\ -1, & \text{if } -1 \le x \le 0 \\ \sin 2\pi x, & \text{if } 0 < x < 1 \\ x, & \text{if } 1 \le x \end{cases}$ at $x = -1,\ 0,\ 1;$

$=$ **when(x<-1, -x–2, when(x<=0, -1, when(x<1, sin(2πx), x)))**

12. $f(x) = \begin{cases} 2, & \text{if } x < 0 \\ 1, & \text{if } 0 \le x < 2 \\ 3, & \text{if } 2 \le x \end{cases}$

at $x = 0,\ 2$

13. $f(x) = \begin{cases} 1, & \text{if } x < -1 \\ |x|, & \text{if } -1 \le x \le 1 \\ 1, & \text{if } 1 < x \end{cases}$

at $x = -1,\ 1$

14. Find $\lim_{h\to 0} \dfrac{(x+h)e^{x+h}-xe^x}{h}$.

15. Find $\lim_{x\to 1} \dfrac{(ax-1)^2-(a-1)^2}{x-1}$.

3.2 The Derivative

The fundamental geometric problem with which we are concerned is that of finding the slope of the graph of a function f at a given point $(a, f(a))$ on that graph. This slope is defined as the slope of the *tangent line* to the graph at the given point.

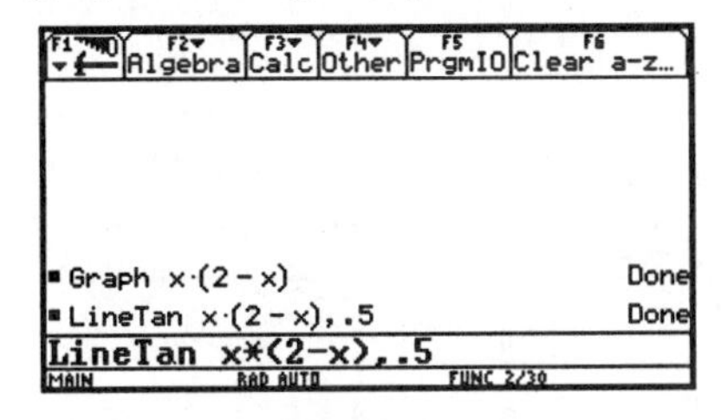

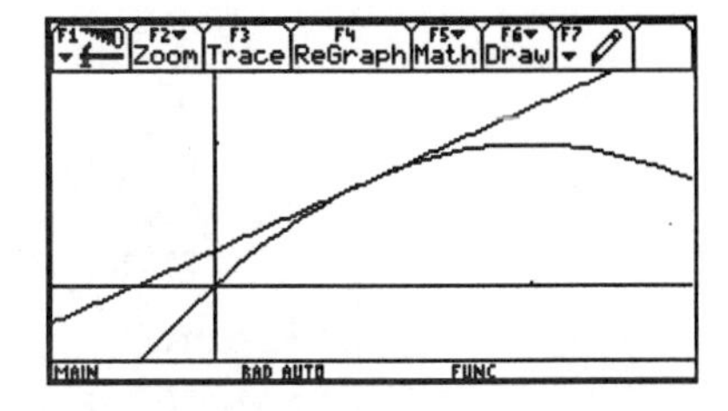

As illustrated above, tangent lines can be plotted from the Home screen with the **LineTan** command. The **Math** menu on the graph screen also has a **Tangent** command (**F5-A**). After selecting **Tangent** from the **Math** menu, you can either type in the desired x-coordinate or use the cursor pad to trace along the curve to the desired point.

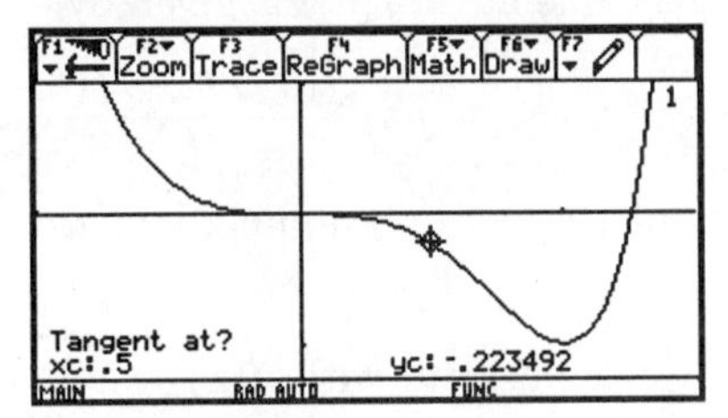

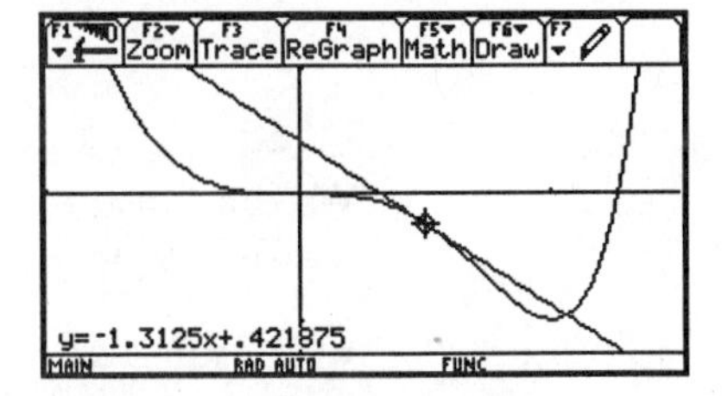

Pressing **ReGraph** (**F4**) will erase the tangent line(s) you've drawn.

Slopes of lines are easy to compute by means of the slope formula

$$m = \frac{y_2 - y_1}{x_2 - x_1},$$

provided we know two points on the line. The difficulty that arises in finding the slope of the tangent line to a graph is that only one point on it is known, namely $(a, f(a))$. So we take an approach that is the essence of calculus: *We approximate the quantity and take the limit as we refine our approximations.*

The slope of the tangent line at $(a, f(a))$ can be approximated by the slope of the *secant line* through $(a, f(a))$ and a nearby point $(a+h, f(a+h))$ on the graph, where h is some small number. The slope of such a secant line may be viewed as a function of h and is easily calculated as

$$m_{PQ_h} = \frac{f(a+h) - f(a)}{h}.$$

The equation of the secant line is then

$$y = f(a) + m_{PQ_h}(x - a).$$

• Example 1a. Let's graph $f(x) = x(2-x)$ along with the secant line through $(.5, f(.5))$ and $(1, f(1))$. We'll first define functions **m(h)** and **secline(x)** by entering

(f(a+h)–f(a))/h →m(h)

and then

f(a) + m(h)*(x–a) →secline(x) .

Then we'll enter **x*(2–x) →f(x)**, **.5 →a** and **.5 →h**. Finally after entering **y1=f(x)** and **y2=secline(x)** in the **Y=** Editor, we plot the graph.

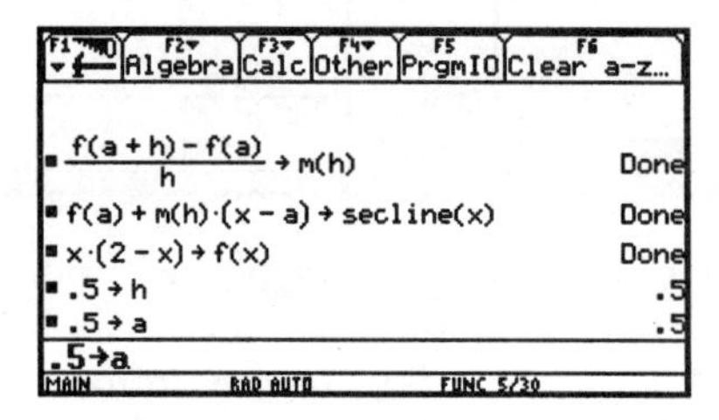

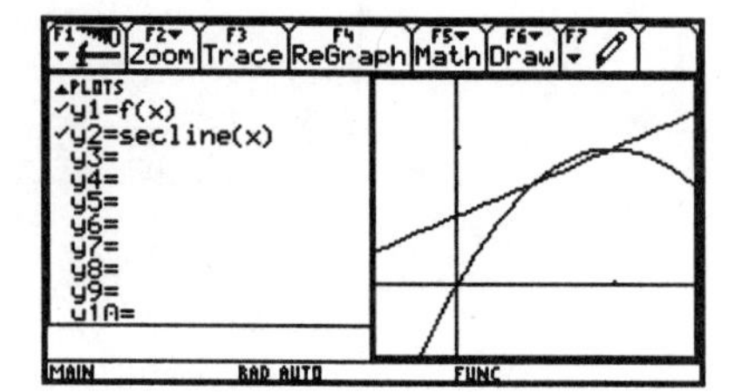

The slope of the tangent line at $(a, f(a))$ is given by

$$m_{\tan} = \lim_{h \to 0} m_{PQ_h} = \lim_{h \to 0} \frac{f(a+h) - f(a)}{h} .$$

We can illustrate this limit process by graphing secant lines for a few decreasing values of h.

• Example 1b. Again we'll illustrate with $f(x) = x(2-x)$ and $a = .5$, plotting secant lines corresponding to $h = .5$, $.25$, and $.05$.

```
f(a + h) - f(a) / h → m(h)              Done
f(a) + m(.5)·(x - a) → secline1(x)      Done
f(a) + m(.25)·(x - a) → secline2(x)     Done
f(a) + m(.05)·(x - a) → secline3(x)     Done
x·(2 - x) → f(x)                        Done
.5 → a                                    .5
.5→a
MAIN      RAD AUTO      FUNC 11/30
```

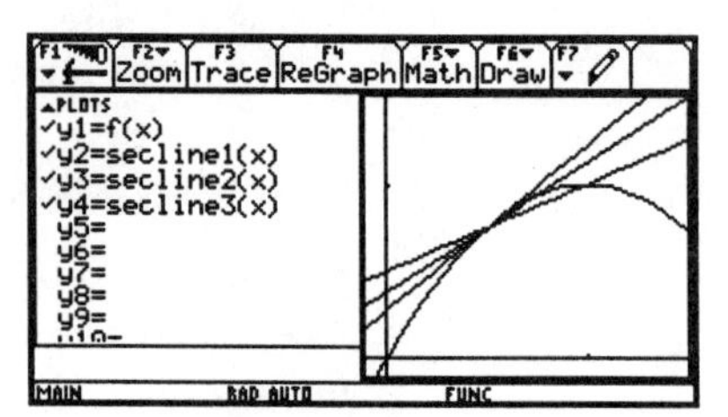

This development motivates the definition of a function f', the *derivative* of f. The definition is

$$f'(x) = \lim_{h \to 0} \frac{f(x+h) - f(x)}{h}$$

for all x at which the defining limit exists. Note that values of f' give tangent line slopes on the graph of $y = f(x)$; that is,

$$f'(a) = m_{\tan} \text{ at } (a, f(a)).$$

• EXAMPLE 2. Let $f(x) = x^3 - x$. Then

$$f'(x) = \lim_{h \to 0} \frac{\left((x+h)^3 - (x+h)\right) - (x^3 - x)}{h}.$$

One quirk of the **TI-89/92** is that if we define $f(x)$, then entering $f(x+h)$ prompts an error message. However, we can circumvent this problem if we define f in terms of a different symbol, say t.

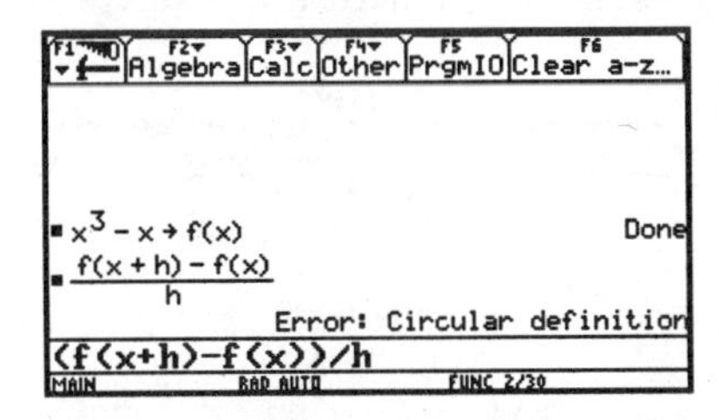


```
F1 F2 Algebra F3 Calc F4 Other F5 PrgmIO F6 Clear a-z...
Error: Circular definition
■ t^3 - t → f(t)                     Done
■ (f(x + h) - f(x))/h    3·x^2 + 3·h·x + h^2 - 1
■ lim (3·x^2 + 3·h·x + h^2 - 1)     3·x^2 - 1
  h→0
■ 3·x^2 - 1 → fprime(x)              Done
ans(1)→fprime(x)
MAIN      RAD AUTO      FUNC 6/30
```

We have found that $f'(x) = 3x^2 - 1$ is the derivative of $f(x) = x^3 - x$. Now the slope of the tangent line to the graph of f at $(1, 0)$, for example, is $f'(1) = 2$, and so the equation of the tangent line there is $y = 2(x - 1)$.

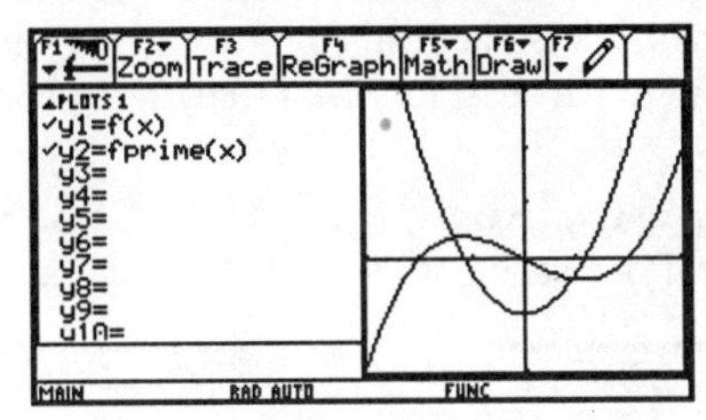

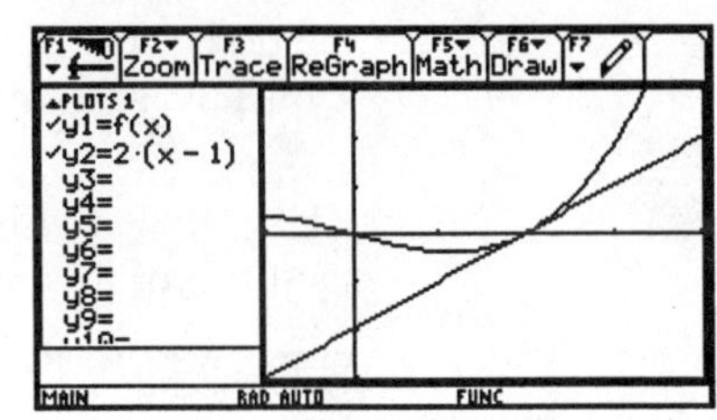

Numerical approximations of derivative values are often useful. A reasonably good approximation to $f'(a)$ can typically be obtained by computing $m_{PQ_h} = \left(f(a+h) - f(a)\right)/h$ for some small value of h, say, $h = .001$. However, a better approximation is obtained from the *central difference formula*

$$\frac{f(a+h) - f(a-h)}{2h}$$

with no additional computational effort. This is in fact what is done by the **nDeriv()** function, found in the **Calc** menu (**F3-A** from the Home screen). The default value of h used by **nDeriv()** is .001. (h may also be specified as a third argument.)

```
F1 F2 Algebra F3 Calc F4 Other F5 PrgmIO F6 Clear a-z...
■ nDeriv(sin(x), x)
        500.·(sin(x + .001) - sin(x - .001))
■ nDeriv(sin(x), x) | x = 0              1.
■ nDeriv(sin(x), x) | x = π/4       .707107
■ nDeriv(sin(x), x) | x = π/3            .5
nDeriv(sin(x),x)|x=π/3
MAIN      RAD AUTO      FUNC 4/30
```

```
F1 F2 Algebra F3 Calc F4 Other F5 PrgmIO F6 Clear a-z...
■ nDeriv(x^3, x)        3.·(x^2 + 3.33333E-7)
■ nDeriv(x^3, x) | x = 0              1.E-6
■ nDeriv(x^3, x) | x = .25          .187501
■ nDeriv(x^3, x) | x = .5           .750001
nDeriv(x^3,x)|x=.5
MAIN      RAD AUTO      FUNC 4/30
```

Also note that **nDeriv()** returns the "numerical derivative" in the form of an expression.

It is also interesting to plot the result of **nDeriv()** together with the original function to which it was applied.

• EXAMPLE 3. Let's plot the graph of $y = \sin x$ together with its numerical derivative. (Does the graph of the numerical derivative look like a familiar function?)

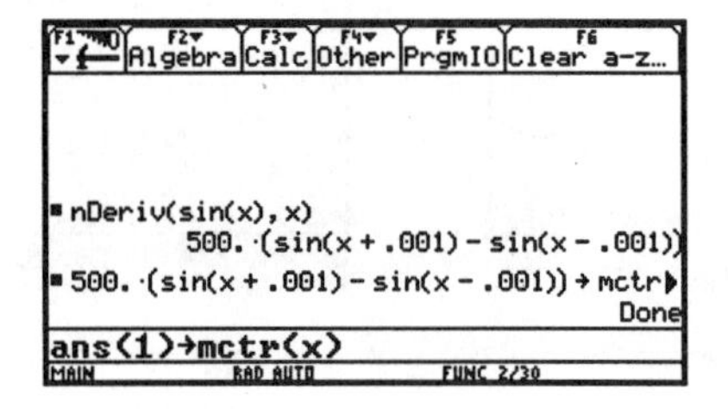

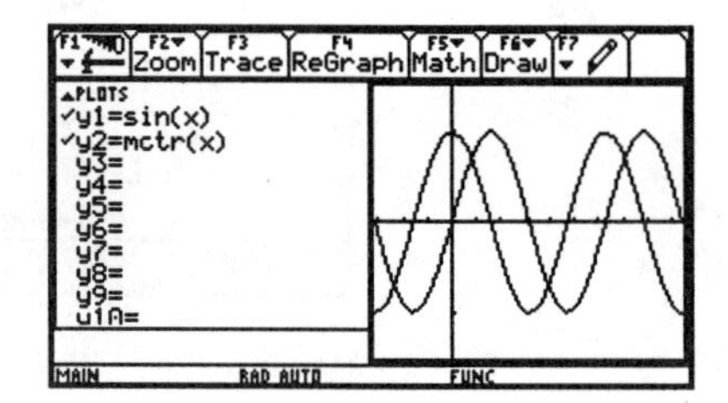

Remember, though, that **nDeriv()** produces only an approximation to the derivative. The derivative is the limit of such approximations as $h \to 0$. Typically, such a limit is quite easy for the **TI-89/92** to compute.

• EXAMPLE 4. Find the derivative of $f(x) = x^3 - x^2 - 3x + 5$ by computing the limit of its numerical derivative as $h \to 0$. Plot the graphs of f and f'.

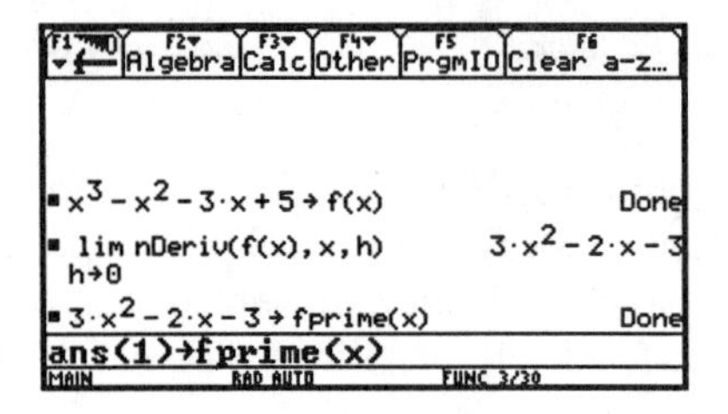

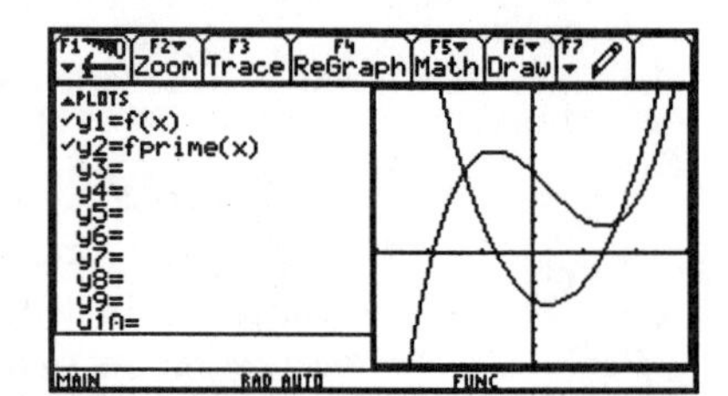

Derivative values (i.e., tangent line slopes) can also be found on the Graph screen by selecting **Derivatives** from the **Math** menu (**F5-6**). After selecting **Derivatives** from the **Math** menu, you can either type in the desired x-coordinate or use the cursor pad to trace along the curve to the desired point.

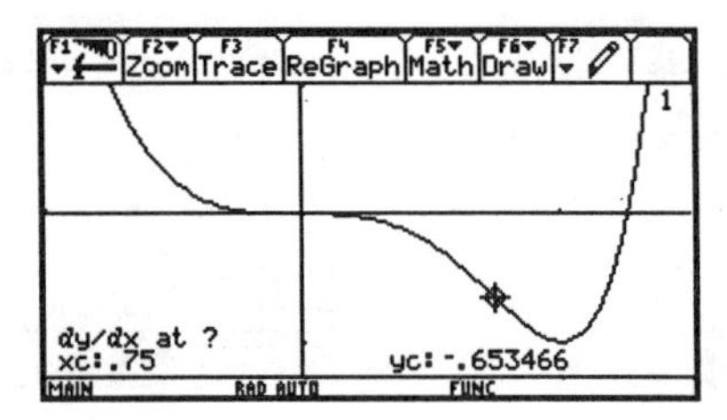

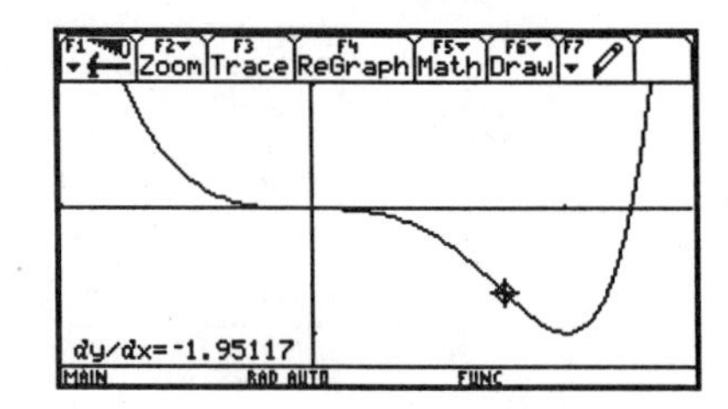

Exercises

1. Plot the function $f(x) = x^2 - 1$ in the window $[-2, 2] \times [-2, 3]$. Then plot its tangent lines at $x = -1,\ 0,$ and 1.

2. Plot the function $f(x) = \sin x$ in the window $[-1, 7] \times [-1.5, 1.5]$. Then plot its tangent lines at $x = 0,\ \frac{\pi}{2},\ \pi,\ \frac{3\pi}{2}$, and 2π.

3. Plot the function $f(x) = \sin\left(\frac{\pi}{2}x\right)$ in the window $[0, 4] \times [-1.5, 1.5]$ along with secant lines through $(1, 1)$ and each of the points $(8/3, -1/2)$, $(2, 0)$, $(5/3, 1/2)$, and $(4/3, \sqrt{3}/2)$.

In Exercises 4–7, compute for the given function f its difference quotient

$$m_{PQ_h} = \frac{f(x+h) - f(x)}{h}.$$

Then compute the limit of m_{PQ_h} as $h \to 0$ to obtain the derivative of f.

4. $f(x) = 3x^2 - x + 2$

5. $f(x) = e^{-x}$

6. $f(x) = 1/x$

7. $f(x) = \sin^2 x$

In Exercises 8–10, for the given function f compare the values of:

a) the difference quotient

$$m_{PQ.1} = \frac{f(a + .1) - f(a)}{.1}$$

at the prescribed point $x = a$

b) **nDeriv(f(x),x,.1)** at $x = a$

c) $f'(a) =$ **limit(nDeriv(f(x),x,h)|x=a, h, 0)**

8. $f(x) = 3x^2 - x + 2,\ a = 1$

9. $f(x) = e^x,\ a = 1$

10. $f(x) = \sin x,\ a = \pi/3$

11. Plot the graph of $f(x) = x^3/3$ along with the graph of **nDeriv(f(x),x)**.

12. Plot the graph of $f(x) = x^3 - x^2 - x$ along with the graph of **nDeriv(f(x), x)**. Use **Trace** to estimate the values of x where **nDeriv(f(x), x)** $= 0$. What is happening with the graph of f at these values of x?

13. For each of the following functions, plot the function along with **nDeriv(f(x),x)** on the indicated interval. Then, from what you observe in the graph, hazard a guess as to exactly what $f'(x)$ is.

    a) $f(x) = \sin x,\quad 0 \le x \le 2\pi$

b) $f(x) = \sin 2x, \quad 0 \leq x \leq 2\pi$

c) $f(x) = \cos x, \quad 0 \leq x \leq 2\pi$

d) $f(x) = e^{-x}, \quad -2 \leq x \leq 2$

14. Find the value of x in the interval $[0, 2]$ where the tangent line to the graph of $f(x) = x^3 - x^2 - x$ is parallel to the secant line through $(0, 0)$ and $(2, 2)$. Plot the graph of f and both lines in the window $[0, 2] \times [-2, 2]$.

3.3 Higher-order derivatives

The **TI-89/92** has a built-in operator for symbolic computation of derivatives. It is the ***d*()** operator located both in the **Calc** menu (**F3-1**) and on the keyboard as **[2nd]-[8]**. Note that this is *not* the same as the letter "**d**." The syntax for using ***d*()** to compute the first derivative of an expression f with respect to the variable *var* is

$$d(f,\ var).$$

The ***d*()** operator is very convenient for computing messy derivatives. To illustrate, let's use ***d*()** to compute the derivatives of

$$f(x) = \frac{x^2\sqrt{x^3 - 2}}{x^2 - 3} \quad \text{and} \quad f(x) = \frac{\sin(3x)}{x^2 + 1}.$$

Notice the alternative "Leibniz" notation $\frac{d}{dx}[f(x)]$ for $f'(x)$.

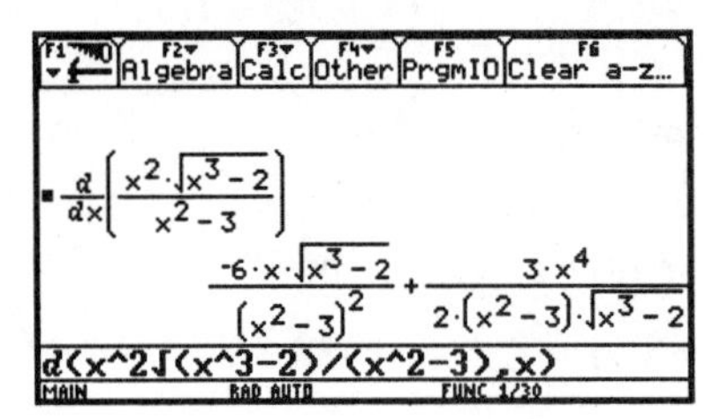

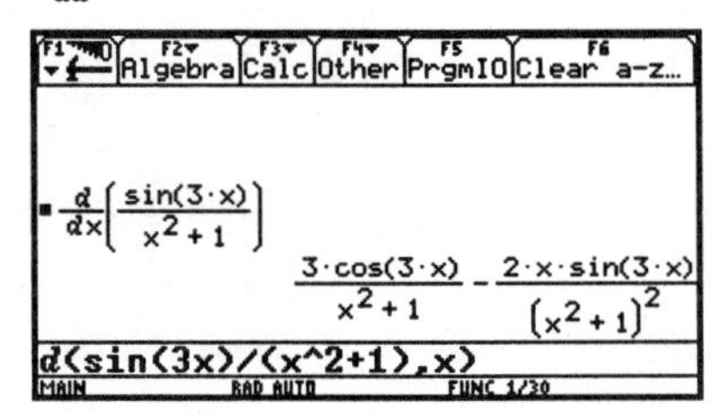

The ***d*()** operator is especially handy for computing higher order derivatives. The second derivative of a function f is, by definition, the derivative of f' and is denoted by f''. The third derivative of a function f is, by definition, the derivative of f'' and is denoted by f''', and so on. In Leibniz notation, higher order derivatives are denoted by

$$\frac{d^2}{dx^2}\Big[f(x)\Big] = f''(x), \quad \frac{d^3}{dx^3}\Big[f(x)\Big] = f'''(x), \text{ and so on.}$$

The syntax for using ***d*()** to compute the nth derivative of an expression f with respect to the variable *var* is

$$d(f,\ var,\ n).$$

The following shows **d()** used to find $\frac{d^2}{dx^2}[x^3]$ and the first four derivatives of $x \cos x$.

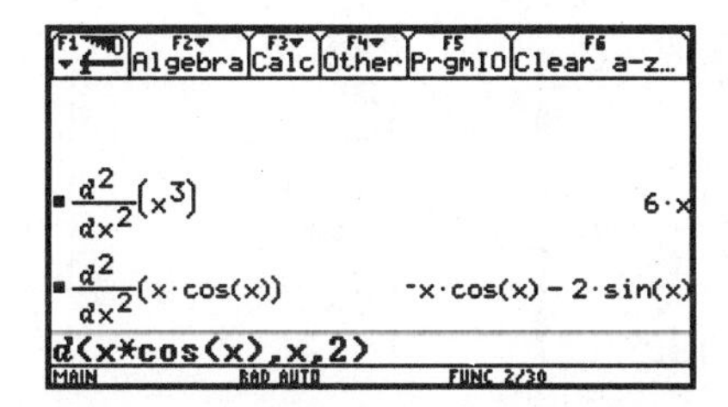

The second-screen output (not detected as a separate image):

The second derivative and concavity. The first and second derivatives of a function f have simple interpretations in terms of the graph of f. The sign of $f'(x)$ determines where the function f is increasing or decreasing. The sign of $f''(x)$ determines where the graph of f is concave up or concave down. This is illustrated as follows with the function $f(x) = x^4 - 2x^2$. First we'll graph f together with f'.

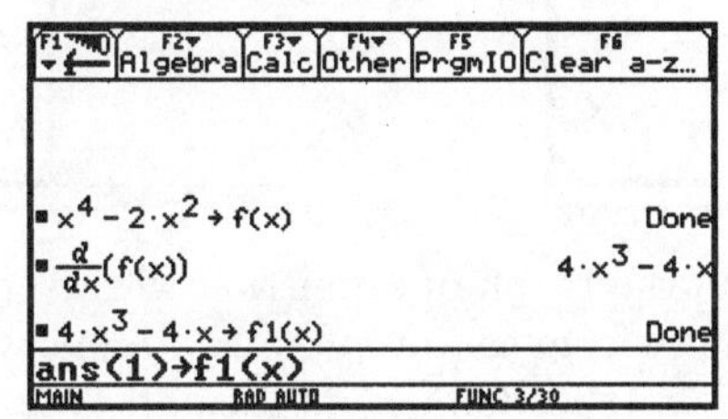

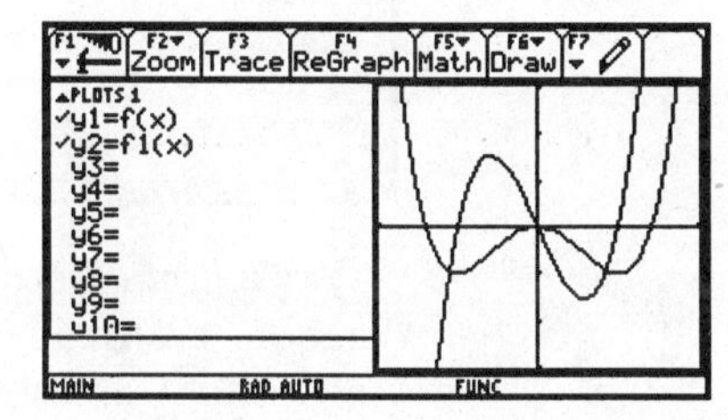

Notice that $f'(x) > 0$ wherever $f(x)$ is increasing, $f'(x) < 0$ wherever $f(x)$ is decreasing, and $f'(x) = 0$ where each local maximum or minimum value of f occurs. These facts are not at all surprising in light of the fact that f' gives the slope of the graph of f. Now let's look at the graph of f together with the graph of f''.

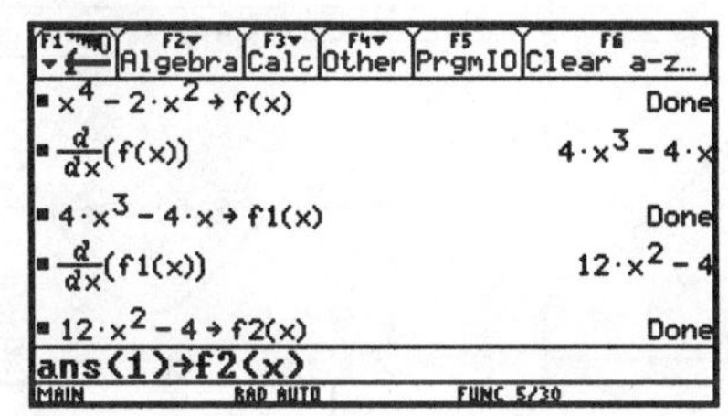

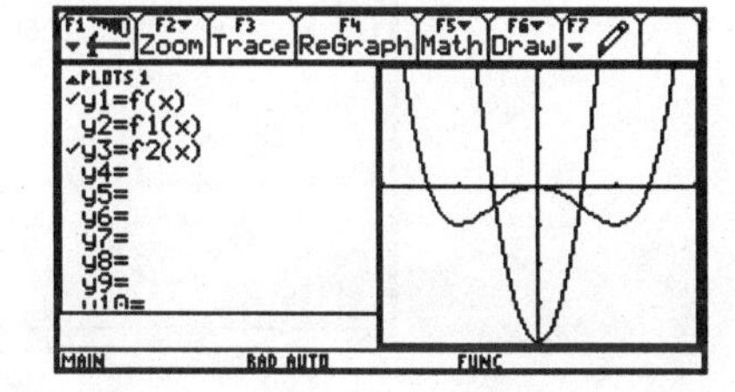

Notice that $f''(x) > 0$ wherever the graph of f is concave up, $f''(x) < 0$ wherever the graph of f is concave down. The reason for this is that the concavity of the graph of f is determined by whether the slope $f'(x)$ is increasing or decreasing. (See Exercise 7.)

It is also easy to perform such investigations using only the **Y=** Editor (although the plotting of derivatives can be considerably slower). Let's look, for example, at $f(x) = x^5 - 2x^3$. We'll enter this function as **y1**, **d(y1(x),x)** as **y2**, and then ***d*(y1(x),x,2)** as **y3**. (Note that in the second picture only **y1** and **y3** are plotted, since **y2** is "unchecked" (**F4**).)

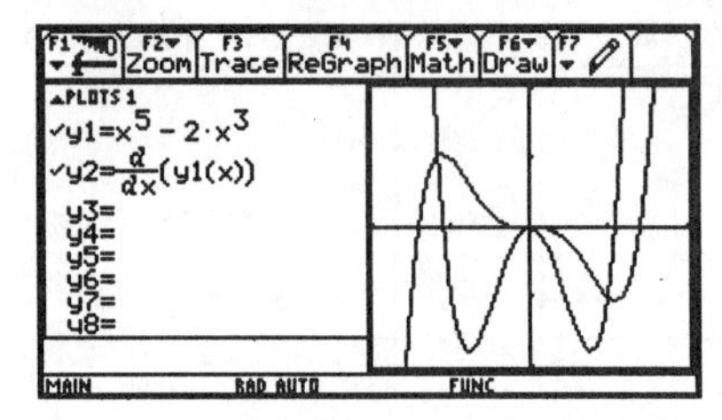

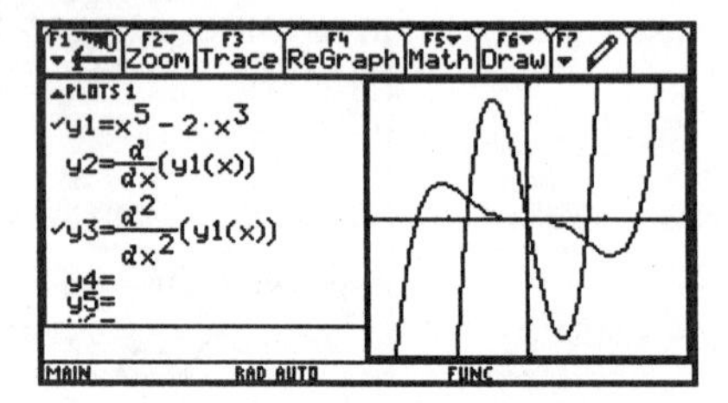

The graph of $f(x) = x^{2/5}(2-x)^{1/3}$ has a *cusp* at $(0,0)$ and a vertical tangent at $(2,0)$. Such features are related to the fact that the graph of f' has vertical asymptote at $x = 0$ and $x = 2$.

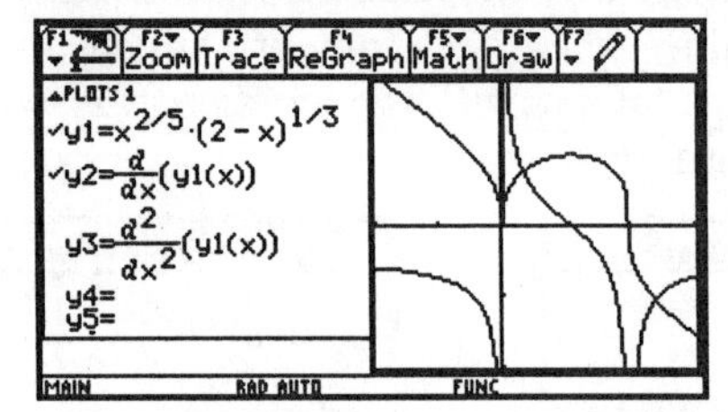

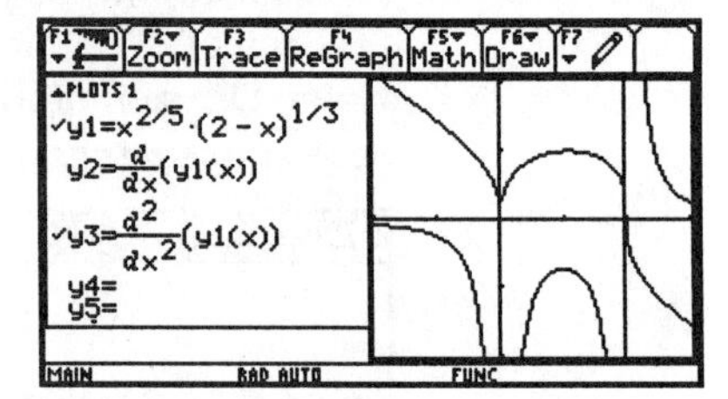

An *inflection* occurs on the graph of a function wherever the graph undergoes a change in concavity from concave upward to concave downward or vice-versa. A point where an inflection occurs (i.e., an *inflection point*) can be located with the **Inflection** command in the Graph screen **Math** menu (**F5-8**). Let's look at the example of $f(x) = \sin x \cos 3x$ on the interval $0 \leq x \leq \pi$. We'll plot the function and then select **Inflection** by pressing **F5-8**. (We are prompted for lower and upper bounds on the x-coordinate of the desired inflection point.)

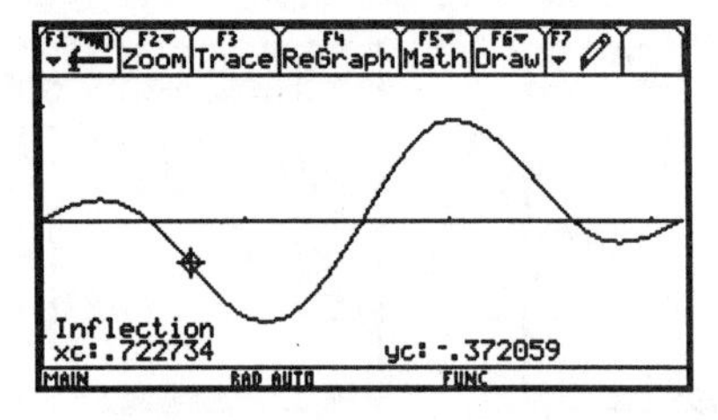

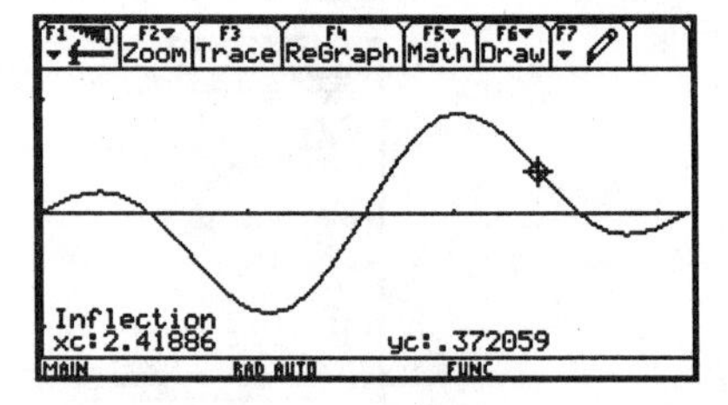

(This graph obviously also has a third point of inflection at $(\pi/2, 0)$.

For more details and examples concerning the effect of derivatives on the shape of the graph of a function, see Sections 4.3–4.6 in Stewart's *Calculus*.

Exercises

1. Plot the function $f(x) = x(1+x^6)^{-1/2}$ along with its first and second derivatives in a $[-2,2] \times [-4,4]$ window using each of the following methods. Notice and explain the difference in speed.

   a) In the **Y=** Editor, define **y1=x/√(1+x^6)**, **y2=*d*(y1(x),x)**, and **y3=*d*(y2(x),x)**. Set window variables and press **◇GRAPH**.

   b) In the **Y=** Editor, define **y1=x/√(1+x^6)**, **y2=dy1(x)**, and **y3=dy2(x)**. From the Home screen, enter ***d*(y1(x),x)**, followed by **ans(1) →dy1(x)**, and enter ***d*(dy1(x),x)**, followed by **ans(1) →dy2(x)**. Set window variables and press **◇GRAPH**.

For the functions in Exercises 2–6, first plot the function together with its derivative and note the correspondence between the sign of the derivative and the increasing/decreasing nature of $f(x)$. Then plot the function together with its second derivative and note the correspondence between the sign of the second derivative and the concavity of the graph of f.

2. $f(x) = \dfrac{1}{x^2+1}$
3. $f(x) = \tan^{-1} x$
4. $f(x) = \dfrac{x}{x^2+1}$
5. $f(x) = \sqrt{x} - x$
6. $f(x) = (x+1)^{2/3}x^{3/5}$

7. For each of the following functions, graph the function in the window $[-2, 2] \times [-2, 2]$. Then use either the **LineTan** command from the Home screen or the **Tangent** item from the Graph screen **Math** menu (**F5-A**) to graph the tangent lines at $x = -1$, 0, and -1. Does the slope of the tangent line increase or decrease? State whether the graph is increasing and concave up, decreasing and concave up, increasing and concave down, or decreasing and concave down.

   a) $f(x) = e^{-2x/3} - 2$
   b) $f(x) = e^{2x/3} - 2$
   c) $f(x) = 2 - e^{2x/3}$
   d) $f(x) = 2 - e^{-2x/3}$

For the functions in 8 and 9, graph the function along with its second derivative. Use **Inflection** from the Graph screen **Math** menu to locate the inflection points on the graph of f. Then use **Zero** from the Graph screen **Math** menu to locate the zeros of f''.

8. $f(x) = x^5 - 2x^3$
9. $f(x) = x^{5/3}(2-x)^2$

For the functions in 10–13, compute $f^{(n)}(0)$ for $n = 1, 2, 3, 4, 5$. ($f^{(n)}(0)$ denotes the nth derivative of f.) By observing a pattern in these numbers, express $f^{(n)}(0)$ in terms of n.

10. $f(x) = xe^{-x}$
11. $f(x) = x\cos x$
12. $f(x) = \dfrac{1}{1+x}$
13. $f(x) = \sqrt{x+1}$

3.4 Programming notes

Secant and tangent lines. The first program that we'll present here, **secgrph()**, will graph a given function together with secant lines through pairs of points in a specified list. The arguments of **secgrph()** are as in

secgrph(f, *xvar*, *xpairs***)**,

where f is an expression in the variable *xvar*, and *xpairs* is a list containing pairs of values of *xvar*.

```
: secgrph(f, xvar, xpairs)
: Prgm: Local i, n, m, dy, a, b
: ClrGraph: ClrDraw: FnOff: PlotsOff
: string(f | xvar=xx) →ff : Graph expr(ff), xx
: dim(xpairs)[1] →n
: For i, 1, n
:    xpairs[i,1] →a: xpairs[i,2] →b
:    (f | xvar=b)–(f | xvar=a) →dy
:    dy/(b–a) →m
:    DrawSlp a, f | xvar=a, m
: EndFor
: DelVar ff
: EndPrgm
```

The following shows several secant lines drawn by **secgrph** on the graph of $y = x^2$ in a $[-2, 2] \times [-1, 4]$ window.

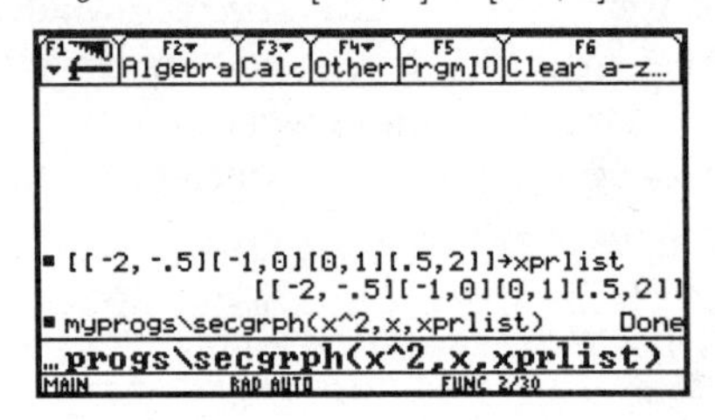

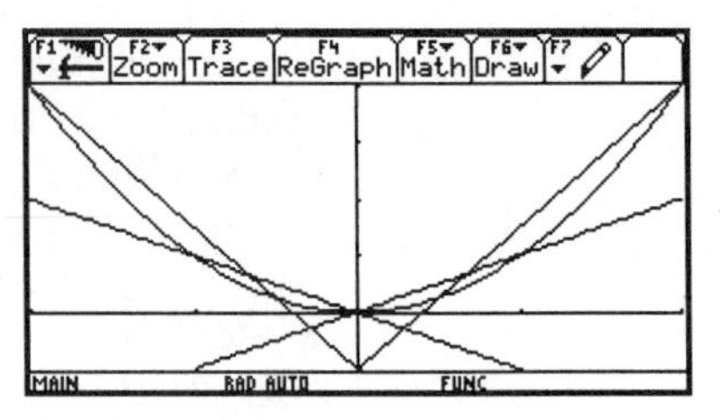

Next we see several secant lines drawn on the graph of $y = x^3 - 2x$ in a $[-2, 2] \times [-2, 2]$ window. All of these secant lines pass through $(0, 0)$.

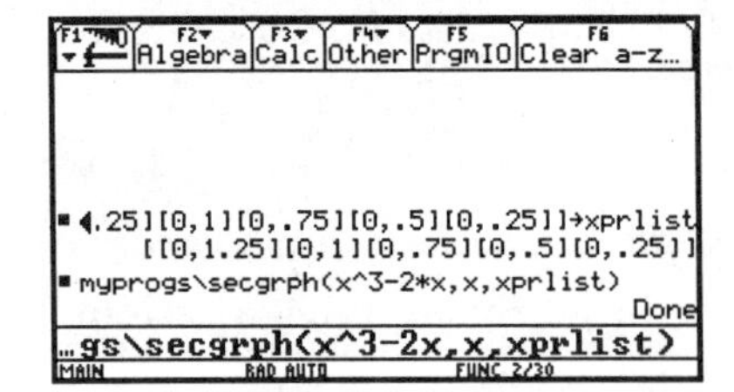

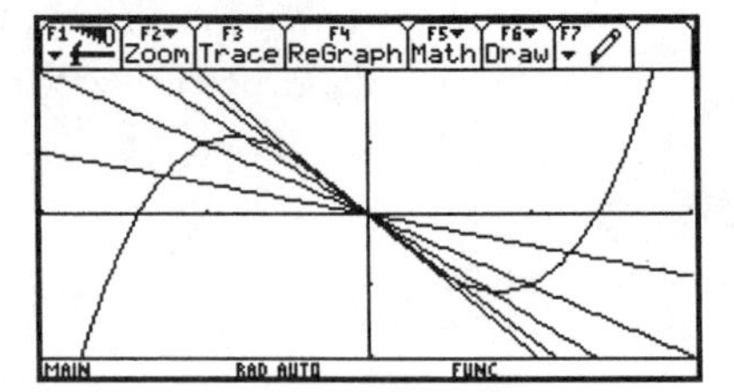

Our second program, **tangrph()**, is one that will graph a given function together with tangent lines at values of the independent variable in a specified list. The arguments of **tangrph()** are as in

tangrph(f, $xvar$, $xlist$**),**

where f is an expression in the variable $xvar$, and $xlist$ is a list containing values of $xvar$.

```
: tangrph(f, xvar, xlist)
: Prgm: Local n, i
: ClrGraph: ClrDraw: FnOff: PlotsOff
: string(f | xvar=xx) →ff
: Graph expr(ff), xx
: dim(xlist) →n
: For i, 1, n
:    LineTan expr(ff) | xx=x, xlist[i]
: EndFor
: DelVar ff
: EndPrgm
```

The following shows a few tangent lines drawn by **tangrph()** on the graph of $y = x^2$ in a $[-2, 2] \times [-1, 4]$ window.

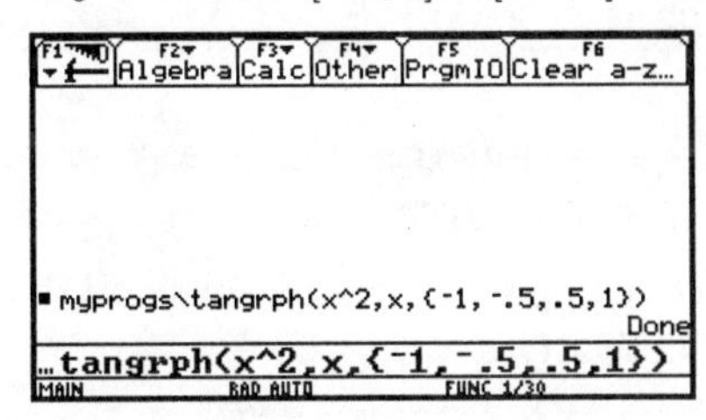

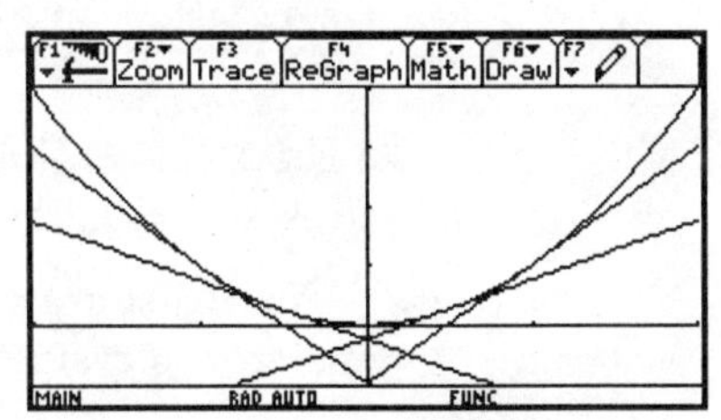

Higher-order derivatives. The following function, **derlist()**, computes a list of the first n derivatives of a given expression. The arguments of **derlist()** are as in

derlist(f, $xvar$, n**),**

where f is an expression in the variable $xvar$, and n is the desired number of derivatives.

```
: derlist(f, xvar, n)
: Func: Local i, thelist, fi, nxt
: {f} →thelist
: f →fi
: For i, 1, n
:    d(fi,xvar) →nxt
:    augment(thelist, {nxt}) →thelist
:    nxt →fi
: EndFor
: Return thelist
: EndFunc
```

Note that the list returned by **derlist()** has $n+1$ elements, the first of which is the expression f. The nth derivative itself is the $(n+1)^{\text{st}}$ member of the list:

derlist(f, $xvar$, n**)[**n**+1].**

The following examples illustrate uses of **derlist()**.

• EXAMPLE 1. *Plot $f(x) = x\sin x$ together with its first, second, and third derivatives.*

We'll use **derlist()** to compute the derivatives and store the resulting list as **fncts**. Once this is done, it is easy to enter the expressions in the **Y=** Editor and then plot the graphs. The plot shown here is in a $[-2\pi, 2\pi] \times [-10, 10]$ window.

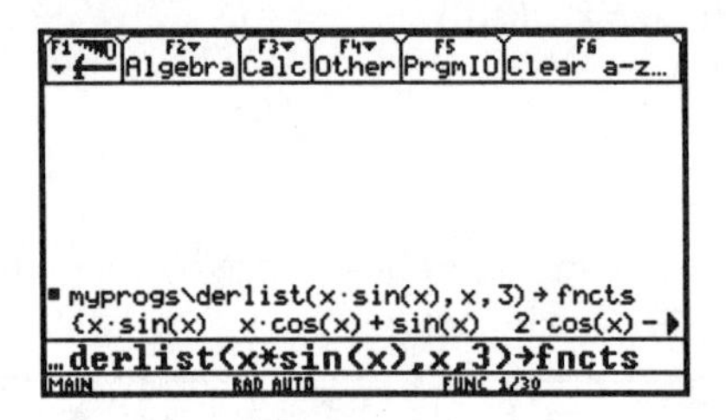

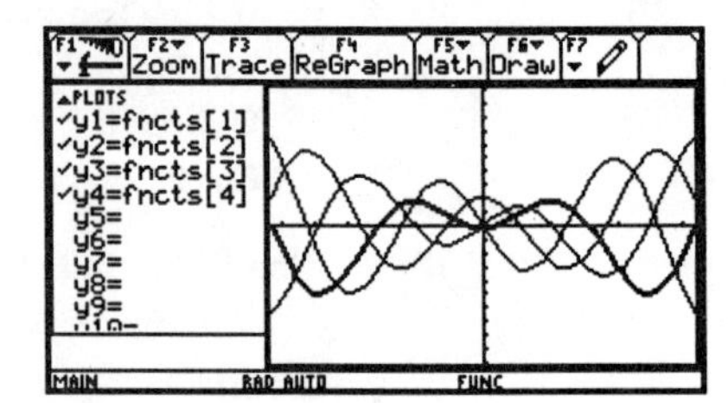

• EXAMPLE 2. *Evaluate $f(x) = \sin(x)\cos(x)$ and each of its first ten derivatives at $x = 0$. Try to observe a pattern in the these values.*

We'll first use **derlist()** to compute the derivatives and then use the "with" operator to evaluate the derivatives at $x = 0$.

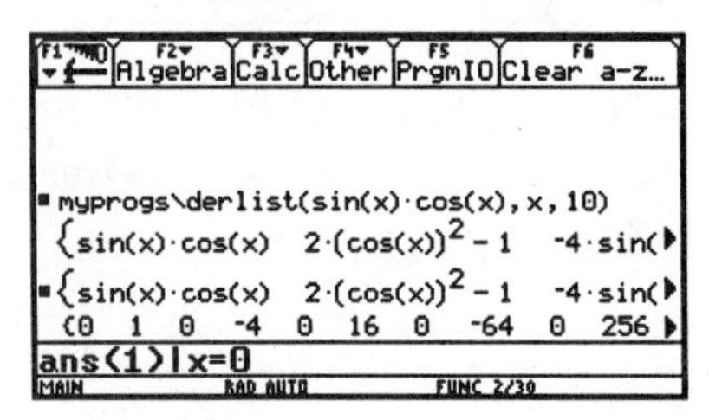

So, assuming that the apparent pattern continues, all even-order derivative are zero at $x = 0$, and odd-order derivatives are powers of 2 with alternating sign; that is,

$$f^{(k)}(0) = \begin{cases} 0, & \text{if } k \text{ is even;} \\ (-2)^{k-1}, & \text{if } k \text{ is odd.} \end{cases}$$

Exercises

1. Use **secgrph()** to plot, in a $[0, 5] \times [0, 8]$ window, the graph of
$$f(x) = x(5 - x)$$
and secant lines through pairs of points consisting of $(1, f(1))$ and each of
$$(3, f(3)),\ (2.5, f(2.5)),\ (2, f(2)),\ (1.5, f(1.5)),\ \text{and } (1.1, f(1.1)).$$
Do this by entering
```
myprogs\secgrph(x*(5-x),x,{{1,3},{1,2.5},{1,2}, {1,1.5},{1,1.1}})
```
2. Use **secgrph()** to plot, in a $[0, 2] \times [0, 2]$ window, the graph of $y = \sqrt{x}$ and secant lines through pairs of points consisting of $(0, 0)$ and each of $(1, 1)$, $(.5, \sqrt{.5})$, $(.25, .5)$, $(.1, \sqrt{.1})$, and $(.01, .1)$. What is happening to the slopes of these secant lines as the second point becomes closer to $(0, 0)$?
3. Use **secgrph()** to plot, in a $[0, \pi/2] \times [0, 1.5]$ window, the graph of $y = \sin x$ and the secant line through $(\pi/4, \sqrt{2}/2)$ and $(.71, \sin(.71))$. Then use **tangrph()** to plot, in the same window, the graph of $y = \sin x$ and the tangent line at $(\pi/4, \sqrt{2}/2)$. Is there any discernible difference between the two pictures?
4. Use **derlist()** to find the third derivative of $f(x) = \sin x^2$.
5. Use **derlist()** to find the first ten derivatives of $f(x) = \sin^2 x$. Then evaluate the list at $x = 0$. What pattern is apparent in the numbers?
6. Use **derlist()** to find the first ten derivatives of $f(x) = (1 - x)^{-1}$. Then evaluate the list at $x = 0$. What pattern is apparent in the numbers?
7. Read the descriptions of **Graph, DrawSlp, LineTan**, and **augment** in Appendix A of your **TI-89/92 Guidebook.**

4 Applications of the Derivative

In this chapter we will explore some of the many applications of the derivative, with the help of the **TI-89/92** and in conjunction with much of the material in Chapters 3 and 4 of Stewart's *Calculus*. We have seen that the derivative of f is a function whose value at $x = a$ gives the slope of the graph of f at the point $(a,\ f(a))$. This geometric idea has numerous variations and lends itself to many applications.

4.1 Velocity, acceleration, and rectilinear motion

Let t be a variable representing time elapsed since some reference time $t = 0$, and imagine a particle moving along a straight-line path in some way. Our interest here is in the function $s(t)$ that gives the position at time t of the moving particle.

Average velocity over a time interval $a \le t \le b$ is defined to be the change in position divided by the change in time:

$$v_{\text{av}} = \frac{s(b) - s(a)}{b - a}.$$

Notice that v_{av} is simply the slope of the secant line through $(a, s(a))$ and $(b, s(b))$.

• EXAMPLE 1. Consider a particle moving along a straight line with position

$$s(t) = (t - 2)^3 + t + 8 \quad \text{for } 0 \le t \le 4.$$

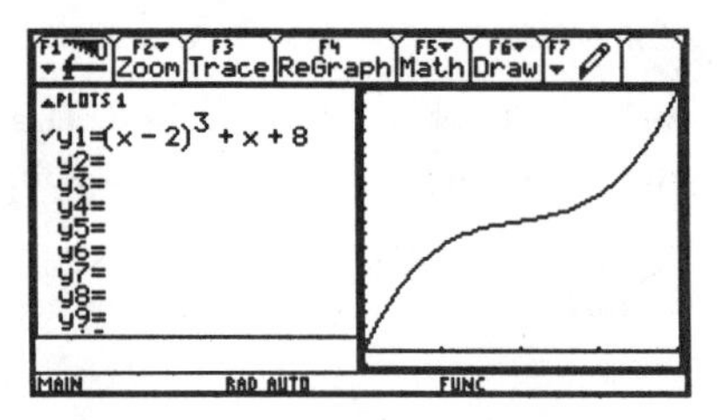

Let's plot secant lines for the time intervals $0 \le t \le 1$ and $0 \le t \le 4$. The slopes of these secant lines are the average velocities of the particle over the respective time intervals.

PLOTS 1
y1=(x - 2)^3 + x + 8
y2=y1(0) + (y1(4) - y1(0))/(4 - 0)·(x - 0)
y3=y1(0) + (y1(1) - y1(0))/(1 - 0)·(x - 0)
y4=
y5=
y6=
y4(x)=
MAIN RAD AUTO FUNC

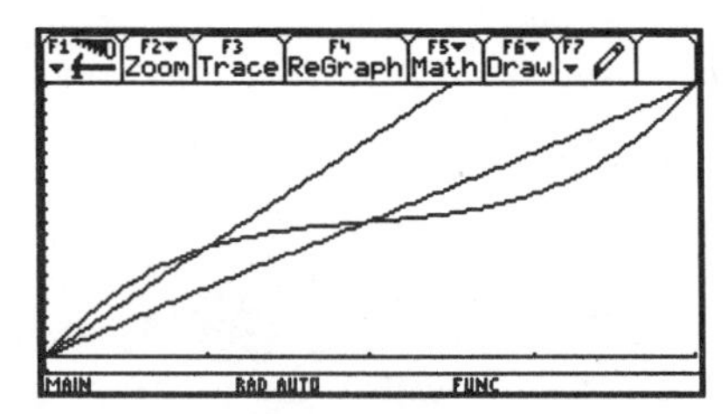

Instantaneous velocity (or simply *velocity*) at a time t is defined to be the limit of the average velocities over intervals $[t, t+h]$ as $h \to 0$. Thus velocity is the derivative of position:

$$v(t) = s'(t) = \lim_{h \to 0} \frac{s(t+h) - s(t)}{h}.$$

We also say that velocity is the (instantaneous) *rate of change* in position.

• EXAMPLE 2. Let's plot $s(t) = (t-2)^3 + t + 8$ along with $v(t) = s'(t)$. We'll do this by entering **y1= (x-2)^3 + x + 8** and **y2=*d*(y1(x))** in the **Y=** Editor.

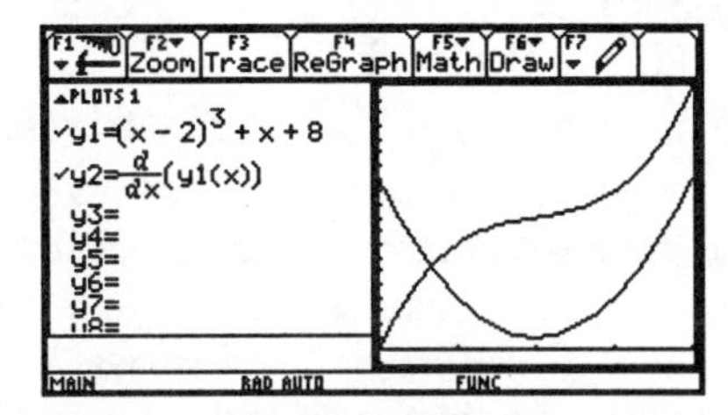

Notice that in this example, $s'(t)$ is always positive, which reflects the fact that the position of the particle is strictly increasing; i.e., the particle is always moving forward.

Just as velocity is the rate of change in position, acceleration is the rate of change in velocity. Thus acceleration is the second derivative of position:

$$a(t) = v'(t) = s''(t).$$

• EXAMPLE 3. Again consider the position function $s(t) = (t-2)^3 + t + 8$, $0 \leq t \leq 4$. In the **Y=** Editor, we'll add **y3=*d*(y1(x),x,2)** (or **y3=*d*(y2(x),x)**) to the list of functions to plot.

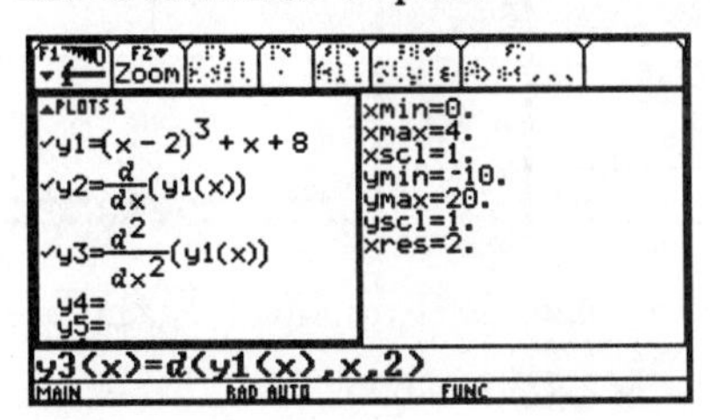

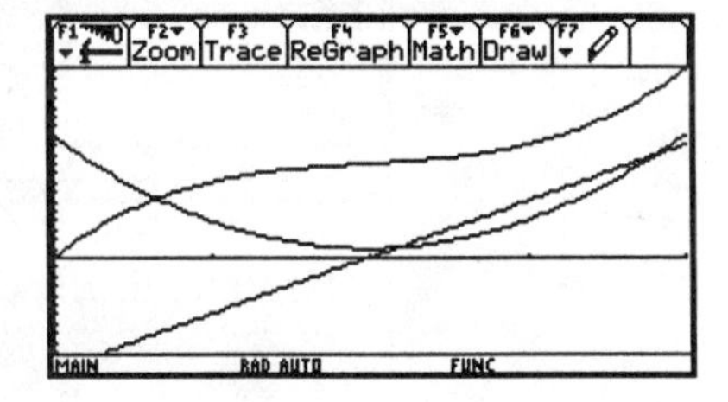

Notice that velocity is decreasing wherever acceleration is negative, and velocity is increasing wherever acceleration is positive.

Simulating motion. There is an interesting way of simulating simple *rectilinear* motion such as this on the **TI-89/92**. It involves creating a parametric plot. By graphing **xt1=** $s(t)$ and **yt1=1**, we cause a horizontal line to be drawn in such a way that the position of the end of the line is at **x1(t)**. By graphing **xt1=1** and **yt1=** $s(t)$, we cause a vertical line to be drawn in such a way that the position of the end of the line is at **y1(t)**. Also, setting either **Leading Cursor** to **ON** in Graph **Formats** (**F1-9**) or setting the Graph **Style** to **Path** (press **F6-6** in the Y= Editor) provides an image of the moving object itself.

A summary of this procedure for simulating motion is as follows:

1) While in **FUNCTION** Graph **MODE**, define the position function as **y1(x)** in the **Y=** Editor. Plot the position, velocity, and/or acceleration if desired.
2) Change the Graph **MODE** to **PARAMETRIC**.
3) Set **Leading Cursor** to **ON** in Graph **Formats** (**F1-9**), or Graph **Style** to **Path** in the **Y=** Editor.
4) For a horizontal path, set **xt1=y1(t)** and **yt1=1** in the **Y=** Editor. For a vertical path, set **xt1=1** and **yt1=y1(t)**.
5) Press ◇**WINDOW** and enter appropriate window variables.
   For a horizontal path,
   - set **ymin=0** and **ymax=2**;
   - adjust **xmin** and **xmax** so that the path stays in the window;

   For a vertical path,
   - set **xmin=0** and **xmax=2**;
   - adjust **ymin** and **ymax** so that the path stays in the window;

   Set **tstep**=.05. (Later adjust **tstep** to speed or slow the motion.)
6) Press ◇**GRAPH**.

• EXAMPLE 4. Let's first try this out on the last example where $s(t) = (t-2)^3 + t + 8$, which we already have entered as **y1**.

```
PLOTS 1            tmin=0.
✓xt1=y1(t)         tmax=4.
✓yt1=1             tstep=.05
 xt2=              xmin=0.
 yt2=              xmax=20.
 xt3=              xscl=1.
 yt3=              ymin=0.
 xt4=              ymax=2.
 yt4=              yscl=1.
 xt5=
yt1(t)=1
MAIN    RAD AUTO    PAR
```

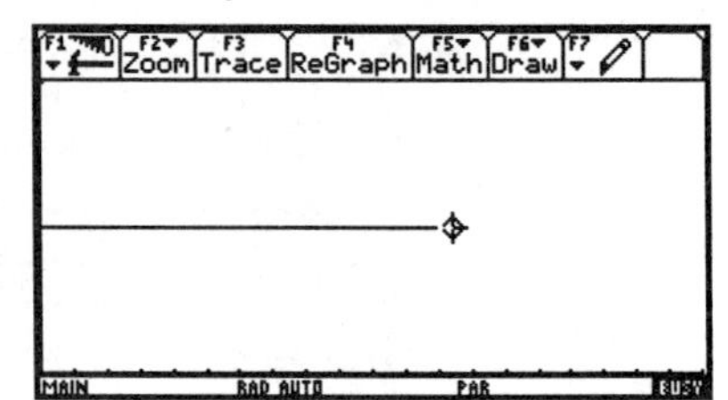

• EXAMPLE 5. Another nice example is provided by the position function

$$s(t) = \sin \pi t.$$

First we'll plot the position and the velocity over the interval $0 \le t \le 4$.

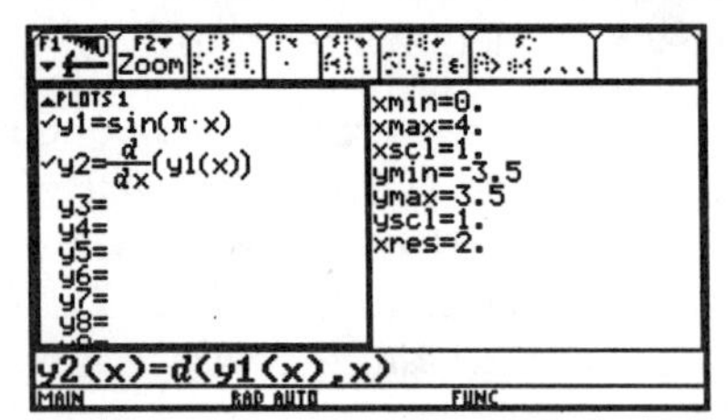

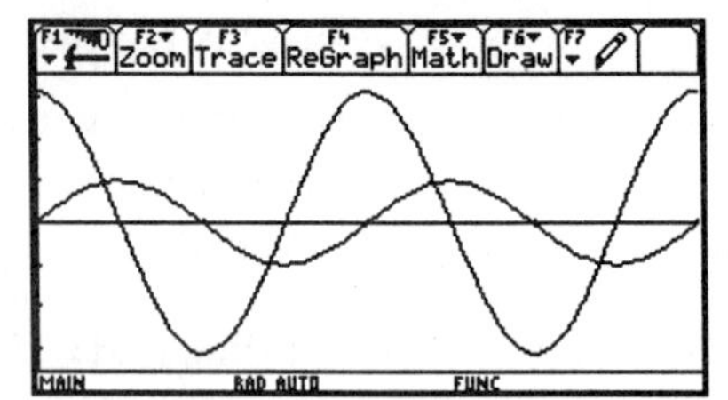

Then we'll switch the Graph **MODE** to **PARAMETRIC**, set appropriate window variables, and simulate the motion in the Graph screen.

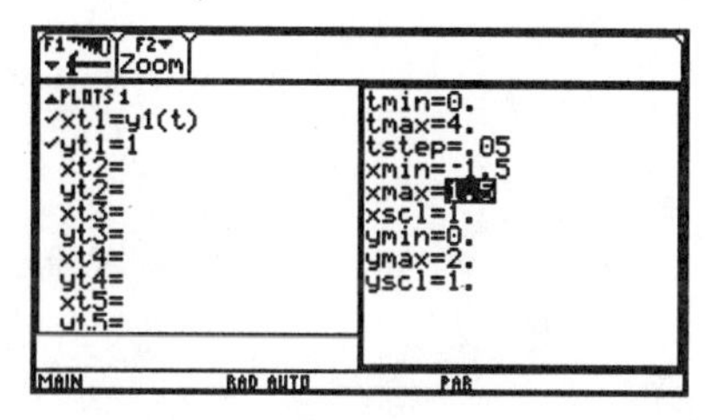

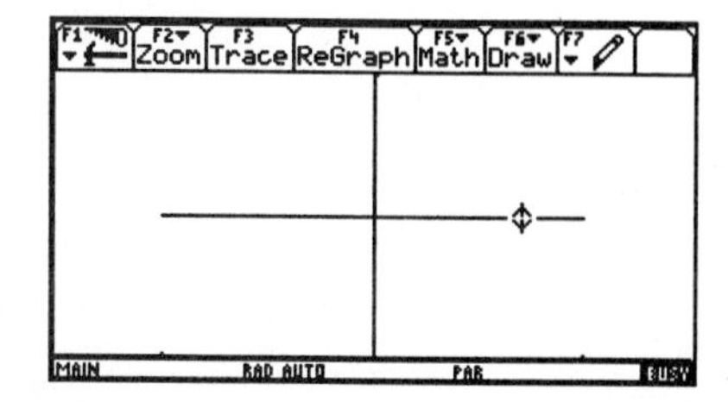

This is an example of *simple harmonic motion.*

Note that even though we have so far chosen to illustrate this procedure "horizontally," many interesting examples of rectilinear motion involve vertical motion, where $s(t)$ (or $y(t)$) represents height. Such examples are best simulated vertically. See Exercises 4–6.

Exercises

For each position function, plot the position, velocity and acceleration on the indicated interval. Then simulate the motion (horizontally) in the manner described above. (*Remark*: Plots of velocity and acceleration are done more quickly if the derivatives are computed on the home screen and then stored as functions.)

1. $s(t) = \dfrac{t-3}{(t-3)^4+1}, \quad 0 \le t \le 6$

2. $s(t) = e^{-t/2} \sin \pi t, \;\; 0 \le t \le 6$

3. $s(t) = \sin(\pi t)\, \cos(5\pi t), \;\; 0 \le t \le 2$

Provided we ignore air resistance, the height (in feet) of a free-falling object, under the influence of gravity, is described approximately by

$$h(t) = -16t^2 + v_0 t + h_0,$$

where v_0 is the velocity at time $t = 0$ (or *initial velocity*), and h_0 is the height at time $t = 0$ (or *initial height*). For each of the following combinations of initial velocity and initial height,

a) plot the height and velocity for $0 \le t \le T$, where T is the positive time at which the height becomes 0;
b) find the maximum height that the object attains;
c) simulate the motion vertically in the manner described above.

4. $v_0 = 100$ ft/sec, $h_0 = 0$

5. $v_0 = 64$ ft/sec, $h_0 = 25$ feet

6. $v_0 = 0$ ft/sec, $h_0 = 100$ feet

4.2 Implicit differentiation and related rates

Suppose that we are interested in studying the graph of an equation such as

$$x^3 - 2x + y^2 = 1,$$

which perhaps defines neither variable as a function of the other. For this particular example, it is not difficult to piece together the graph by plotting each of

$$y = \pm\sqrt{1 + 2x - x^3}.$$

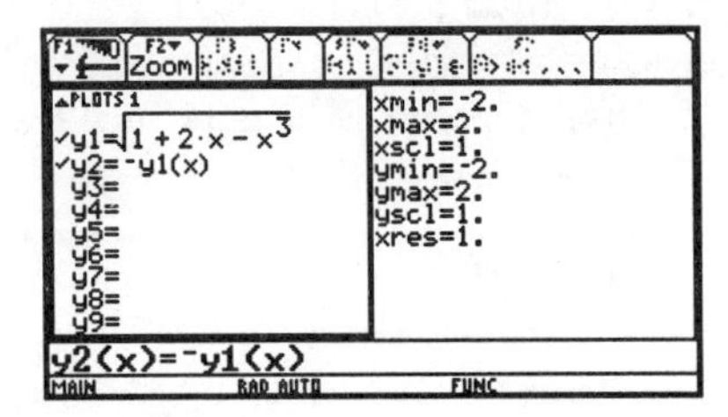

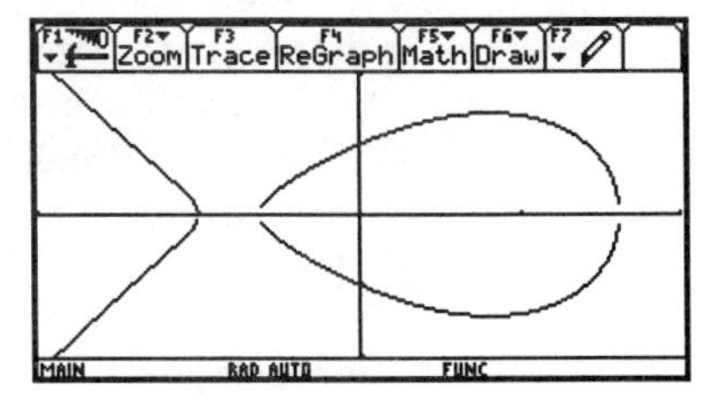

In general, however, it may not be possible to solve for either variable explicitly. But finding either of the derivatives $\frac{dy}{dx}$ or $\frac{dx}{dy}$ is typically not difficult. We can find either of these derivatives, directly from the equation, by *implicit differentiation.* The process is this for finding $\frac{dy}{dx}$:

1) Find the derivative of each side of the equation with respect to x.

2) Solve the resulting equation for $\frac{dy}{dx}$ in terms of x and y.

In our current example, step 1 results in the equation

$$3x^2 - 2 + 2y\frac{dy}{dx} = 0,$$

which is easily solved for $\frac{dy}{dx}$, producing

$$\frac{dy}{dx} = \frac{2 - 3x^2}{2y}.$$

Note that we can now easily see, for example, that $\frac{dy}{dx} = 0$ at each point on the graph where $x = \pm\sqrt{2/3}$, and that $\frac{dy}{dx}$ is undefined at each point on the graph where $y = 0$ (resulting, for this example, in vertical tangent lines).

Now let's walk through this procedure on the **TI-89/92**. First we'll enter the equation and store it in the variable "eqn." Notice that we define the equation in terms of **y(x)** rather than **y**. Then the ***d*()** operator does the implicit differentiation for us.

$x^3 - 2\cdot x + (y(x))^2 = 1 \rightarrow eqn$ $(y(x))^2 + x^3 - 2\cdot x = 1$
$\frac{d}{dx}(eqn)$ $2\cdot y(x)\cdot\frac{d}{dx}(y(x)) + 3\cdot x^2 - 2 = 0$
d(eqn,x)

Now, before we can solve for $\frac{dy}{dx}$, we have to substitute some symbol for it using the "with" operator (|). Then we can use the **solve()** command to solve for that symbol.

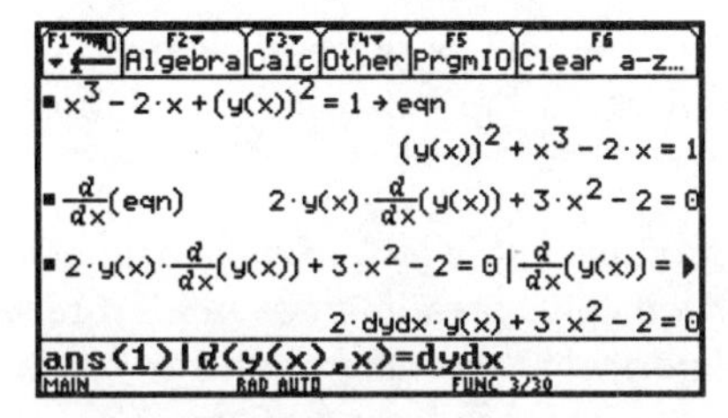

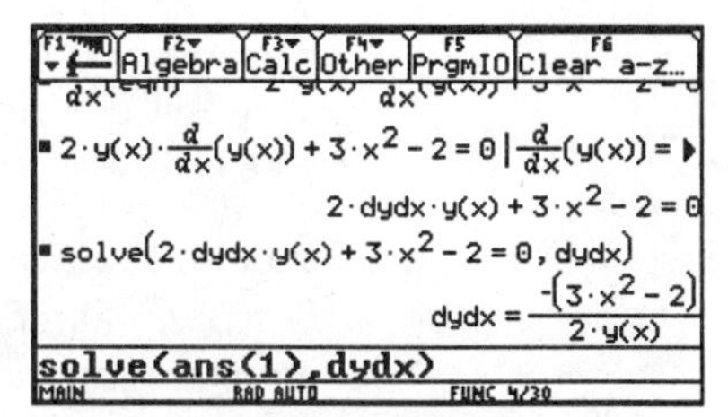

Related rates. When two or more quantities that are themselves functions of time are related through some equation, a relationship among the rates of change in those quantities can be obtained by implicit differentiation with respect to t.

• EXAMPLE 1. *A particle is moving along the curve*

$$x^3 + x - y^3 - y = 8$$

in such a way that the x-coordinate of the particle's position is changing at a constant rate $\frac{dx}{dt} = 1$. How fast is the y-coordinate of the particle's position changing when the particle is at the point $(2, 1)$?

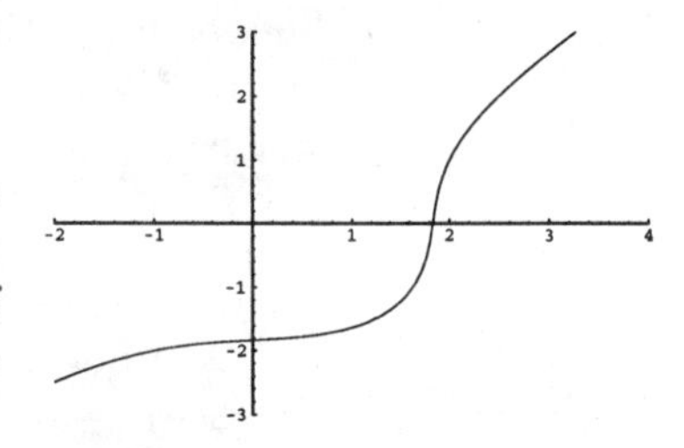

To solve this problem, we first enter the equation of the path, using $x(t)$ and $y(t)$ to indicate that x and y are to be treated as functions of t. Then we'll differentiate implicitly with respect to t, substituting the symbol "**ry**" for $\frac{dy}{dt}$.

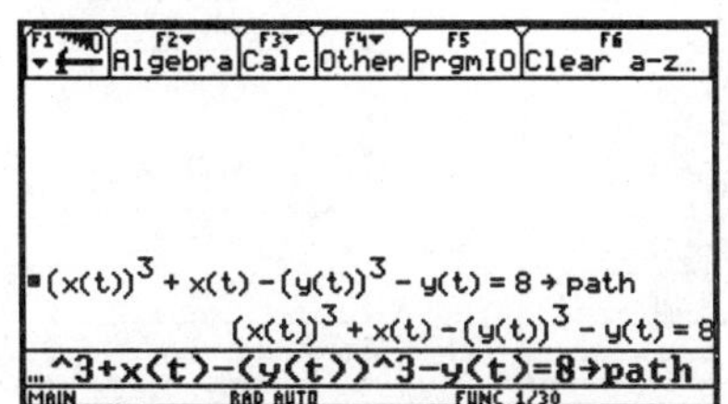

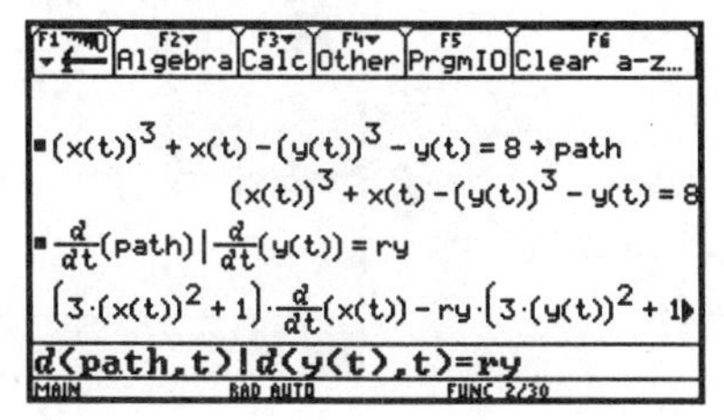

Next we solve for **ry**, substituting the rate $\frac{dx}{dt} = 1$. The final step then is to substitute in the coordinates $x(t) = 2$ and $y(t) = 1$.

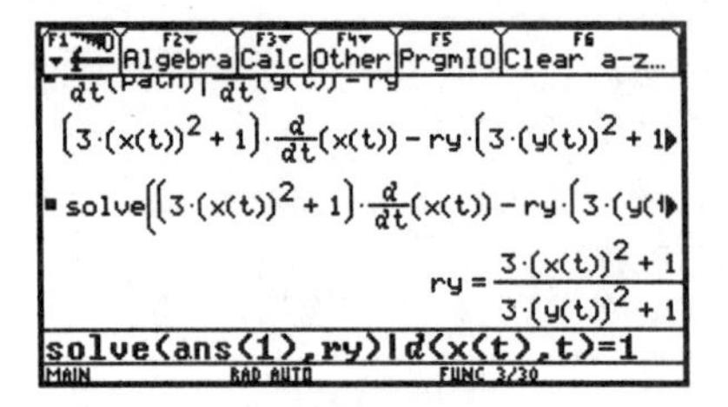

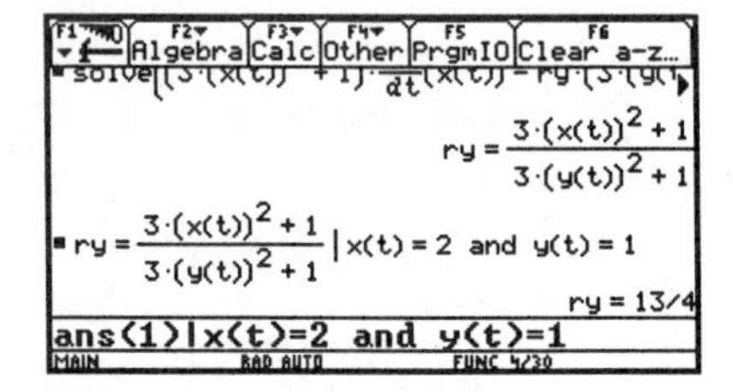

Thus our final answer is that $\frac{dy}{dt} = \frac{13}{4}$ when the particle's position is $(2, 1)$.

Formulas from geometry (as well as physics, chemistry, and many other areas) often describe the relationship between two or more quantities that may be changing with time.

• EXAMPLE 2. *One tenth of one cubic inch of oil is dropped gently onto the surface of a pan of water, quickly spreading out in all directions and taking on an approximately cylindrical shape. The radius of the oil is observed to be increasing at a rate of 1/2 inch per second at the instant when the radius is 5 inches. Find the rate of change in the thickness of the oil at that instant.*

The relationship between radius r and thickness y is given by the formula for the volume of a cylinder:

$$\pi\, r(t)^2 y(t) = 1.$$

So we'll first enter this equation and differentiate with respect to t. Because of the volume formula, we will then substitute $1/(\pi\, r(t)^2)$ for $y(t)$.

```
d(eqn,t)|d(y(t),t)=ry
MAIN RAD AUTO FUNC 2/30
```

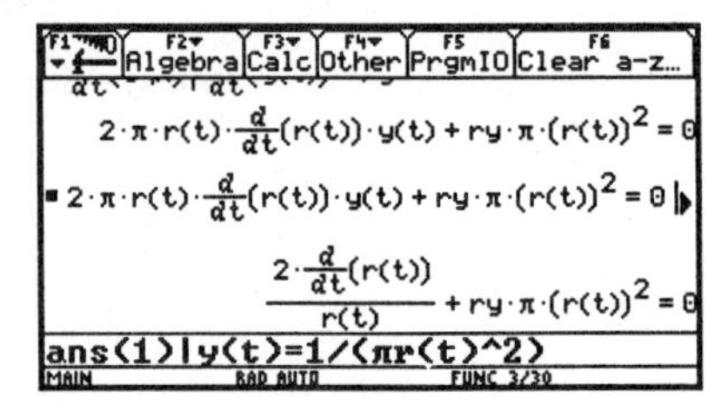

Now, finally, we'll solve for **ry** $= \frac{dy}{dt}$, and then substitute in $r(t) = 5$ and $\frac{dr}{dt} = 1/2$ to produce the final result.

```
solve(ans(1)*r(t),ry)
MAIN RAD AUTO FUNC 4/30
```

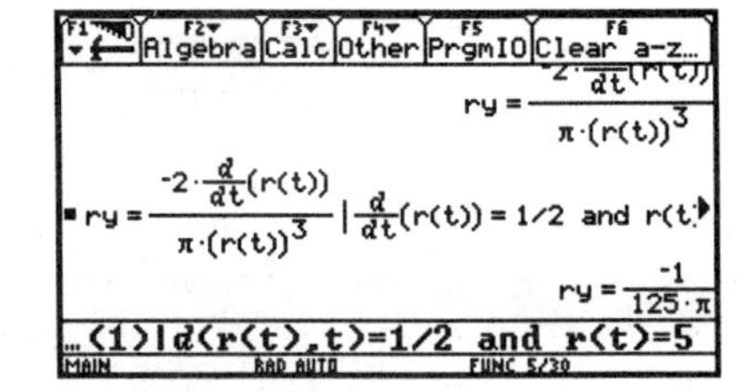

Thus the thickness is *decreasing* at $1/(125\pi) \approx .0025$ inches per second at the instant when $r = 5$.

Exercises

1. For each of the following equations, find the slope of the tangent line to the graph at the indicated point.

   a) $x\,y^2 - y\,x^3 = 2$ at $(0,\ 2)$

   b) $2\sin\big(\pi y\cos(\pi x)\big) + 3xy = 2$ at $(1/3,\ 1)$

   c) $x\sin\pi y - y\cos\pi x = 0$ at $(1/4,\ 1/4)$

2. A particle is moving along the parabola $y = x^2$ in such a way that the x-coordinate of the particle's position changes at a constant rate $\frac{dx}{dt} = 2$. Find the rate of change in the y-coordinate of the particle's position at the instant when the position of the particle is:

   a) $(-1, 1)$ b) $(0, 0)$ c) $(1, 1)$ d) $(2, 4)$

3. A particle is moving along the top half of the circle $x^2 + y^2 = 1$ in such a way that the x-coordinate of the particle's position changes at a constant rate $\frac{dx}{dt} = 1$. Find the rate of change in the y-coordinate of the particle's position at the instant when the position of the particle is:

   a) $(-\sqrt{2}/2,\ \sqrt{2}/2)$ b) $(0, 1)$ c) $(\sqrt{3}/2,\ 1/2)$

4. For each of the following equations, differentiate implicitly and then solve for the indicated rate. The symbol "k" is always a constant.

   a) $V(t) = \frac{4\pi}{3}\,r(t)^3,\ \dfrac{dr}{dt}$

   b) $A(t) = 2\pi r(t)\big(r(t) + h(t)\big),\ \dfrac{dr}{dt}$

   c) $V(t) = \frac{\pi}{3}\,r(t)^2 h(t),\ \dfrac{dh}{dt}$

   d) $\dfrac{\sin\theta(t)}{\sin\phi(t)} = k,\ \dfrac{d\theta}{dt}$

   e) $p(t)V(t) = k\,T(t),\ \dfrac{dp}{dt}$

4.3 Linear and quadratic approximation

Note: This section is closely related to the *Laboratory Project: Taylor Polynomials* that follows Section 3.10 in Stewart's *Calculus*.

Consider the following problem.

Given a differentiable function f and a number a in its domain, find the linear function λ_a that best approximates f near a in the sense that

$$\lambda_a(a) = f(a) \quad \text{and} \quad \lambda_a'(a) = f'(a).$$

In other words, we're looking for the linear function whose graph passes through $(a, f(a))$ with the same slope as the tangent line to the graph of f there. Since λ_a is linear, we can assume that $\lambda_a(x) = mx + b$, and so our requirements on λ_a become

$$ma + b = f(a) \quad \text{and} \quad m = f'(a).$$

Now we solve for m and a to get $m = f'(a)$ and $b = f(a) - af'(a)$. This gives us, after a little rearranging, the *linearization* of f at $x = a$:

$$\lambda_a(x) = f(a) + f'(a)(x - a).$$

Of course, this is nothing more than the function whose graph is the tangent line to the graph of f at $(a, f(a))$.

• EXAMPLE 1. Let's find the linearization of $f(x) = x^3$ at $x = 3/4$.

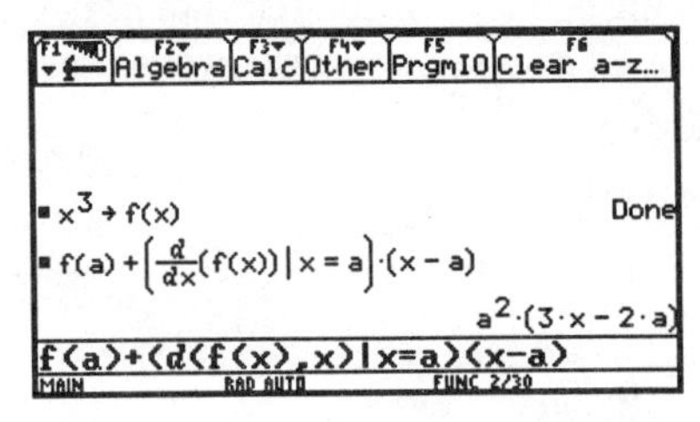


```
Algebra Calc Other PrgmIO Clear a-z...
■ f(a) + (d/dx(f(x)) | x = a)·(x − a)
                                a^2·(3·x − 2·a)
■ a^2·(3·x − 2·a) → λ(x)                   Done
■ 3/4 → a                                   3/4
■ λ(x)                          27·(2·x − 1)/32
λ(x)
MAIN        RAD AUTO        FUNC 5/30
```

To get an idea of the range of values of x for which $\lambda_{3/4}(x)$ gives a good approximation to $f(x)$, let's plot both graphs on the interval $[.5, 1]$ and make a table of values for x near $3/4$.

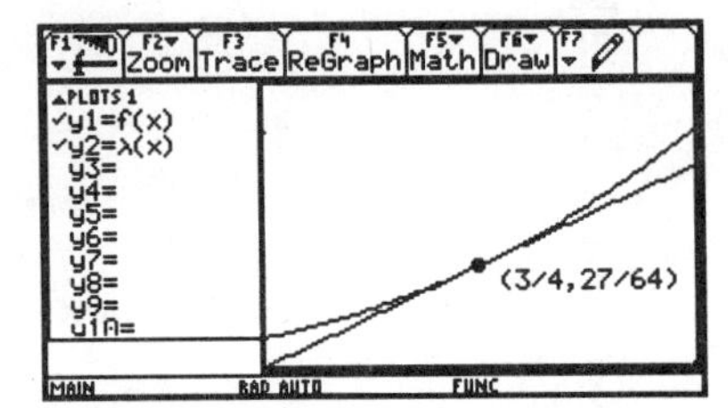

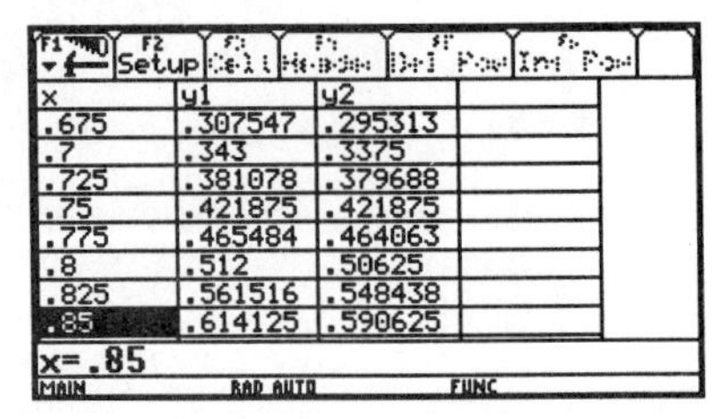

x	y1	y2
.675	.307547	.295313
.7	.343	.3375
.725	.381078	.379688
.75	.421875	.421875
.775	.465484	.464063
.8	.512	.50625
.825	.561516	.548438
.85	.614125	.590625

x=.85

So we see that $\lambda_{3/4}(x)$ gives at least one-place accuracy, roughly, when $|x - 3/4| \le 0.125$ and at least two-place accuracy when $|x - 3/4| \le 0.05$.

It should not be difficult to convince yourself that how well $\lambda_a(x)$ approximates $f(x)$ near $x = a$ depends very much upon the shape of the graph of f near $(a, f(a))$.

Quadratic approximation. Our approach here will parallel our approach to the linear approximation. Consider the following problem.

> *Given a differentiable function f and a number a in its domain, find the quadratic function q_a that best approximates f near a in the sense that*
>
> $$q_a(a) = f(a), \quad q_a'(a) = f'(a) \text{ and } q_a''(a) = f''(a).$$

To make the derivation proceed more smoothly, we'll look for $q_a(x)$ in the form $q_a(x) = c_0 + c_1(x-a) + c_2(x-a)^2$. Because of this, we have

$$q_a(a) = c_0, \quad q_a'(a) = c_1, \quad \text{and} \quad q_a''(a) = 2c_2.$$

As a result, the coefficients we want are $c_0 = f(a)$, $c_1 = f'(a)$, and $c_2 = f''(a)/2$, and so we arrive at

$$q_a(x) = f(a) + f'(a)(x-a) + \frac{1}{2}f'(a)(x-a)^2.$$

Notice that because of the form in which we chose to write this, the first two terms are precisely the linear approximation $\lambda_a(x)$.

• EXAMPLE 2. Let's find the quadratic approximation of $f(x) = \cos x$ at $a = 0$.

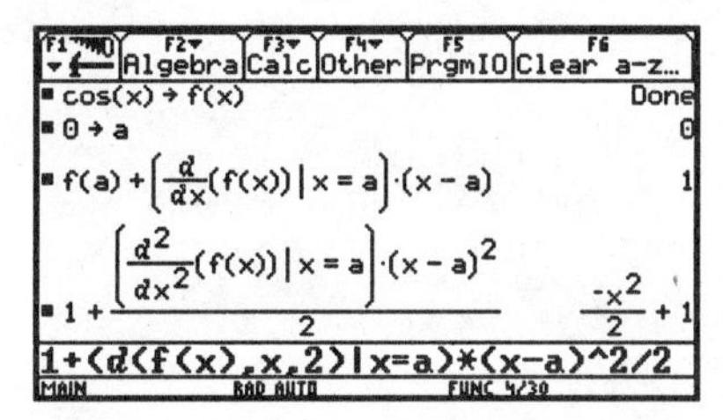

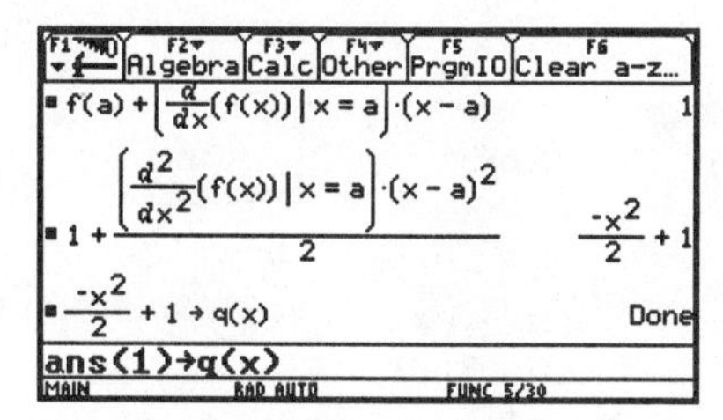

Now we'll graph this quadratic approximation together with $\cos x$. Then we'll zoom in a bit closer and look at the graphs on the interval $[-1.5, 1.5]$ along with the linear approximation (**F5-A**) at the same point for comparison.

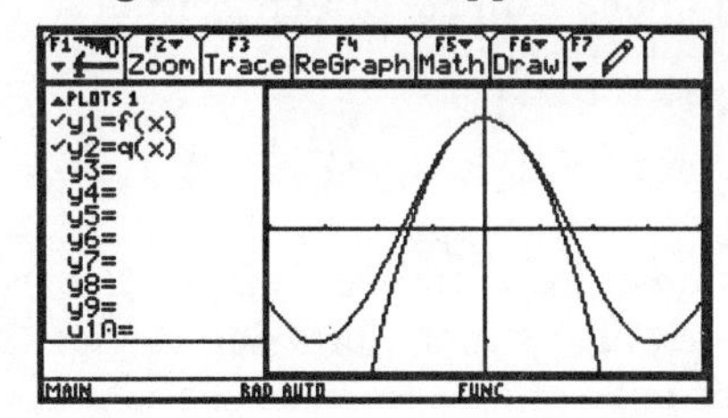

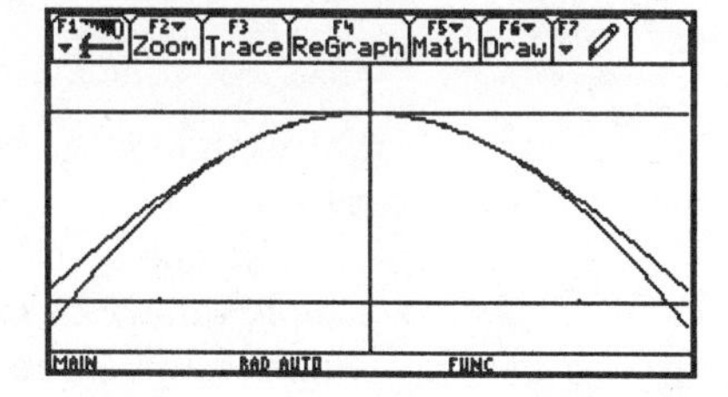

So we see that the quadratic approximation does a reasonably good job of approximating $\cos x$ over a fairly wide range of values of x and is far superior to the linear approximation.

Exercises

1. a) Find the linear approximation to $f(x) = \sqrt[3]{x}$ at $x = 8$. Then use it to approximate $\sqrt[3]{9}$. Compare the approximation with the correct value.
   b) Find the quadratic approximation to $f(x) = \sqrt[3]{x}$ at $x = 8$. Then use it to approximate $\sqrt[3]{9}$. Compare the approximation with the correct value.

For each function in Exercises 2–5, graph the function along with both its linear and quadratic approximations at the specified point.

2. $f(x) = e^{-x}$ at $x = 0$
3. $f(x) = \dfrac{1}{1-x}$ at $x = 0$
4. $f(x) = \cos x$ at $x = \pi/6$
5. $f(x) = \frac{1}{3}x^3$ at $x = 1$

6. Find the equation of the parabola that best approximates the circle $x^2 + y^2 = R^2$ at the point $(0, -R)$.

7. Use the quadratic approximation to $\cos x$ at $x = 0$ to derive an approximate formula for the solution of $\cos x = k\,x$. (*Hint*: *The quadratic formula is your friend.* ☺) For what values of k does the formula give a reasonably accurate result? Base your answer on the graphs of $y = \cos x$ and $y = k\,x$.

8. By analogy with the derivation of the quadratic approximation, derive the *cubic* approximation to a function f about $x = a$. Then find the cubic approximation to $f(x) = \sin x$ at $x = 0$.

4.4 Newton's Method

This section is about solving equations, or equivalently, finding zeros of functions. Any equation in one variable can be written in the form

$$f(x) = 0.$$

For example, the equation $x^2 = 5$ is equivalent to $f(x) = 0$ where $f(x) = x^2 - 5$.

It it not at all uncommon to encounter an equation whose solution(s) simply cannot be expressed in any exact form and so must be approximated numerically. A simple example of such an equation is

$$x^2 = \cos \pi x.$$

To view this as a problem of finding the zeros of a function, let $f(x) = x^2 - \cos \pi x$. From the graph of this function, we see that there are two solutions, each of which is the negative of the other. So let's concentrate on finding the positive solution, which clearly lies in the interval $[0,1]$.

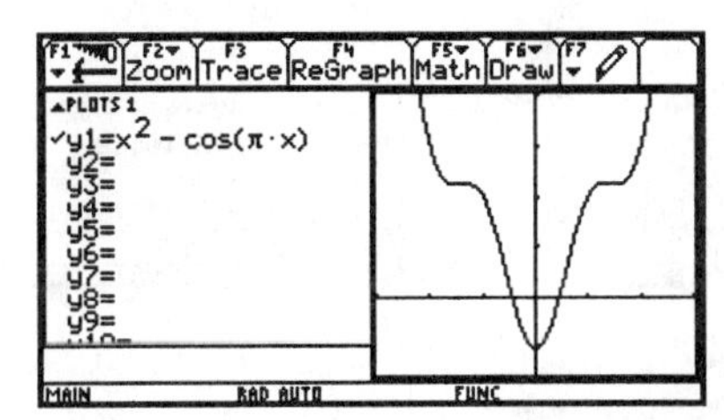

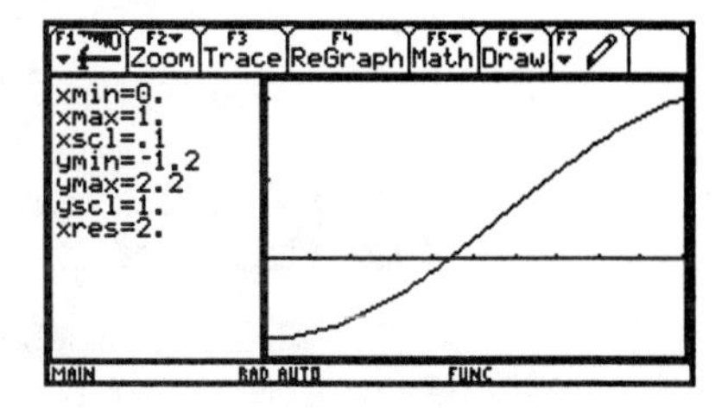

The idea behind Newton's Method is that an approximation x_0 to the solution of $f(x) = 0$ can be improved by solving the *linearized problem*

$$f(x_0) + f'(x_0)(x - x_0) = 0$$

for a new approximation x_1. This amounts to finding the point x_1 where the tangent line to the graph of f at $(x_0, f(x_0))$ crosses the x-axis.

So suppose in our current example that we take $x_0 = .25$ as a first approximation to the solution of $x^2 - \cos \pi x = 0$. Let's graph the tangent line at $(.25, f(.25))$. Note that the x-intercept of this tangent line, which is approximately .49, is much closer to the solution than was x_0. If we were to repeat this process, using the tangent line at $(.49, f(.49))$, we would get still closer to the solution in a rather dramatic fashion.

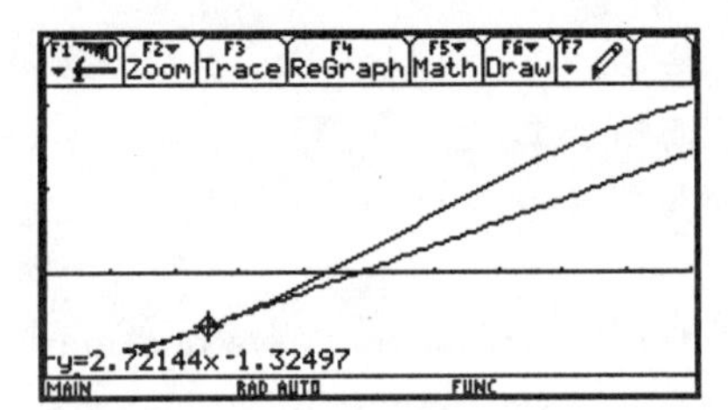

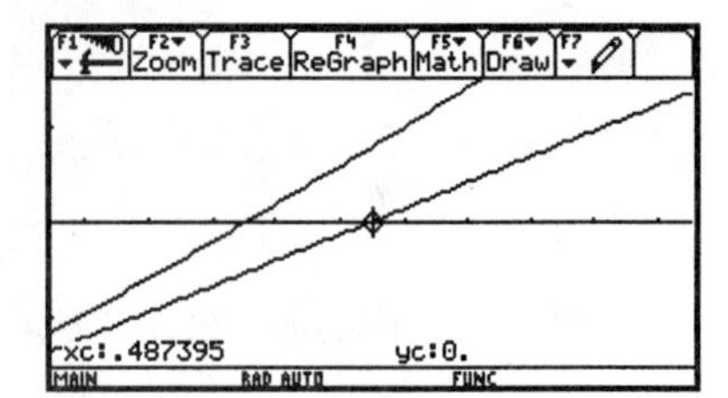

Newton's Method is an iterative procedure that (sometimes) generates successively improved approximations to a solution of $f(x) = 0$. From a current approximation x_k we solve the linearized problem

$$f(x_k) + f'(x_k)(x - x_k) = 0$$

to obtain a new approximation x_{k+1}. So the formula for x_{k+1} is

$$x_{k+1} = x_k - \frac{f(x_k)}{f'(x_k)}.$$

An easy way to carry out this iterative calculation is to define a function **newt**(x), based the right side of this formula. Then once $f(x)$ is defined, we can successive apply **newt()**, after beginning with some initial approximation x_0, by repeatedly entering **newt(ans(1))**.

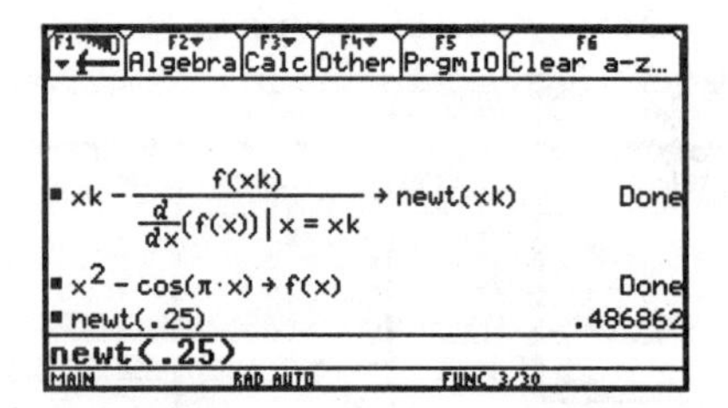


```
F1 F2 F3 F4 F5 F6
Algebra Calc Other PrgmIO Clear a-z...
■ x^2 - cos(π·x) → f(x)                Done
■ newt(.25)                   .486862261603
■ newt(.4868622616031)        .439259486983
■ newt(.43925948698324)        .43843111569
■ newt(.43843111568971)       .438430779482
■ newt(.43843077948157)       .438430779482
■ newt(.43843077948153)       .438430779482
newt(ans(1))
MAIN          RAD AUTO          FUNC 8/30
```

(For the display above, we have changed the setting of the **Display Digits MODE** to **FLOAT 12**.) Notice how rapidly the successive approximations converge to the exact value of the solution. Notice in particular that after only three steps the approximation was already correct to six decimal places, and the fourth step then resulted in 12-place accuracy!

We should point out that the **TI-89/92**'s **nSolve()** function uses (a modified) Newton's Method combined with a search strategy for finding a suitable initial approximation.

Exercises

In Exercises 1–5, plot the given function and determine crude initial approximations to each of the zeros of the function. Then apply Newton's Method to find each zero, accurate to at least eight decimal places.

1. $f(x) = x^3 - x^2 - 2x + 1$
2. $f(x) = 2x^5 - 5x^3 + 2x^2 + 3x - 1$
3. $f(x) = x^2 e^{-x/2} - 1$
4. $f(x) = 2.95 \cos x - x$
5. $f(x) = \tan^{-1} x - x^2$

Exercises 6 and 7 demonstrate how Newton's Method may perform poorly or fail to find a solution at all in certain circumstances.

6. The only positive zero of $f(x) = x^3 - 3x^2 + 4$ is $x = 2$. Try to find this by Newton's Method, starting with an initial approximation $x_0 = 2.1$. Then graph the function. To what property of the function might we attribute the behavior of Newton's Method on this problem?

7. The function $f(x) = \tan x - x$ has a zero in the interval $29 \leq x \leq 30$. Try to find it with Newton's Method, beginning with an initial approximation of:

   a) $x_0 = 29.75$ b) $x_0 = 29.85$ c) $x_0 = 29.80$

4.5 Optimization

One of the most important applications of calculus is optimization. Optimization is about finding maximum and minimum values (i.e., *maxima* and *minima*) of functions. Maxima and minima collectively are referred to as extreme values, or *extrema.* In this section we will look at some general issues related to finding extrema of functions. The next section will be devoted to applied problems. The material here supplements Sections 4.1 and 4.7 of Stewart's *Calculus*.

Local extrema. A function can have numerous local maxima and local minima. For example, consider the function

$$f(x) = x^5 - 4x^4 - x^3 + 16x^2 - 12x.$$

The graph of this function indicates two local minima and two local maxima. The values of x where these extrema occur can be located by finding the *critical points* of the function, which, in this case, are just the zeros of the derivative,

$$f'(x) = 5x^4 - 16x^3 - 3x^2 + 32x - 12.$$

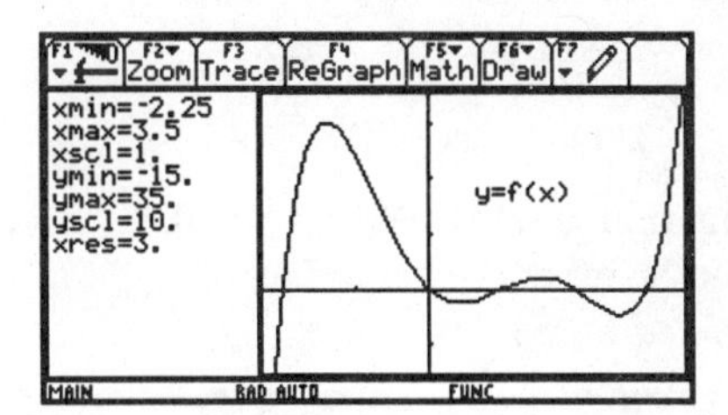

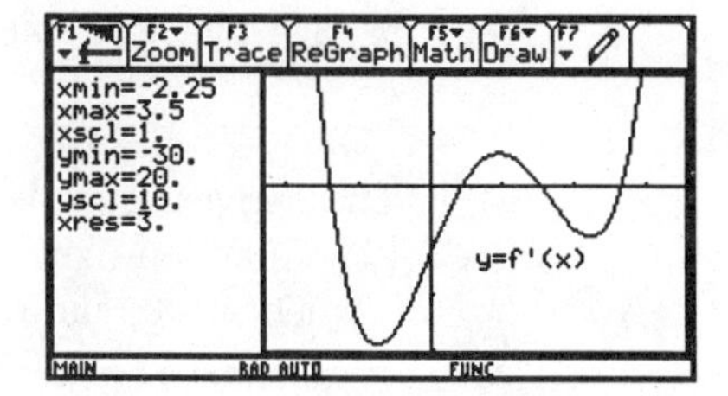

So let's plot f and f' together and then find the zeros of f' and the value of f at each of those zeros. We'll find the zeros of f' by using the **Zero** command from the Graph screen **Math** menu (**F5-2**) and the values of f at those points by using either the **Minimum** (**F5-3**) or **Maximum** (**F5-4**) command from the same menu. Each of these commands requires us to choose between the two functions (with **[⇑]** or **[⇓]**) and to input lower and upper bounds on the value of x we're interested in.

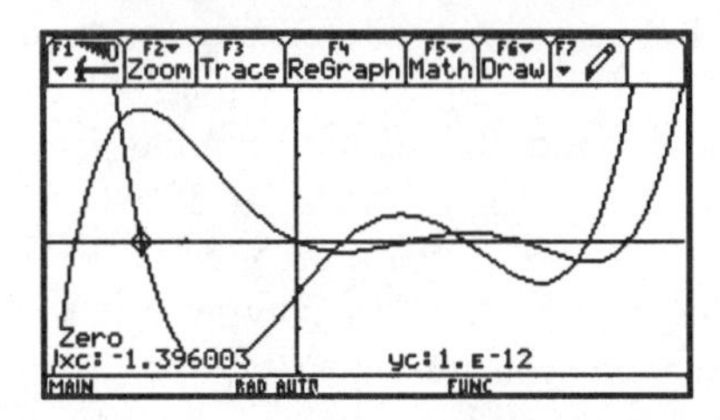

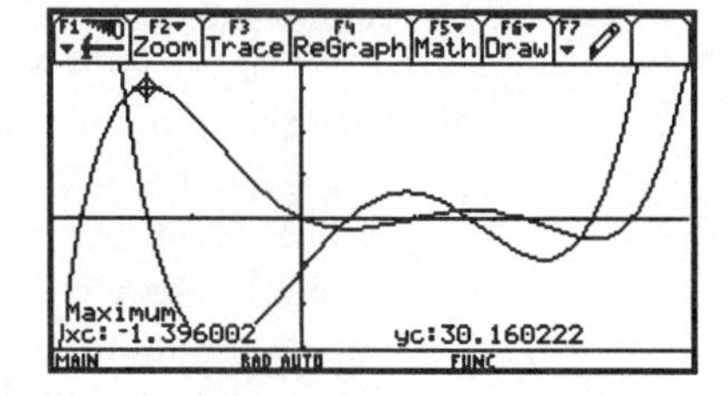

For the remainder of this investigation, we'll concentrate on the portion of the graphs where $x \geq 0$.

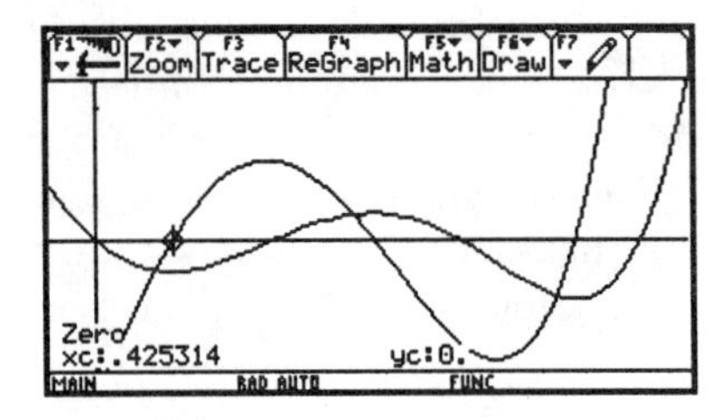

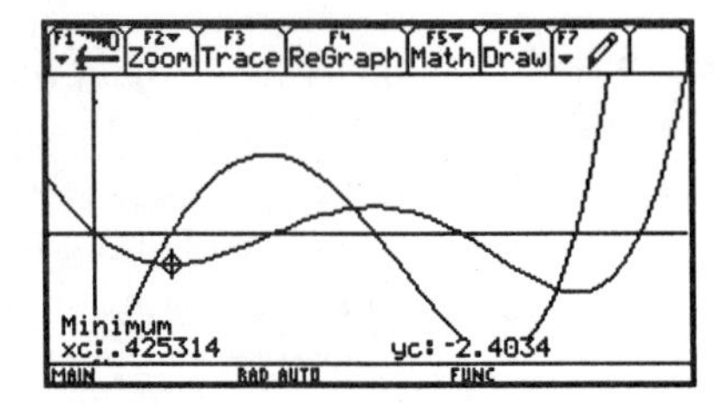

It is left to you to complete this investigation by finding the two remaining extrema.

We must be careful at this point to emphasize the fact that although local extrema often occur at critical points where $f'(x) = 0$, such a critical point may just as well produce no local extremum at all. A nearly trivial example is $f(x) = x^3$. The derivative, $f'(x) = 3x^2$, is zero at $x = 0$; yet neither a local maximum nor local minimum occurs there.

Other kinds of critical points. In addition to the zeros of f', values of x (in the domain of f) where $f'(x)$ does not exist are also critical points of f. A nice example is provided by

$$f(x) = (|x| - 2)^{1/3}.$$

The derivative of this function is undefined at $x = 0, \pm 2$ and nowhere zero. (The function $\text{sign}(x)$ seen below is the derivative of $|x|$; it is -1 when $x < 0$, 1 when $x > 0$, and undefined at $x = 0$.)

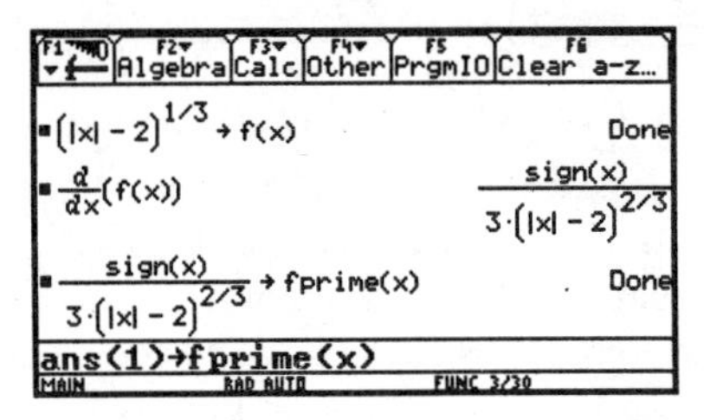

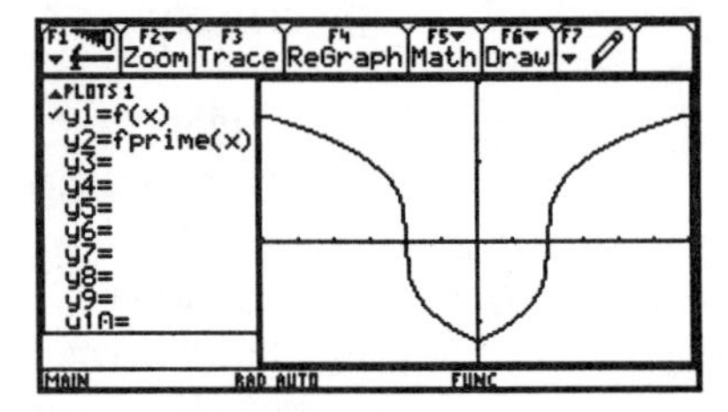

At $x = \pm 2$ the derivative does not exist because the tangent line to the graph is vertical; at $x = 0$ the derivative does not exist because the slope changes instantaneously from -1 to 1, causing a corner-point in the graph. Notice that the function has a local (and global) minimum at $x = 0$, while neither $x = -2$ nor $x = 2$ produce any kind of extremum. If we ask the **TI-89/92** to locate the minimum (**F5-3**), it takes several seconds to do the computation, because it cannot find the critical point by solving $f'(x) = 0$ and so must resort to a more basic and time-consuming search for the minimum. Notice also that graphing the derivative reveals vertical asymptotes at $x = \pm 2$ and a discontinuity at $x = 0$.

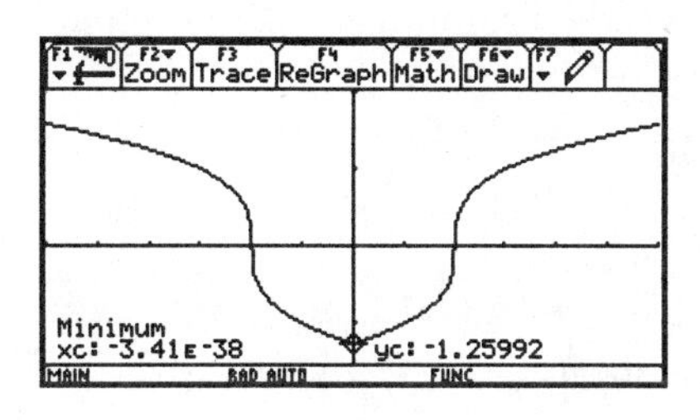

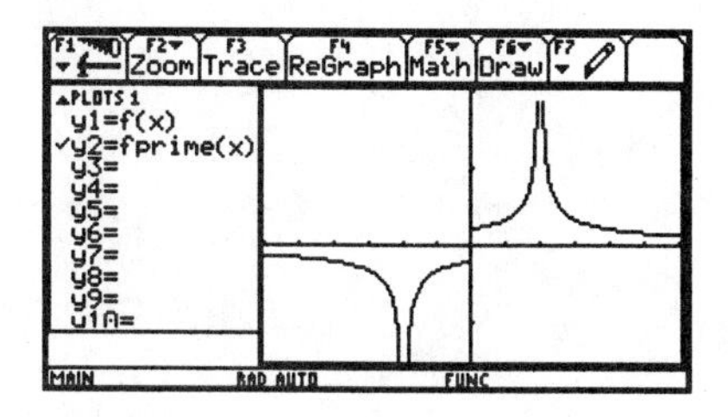

Absolute extrema on closed intervals. One of the most important theorems in calculus states that any continuous function on a closed, bounded interval attains both an absolute minimum and an absolute maximum value on that interval. Moreover, each of these absolute extrema either is a local extremum in the interior of the interval or else occurs at one of the endpoints of the interval. For a simple example, consider

$$f(x) = x^3 - x \quad \text{on the closed interval } [-1, 3/2].$$

Let's graph the function on this interval and then use the **Minimum** and **Maximum** commands from the **Math** menu (**F5-3** and **F5-4**) to find the extreme values, each time entering -1 and 1.5 as the lower bound and upper bound, respectively.

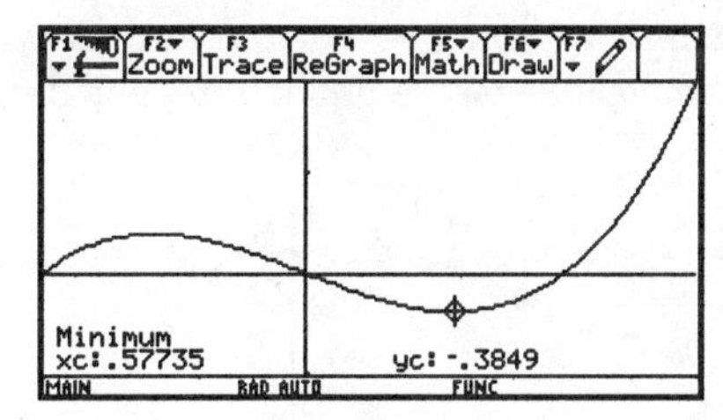

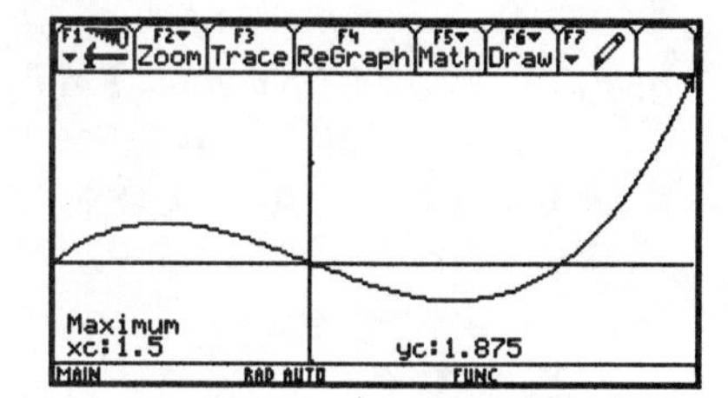

Thus, the minimum value occurs at a critical point in the interior of the interval, while the maximum occurs at an endpoint.

Exercises

For the functions in Exercises 1–4,

a) find all critical points of the function;
b) find each local extremum and the value of x at which it occurs.

1. $f(x) = x^5 - x^3$
2. $f(x) = x^{1/3}(x - 1)$
3. $f(x) = x^3 e^{-x^2}$
4. $f(x) = |x^3 - x^2|$

In Exercises 5–8, find the (absolute) maximum and minimum values of the function on the specified closed interval.

5. $f(x) = \ln x$ on $[1, 3]$

6. $f(x) = |3x^3 - 2x^2|$ on $[0, 1]$

7. $f(x) = x(x-1)(x-2)$ on $[0, 2]$

8. $f(x) = e^{-x} \sin 3x$ on $[0, 5]$

4.6 Applied optimization problems

Optimization is such an important topic because of its applications. Many applications come from business and manufacturing.

• EXAMPLE 1. *An aquarium is to be constructed to hold 20 cubic feet of water. The two ends of the aquarium are to be square, and the aquarium has no top. The glass used for the four sides costs \$.50 per square foot, while cheaper glass used for the bottom costs \$.35 per square foot. Glue and rubber caulking to fasten and seal the joints between pieces of glass costs \$.10 per foot. Finally, framing around the bottom and top perimeters costs \$.05 per foot. Find the dimensions of the aquarium that minimize the total material cost.*

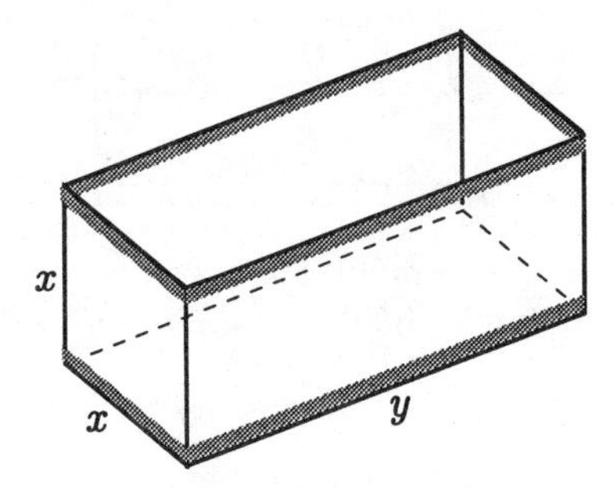

Let x and y be the dimensions indicated in the figure. We will first express the total material cost in terms of x and y. The cost of the glass for the sides and bottom will be $.50(2x^2 + 2xy) + .35xy$. The cost of the glue and caulking for the joints will be $.10(6x + 2y)$. The cost of the framing will be $.05\big(2(2x + 2y)\big)$. Putting all this together gives us the total cost:

$$\begin{aligned} C &= .50(2x^2 + 2xy) + .35xy + .10(6x + 2y) + .05(2(2x + 2y)) \\ &= x^2 + 1.35xy + .8x + .4y. \end{aligned}$$

Now, because the volume is to be 20 cubic feet, we have $x^2y = 20$, and so

$$y = 20/x^2.$$

Substitution of this into the cost function gives the cost as a function of x alone:

$$C(x) = x^2 + 1.35x\left(20/x^2\right) + .8x + .4\left(20/x^2\right)$$
$$= \frac{x^4 + .8x^3 + 27x + 8}{x^2}.$$

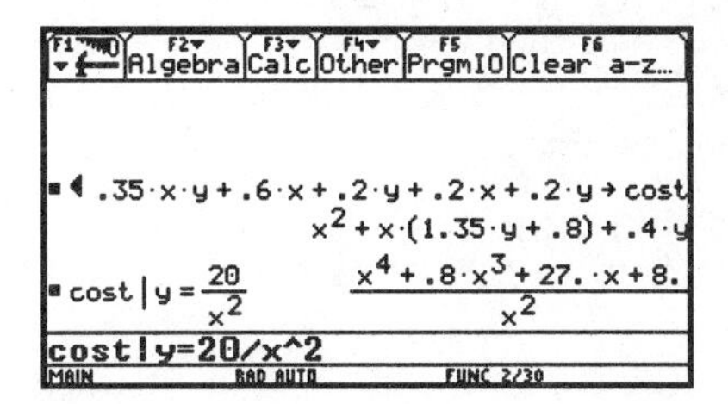

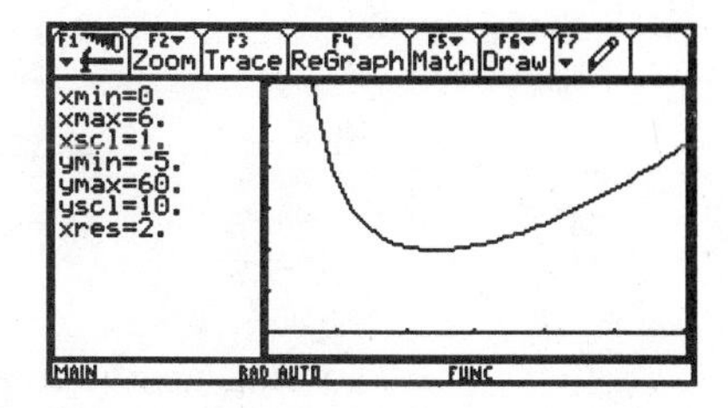

Let's now plot the derivative of the cost and find the critical point that minimizes the cost.

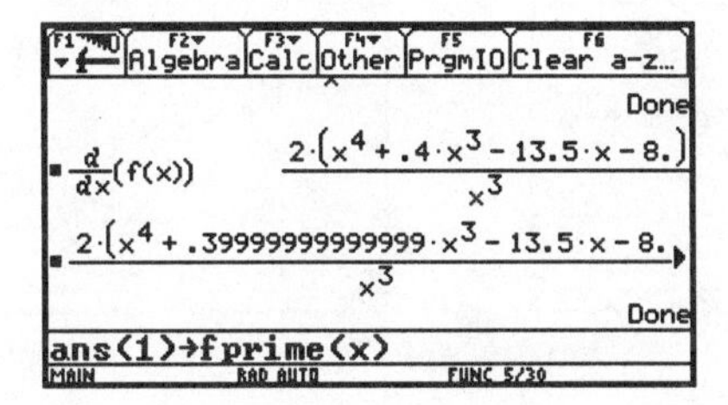

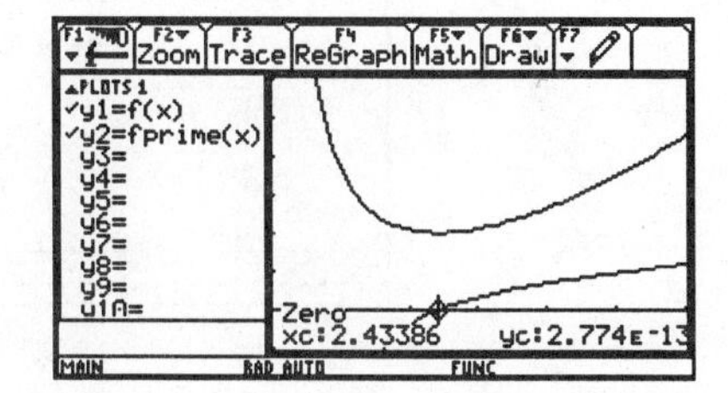

So we conclude finally that the cost is minimized if the base of the aquarium is $x = 2.434$ feet by $y = 20/2.434^2 = 3.376$ feet.

We conclude this section with a classical geometric problem that is posed in such a way that we cannot use a completely graphical approach, as was possible in the previous example. The problem is stated as follows.

• Example 2. *Find the dimensions of the smallest right circular cone that can contain a sphere of radius ρ.*

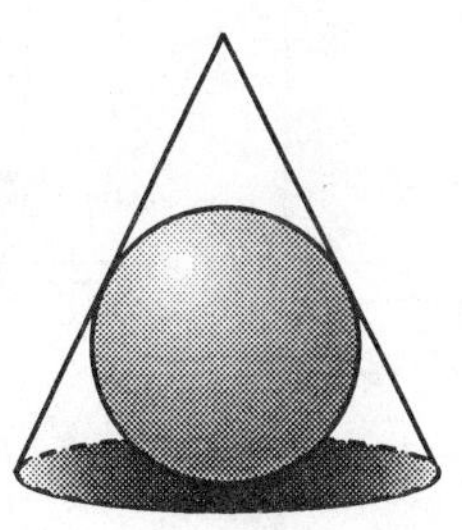

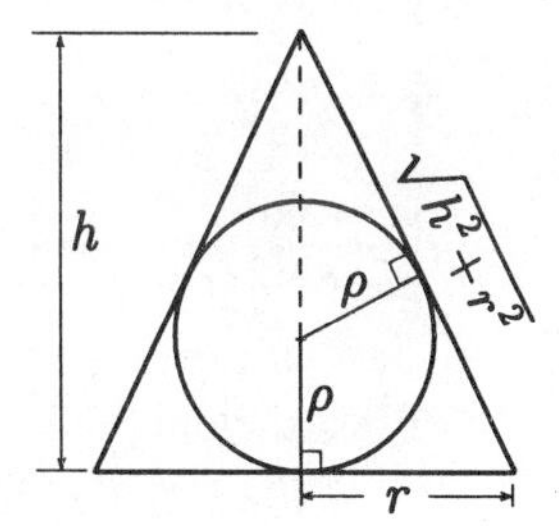

The objective here is to minimize the cone's volume:

$$V = \frac{\pi}{3}r^2h,$$

where r and h are the radius and height of the cone, respectively. However, we must first express this volume in terms of one variable. Examining the cross-section in the figure, we can use the principle of similar triangles to obtain a relationship between r and h, namely,

$$\frac{h-\rho}{\rho} = \frac{\sqrt{h^2+r^2}}{r},$$

which, after we square both sides, becomes

$$\frac{(h-\rho)^2}{\rho^2} = \frac{h^2+r^2}{r^2}.$$

The plan now is to solve this equation for r^2 and substitute the resulting expression into the volume formula to obtain the volume as a function of h.

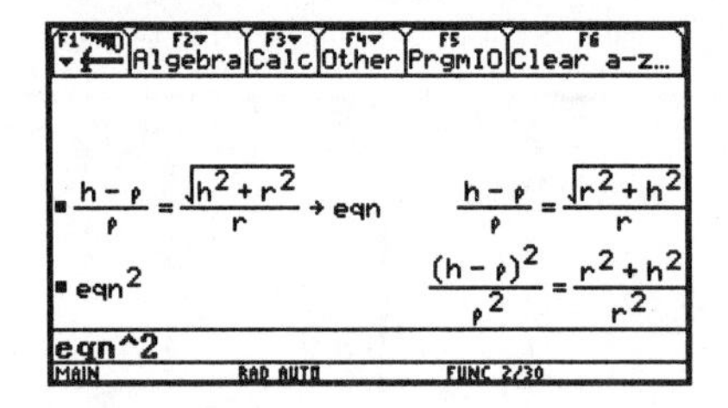

The result is

$$V(h) = \frac{\pi h^2 \rho^2}{3(h-2\rho)}.$$

Note that this form of $V(h)$ makes clear algebraically the fact that the only interesting values of h for this problem are $h > 2\rho$. (This is also obvious from the geometry of the problem.) Now to find the value of h that minimizes $V(h)$, we compute the derivative of V and find its zeros.

Thus the only critical point of interest is $h = 4\rho$. Moreover, closer inspection of $V'(h)$ reveals that $V'(h) < 0$ when $2\rho < h < 4\rho$ and $V'(h) > 0$ when $h > 4\rho$. Therefore, $V(h)$ has a minimum value at $h = 4\rho$. The corresponding value of r is $\sqrt{2}\,\rho$.

To graphically visualize what we've done, let's plot $V(h)$ and $V'(h)$ with $\rho = 1$. The minimum cone volume will occur when $h = 4$.

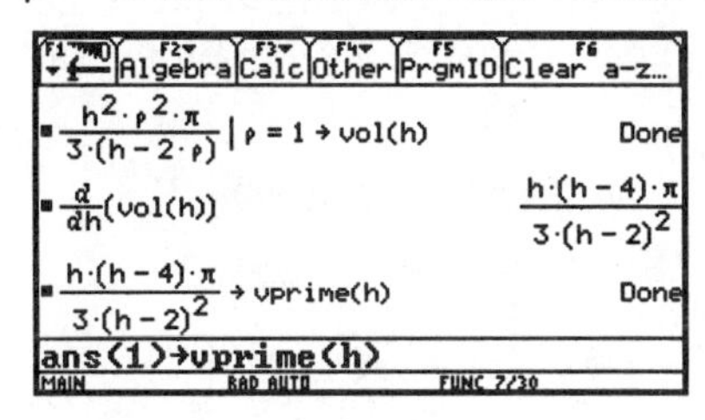

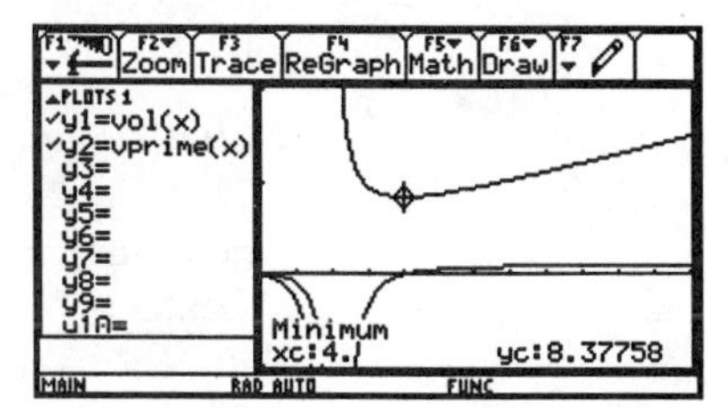

Exercises

1. Rework the aquarium problem ignoring all but the cost of the glass.
2. Rework the aquarium problem given that framing around the bottom and top perimeters costs $.50 per foot.
3. Find the dimensions of the largest right circular cone that can be contained in a sphere of radius ρ.
4. Find the dimensions of the largest right circular cylinder that can be contained in a sphere of radius ρ.
5. Find the area of the largest rectangle (with sides parallel to the coordinate axes) that can be contained in the region in the plane bounded by the graphs of $y = 0$, $y = x^2$, and $x = 1$.
6. Find the area of the largest triangle (with two sides parallel to the coordinate axes) that can be contained in the region in the plane bounded by the graphs of $y = 0$, $y = x^2$, and $x = 1$.
7. Find the point on the graph of $y = 1/x^2$, $x > 0$, that is closest to the origin.
8. Find the minimum surface area of a closed cylindrical can with a volume of 30 cubic inches.
9. Find the maximum volume of a closed cylindrical can with a surface area of 100 square inches.

4.7 Programming notes

No one's life can be complete without writing a program that implements Newton's Method. The following is such a program for the **TI-89/92**.

```
: newt(f, xvar, x0, tol)
: Prgm: Local df, xi, xnew, err, dgts: ClrIO:
: getMode("display digits") →dgts
: setMode("display digits","fix 12")
: d(f,xvar) →df : x0 →xnew: 2*tol →err
: Disp "{      xi            f(xi)      }"
: While err>tol
:    xnew →xi
:    Disp {xi, f|xvar=xi}
:    xvar–f/df|xvar=xi →xnew
:    If xnew≠0 Then : abs((xi–xnew)/xnew) →err
:    Else : abs((xi–xnew)/tol) →err
:    EndIf
: EndWhile
: Disp {xnew, f|xvar=xnew}
: setMode("display digits",dgts)
: EndPrgm
```

This program takes arguments f, *xvar*, $x0$ and *tol*, in which f is an expression in the variable *xvar*, $x0$ is an initial guess at the solution of $f = 0$, and *tol* is the desired "tolerance" for the resulting approximation. (The program stops when an estimate of the relative error becomes less than *tol*.) The program is also designed to display, in tabular form, each iterate together with the corresponding value of the function f.

• EXAMPLE 1. *Use Newton's Method to find the positive solution of*

$$x^2 = \cos x$$

with a relative error of no more than 10^{-8}. *Use an initial guess of* $x_0 = 1$, *and display each iterate and the corresponding function value.*

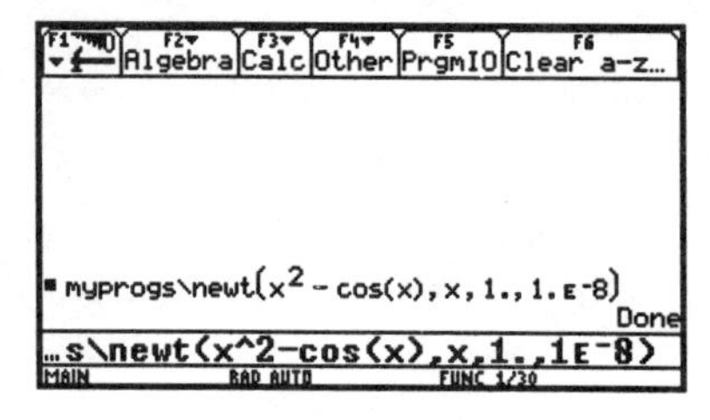


```
{      xi              f(xi)        }
{1.000000000000   .459697694132}
{.838218409905    .033821688041}
{.824241868226    .000261002758}
{.824132319051    .000000016076}
{.824132312303    -1.000000000000E-14}
MAIN          RAD AUTO          FUNC 4/30
```

It is important to note that this program does its computations symbolically or in exact rational arithmetic unless either $x0$ is an "approximate" real number or the **Exact/Approx MODE** is set to **APPROXIMATE**. The simplest way to ensure that the program uses approximate real arithmetic is to enter $x0$ with a decimal point (as in Example 1 above).

• EXAMPLE 2. *Use Newton's Method to approximate* $\sqrt[3]{2}$, *with a relative error of no more than* 10^{-8}. *Use exact rational arithmetic with initial guess* $x_0 = 5/4$, *and display each iterate and the corresponding function value.*

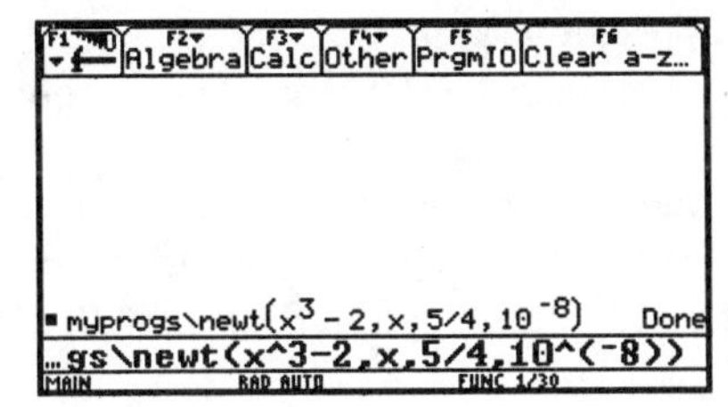

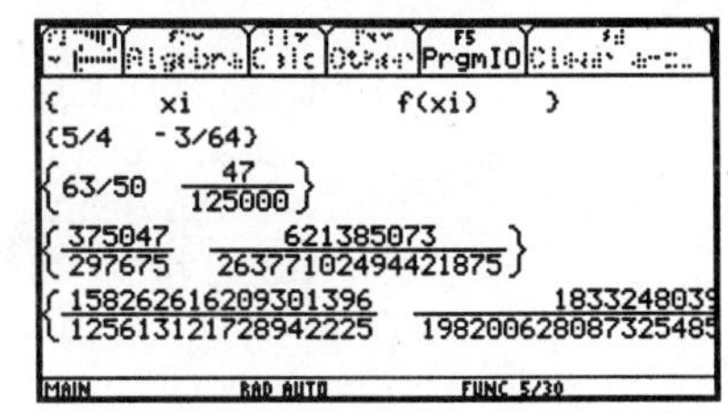

Exercises

1. Compute π by applying **newt()** to the function $f(x) = \tan(.25x) - 1$ with initial guess $x_0 = 3$ and tolerance **tol** $= 10^{-8}$.

2. Compute $\sqrt{3}$ by applying **newt()** to the function $f(x) = x^2 - 3$ with initial guess $x_0 = 2$ and tolerance **tol** $= 10^{-8}$. Do this once using exact rational arithmetic and then again using approximate real arithmetic.

3. Graph the function $f(x) = \sin(x^2) - \sin^2 x$ on the interval $0 \le x \le \pi$, and use **Trace** to find rough initial guesses for each of the zeros of f in this interval. Then use **newt()** to find each of the zeros.

4. The equation $3 + \ln x = \sqrt{x}$ has two solutions. Use the **sgnchng()** function from Section 2.5 to bracket the least of the solutions between multiples of .01 and the greater of the solutions between consecutive integers. Then use **newt()**, with appropriate initial guesses, to find the two solutions.

5. The reciprocal of π can be computed by applying Newton's Method to the function $f(x) = 1/x - \pi$. Experimentally find estimates for the endpoints of the interval containing precisely those initial guesses x_0 for which the Newton iteration converges. Explain your findings with the help of a graph of the function. Can you find both endpoints *exactly*?

6. **Programming challenge.** Create a program, **grphnewt(***f, xvar, x0, tol***)**, that will graphically depict Newton's Method and report the solution as in the following picture.

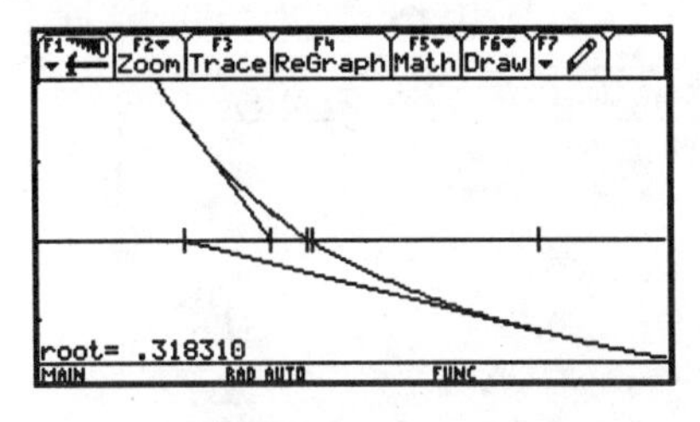

5 Integration

The fundamental theme underlying all of calculus is calculation of the limit of successively improved approximations. This idea, applied to the problem of finding the slope of a curve, leads to the definition of the derivative. The same idea, when applied to the problem of finding the area under a curve, leads to the definition of the definite integral. In this chapter, we will use the **TI-89/92** to explore many of the basic concepts surrounding integration, which are more thoroughly developed in Section 4.10 and Chapter 5 of Stewart's *Calculus*.

5.1 Antiderivatives

Given a function $f(x)$, any function $F(x)$ such that $F'(x) = f(x)$ is called an antiderivative of f. If $F(x)$ is an antiderivative of f, then so is $F(x)+C$ for any constant C, since the derivative of any constant function is zero. In fact, given any antiderivative $F(x)$, *every* antiderivative must have the form $F(x)+C$ for some constant C. The family of all antiderivatives of f is called the *indefinite integral* of f, denoted by

$$\int f(x)dx.$$

So given any antiderivative $F(x)$ of f, we can describe the indefinite integral by writing

$$\int f(x)dx = F(x) + C.$$

For example, since we know that the derivative of x^3 is $3x^2$ and that the derivative of $\sin x$ is $\cos x$, we have

$$\int 3x^2 dx = x^3 + C \quad \text{and} \quad \int \cos x\, dx = \sin x + C.$$

The **TI-89/92** has the ability to antidifferentiate many types of functions. This is done with the $\int$**()** operator (**[2nd]-7**), also found in the Home screen **Calc** menu (**F3-2**). To compute the indefinite integral of a function $f(x)$, we would enter $\int$**(f(x),x,c)**, as seen in the following examples.

F1 F2 Algebra F3 Calc F4 Other F5 PrgmIO F6 Clear a–z...

$\int(12\cdot x^3 - 3\cdot x - 5)dx$ $\quad 3\cdot x^4 - \frac{3\cdot x^2}{2} - 5\cdot x + c$

$\int(x^2\cdot\sqrt{1+x^3})dx$ $\quad \frac{2\cdot(x^3+1)^{3/2}}{9} + c$

∫(x^2√(1+x^3),x,c)

MAIN RAD AUTO FUNC 2/30

F1 F2 Algebra F3 Calc F4 Other F5 PrgmIO F6 Clear a–z...

$\int(x^2\cdot\sqrt{1+x^3})dx$ $\quad \frac{2\cdot(x^3+1)^{3/2}}{9} + c$

$\int((\sin(x))^2)dx$ $\quad \frac{^-\sin(x)\cdot\cos(x)}{2} + \frac{x}{2} + c$

$\int\left(\frac{x+3}{x+1}\right)dx$ $\quad 2\cdot\ln(|x+1|) + x + c$

$\int(x\cdot e^{-x})dx$ $\quad {}^-x\cdot e^{-x} - e^{-x} + c$

∫(xe^(⁻x),x,c)

MAIN RAD AUTO FUNC 5/30

If the third argument is omitted, the $\int()$ operator returns a representative antiderivative with $C = 0$, which suffices in many situations that call for antidifferentiation.

By definition, the derivative of any antiderivative of f is f. This can be illustrated by applying the derivative operator $d()$ to the result of the $\int()$ operator.

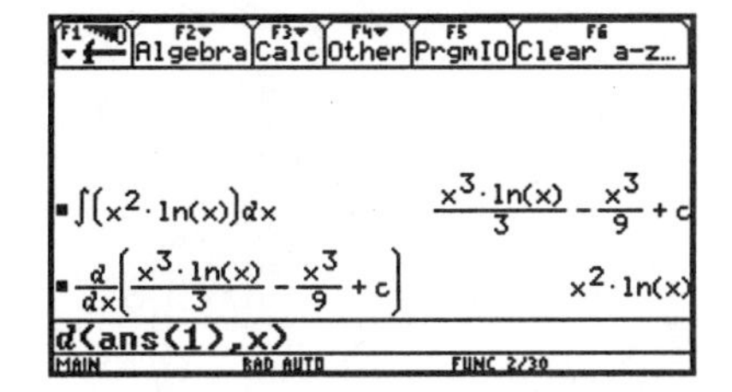

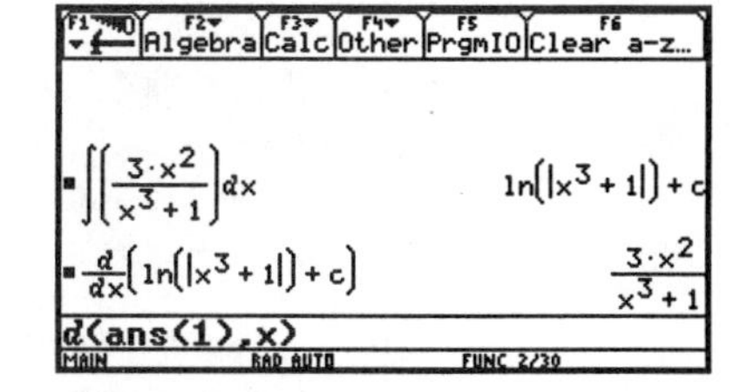

Also, any function f is an antiderivative of its derivative; that is,

$$\int f'(x)dx = f(x) + C.$$

This can be illustrated by applying the $\int()$ operator to the result of the derivative operator $d()$.

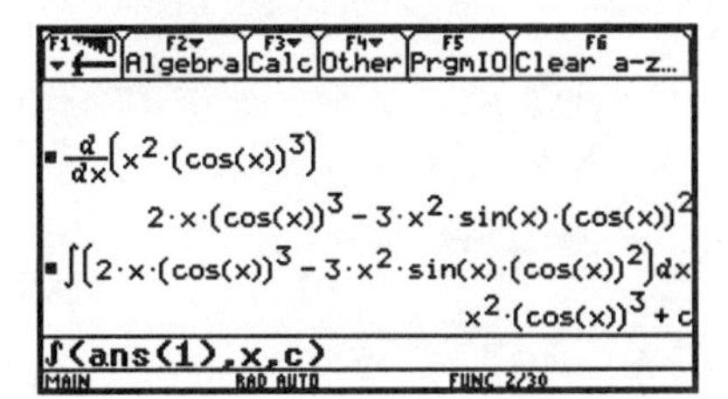

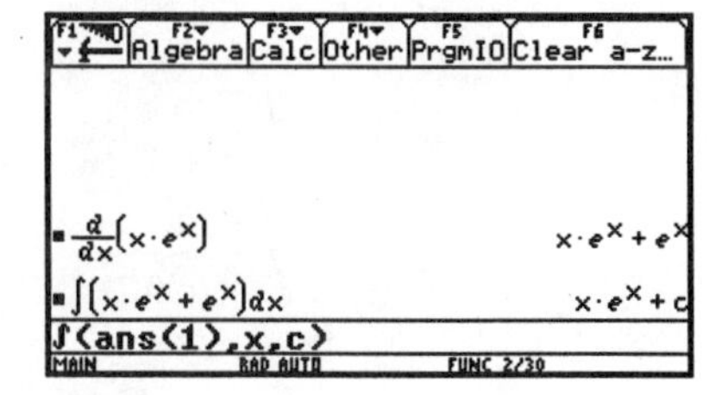

Not every function can be antidifferentiated in this way. For example, let's try to find antiderivatives for $\sin x^2$ and $(1 + x^3)^{-1/2}$.

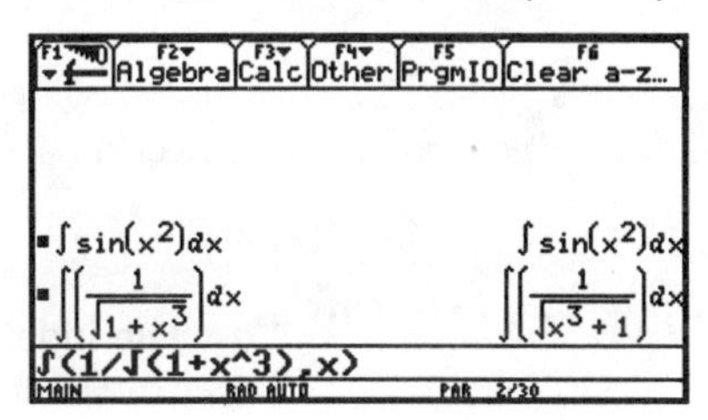

The difficulty here is not that these functions do not have antiderivatives, but that their antiderivatives cannot be expressed in terms of other elementary functions. Soon we will discuss a way of expressing antiderivatives of such functions.

Exercises

In Exercises 1–5, find the indefinite integral by hand, using basic antidifferentiation rules from Section 4.10 in Stewart's *Calculus*. Then check your work by having the **TI-89/92** do the computation.

1. $\int (x^3 - 6x + 3)\,dx$

2. $\int 5\sqrt{x}\,dx$

3. $\int \frac{dx}{7x^2}$

4. $\int (\cos x - \sin x)\,dx$

5. $\int (x + e^x)\,dx$

In Exercises 6–10, use the **TI-89/92** to compute the indefinite integral. Then check the result by computing its derivative.

6. $\int x\sqrt{x+1}\,dx$

7. $\int \frac{x+3}{x+1}\,dx$

8. $\int \frac{dx}{9+4x^2}$

9. $\int e^{-x} \cos 2x\,dx$

10. $\int \sin^3 x\,dx$

In Exercises 11 and 12, find and graph the function f that satisfies the given conditions. (See examples in Section 4.9 of Stewart's *Calculus*.)

11. $f'(x) = x^2 - 4x$, $f(1) = 2$

12. $f'(x) = x\,e^{-x}$, $f(0) = 3$

5.2 Limits of sums and the area under a curve

In this section we will apply the basic idea of calculus—calculation of the limit of successively improved approximations—to the problem of finding the area under the graph of a nonnegative function. So consider the problem of finding the area of the region in the plane bounded by the graph of $y = x^2$, the x-axis, and the line $x = 1$. Let's first get a picture of this region by graphing $y = x^2$ and then shading the region under the graph between $x = 0$ and $x = 1$.

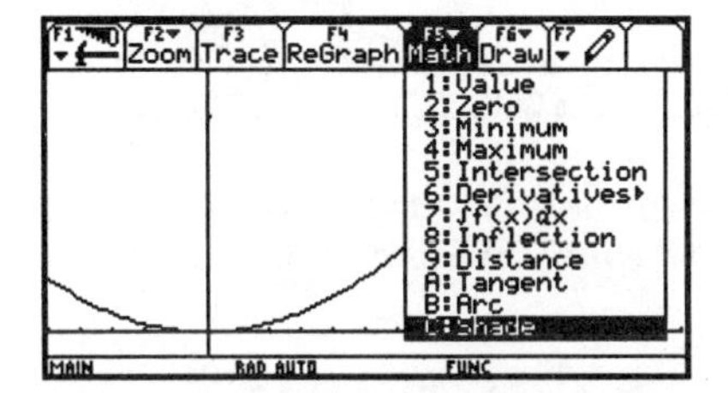

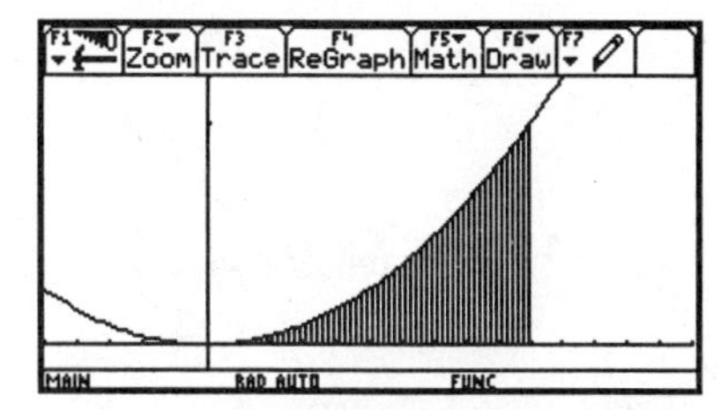

One approach to approximating this area is to approximate the region with a collection of adjacent, non-overlapping rectangles, each of which has its height given by some value of the function $f(x) = x^2$.

If we decide to approximate the region with rectangles whose top-left corners touch the curve, then the resulting area approximation amounts to finding the area of a region such as the one shaded below, which shows ten such rectangles, each with width $\Delta x = 0.1$. (Notice the use of the **floor()** function to graph the "step function" in the picture. See Section 2.1 for more on this.)

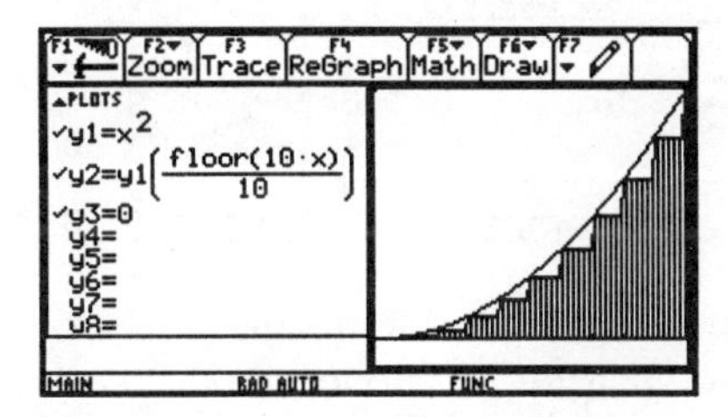

The heights of these rectangles are, respectively,

$$0^2,\ .1^2,\ .2^2,\ .3^2,\ .4^2,\ \ldots,.9^2,$$

or, $\left((i-1)\cdot 0.1\right)^2$ for $i = 1, 2, \ldots, 10$. Thus the areas of the rectangles are, respectively, $\left((i-1)\cdot 0.1\right)^2(0.1)$ for $i = 1, 2, \ldots, 10$, and so the sum of their areas is

$$\sum_{i=1}^{10} \left((i-1)\cdot 0.1\right)^2(0.1) = 0.285.$$

Such an approximation to the area is called a *left-endpoint approximation.* For this particular function, a left-endpoint approximation provides an approximation by *inscribed rectangles,* which is guaranteed to give an underestimate of the true area under the curve.

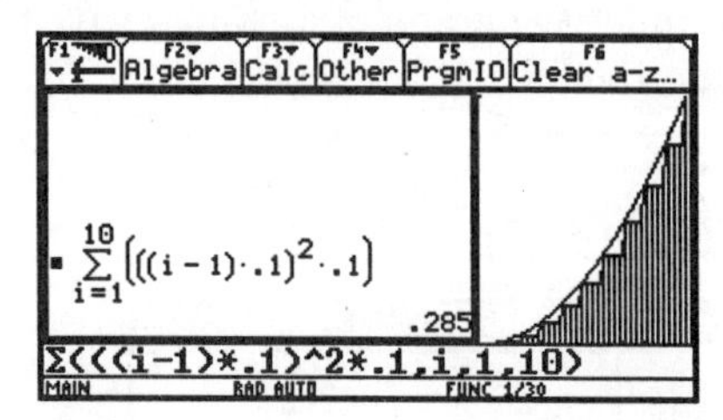

Choosing to have the top-right corner of each rectangle touch the curve results in a *right-endpoint approximation.* The heights of these rectangles are, respectively,

$$.1^2,\ .2^2,\ .3^2,\ .4^2,\ \ldots, 1^2,$$

or, $(i \cdot 0.1)^2$ for $i = 1, 2, \ldots, 10$. Thus the areas of the rectangles are, respectively, $(i \cdot 0.1)^2(0.1)$ for $i = 1, 2, \ldots, 10$, and so the sum of their areas is

$$\sum_{i=1}^{10} (i \cdot 0.1)^2(0.1) = 0.385.$$

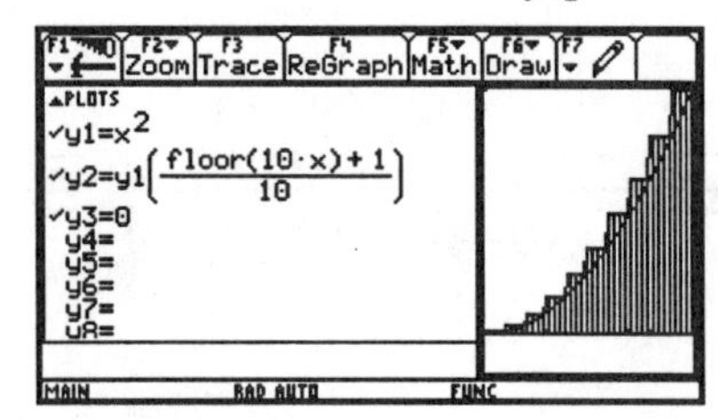

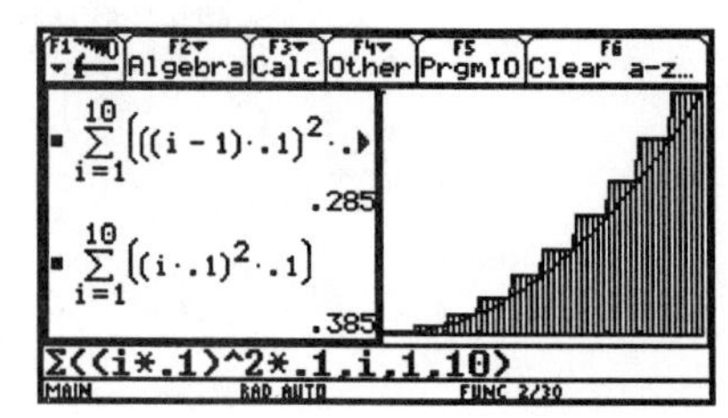

For this particular function, a right-endpoint approximation provides an approximation by *circumscribed rectangles,* which is guaranteed to give an over-estimate of the true area under the curve.

An approximation using ten rectangles that is better than either of those above can be had by letting the midpoint of the top of each rectangle touch the curve. The areas of these rectangles are, respectively, $((i - .5) \cdot 0.1)^2(0.1)$ for $i = 1, 2, \ldots, 10$, and so the sum of their areas is

$$\sum_{i=1}^{10} ((i - .5) \cdot 0.1)^2(0.1) = 0.3325.$$

Such an approximation to the area is called a *midpoint approximation.*

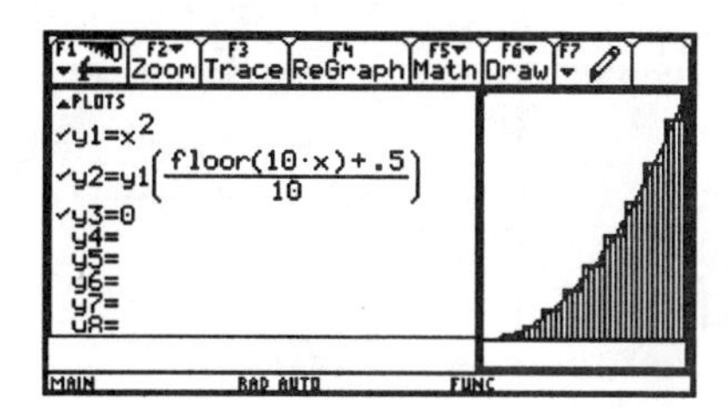

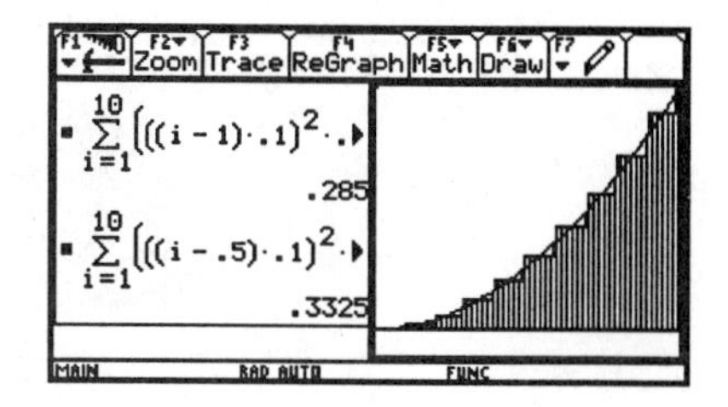

It should be quite obvious that each of the above approximations can be improved by simply using more rectangles. In fact, the exact area can be obtained by taking the limit of any of these types of approximations as the number of rectangles approaches infinity. So let's compute, for simplicity, the right-endpoint approximation with n rectangles, where n may be any positive integer. By analogy with what we saw above, this area is

$$\sum_{i=1}^{n} (i/n)^2 (1/n).$$

Let's now compute a "closed form" of this sum in terms of n, then several values of the sum for increasing values of n, and finally the limit as $n \to \infty$.

```
Σ(i=1..10) ((i·.1)^2·.1)                    .385
Σ(i=1..n) ((i/n)^2·(1/n))     (n+1)·(2·n+1)/(6·n^2)
(n+1)·(2·n+1)/(6·n^2) → area(n)             Done
ans(1)→area(n)
MAIN          RAD AUTO          FUNC 4/30
```

```
(n+1)·(2·n+1)/(6·n^2) → area(n)             Done
area(10.)                                   .385
area(100.)                                .33835
area(1000.)                             .3338334
lim (n→∞) area(n)                            1/3
limit(area(n),n,∞)
MAIN          RAD AUTO          FUNC 8/30
```

So we have determined that the exact value of the area under the curve is $\frac{1}{3}$.

In general, the area under the graph of a continuous, nonnegative function $f(x)$ between $x = a$ and $x = b$ can be found by computing

$$\lim_{n\to\infty} \sum_{i=1}^{n} f(a + i\Delta x)\Delta x,$$

where $\Delta x = (b - a)/n$.

For instance, the area under the graph of $f(x) = \sin x$ between $x = 0$ and $x = \pi$ is

$$\lim_{n\to\infty} \sum_{i=1}^{n} \sin\left(\frac{i\pi}{n}\right)\frac{\pi}{n}.$$

A closed form of this sum is not possible; nevertheless, we can estimate the limit numerically.

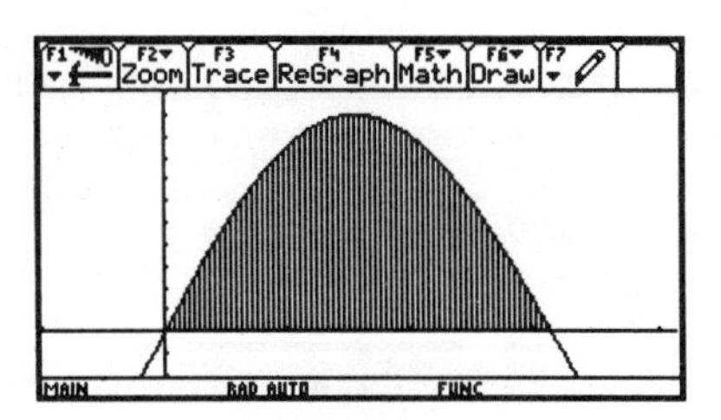

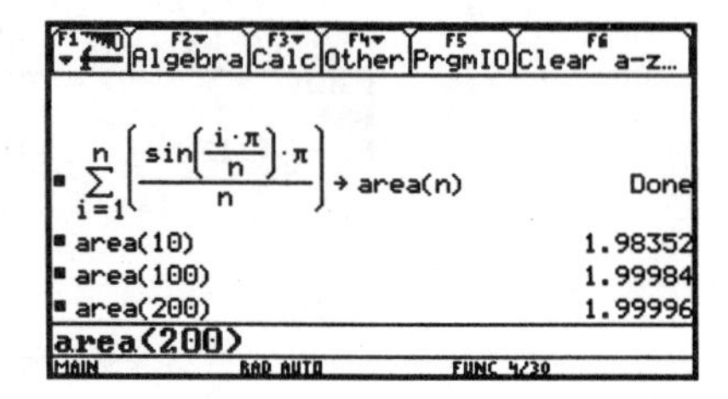

The computations shown suggest that the area under the curve equals 2.

Exercises

In Exercises 1–3, graph the function and shade the area under the curve. Then, using ten rectangles, compute right-endpoint, left-endpoint, and midpoint approximations to the area under the curve.

1. $f(x) = \sqrt{x}, \quad 0 \leq x \leq 1$
2. $f(x) = e^x, \quad -1 \leq x \leq 1$
3. $f(x) = \sin^2 x\ , \quad 0 \leq x \leq \pi$

In Exercises 4 and 5, graph the function and compute a closed form for the sum

$$\sum_{i=1}^{n} f(a + i\Delta x)\Delta x,$$

where $\Delta x = (b - a)/n$. Then compute the limit as $n \to \infty$ to obtain the exact area under the curve.

4. $f(x) = 1 - x^2, \quad -1 \leq x \leq 1$
5. $f(x) = x^2(2 - x)^2, \quad 0 \leq x \leq 2$

In Exercises 6 and 7, graph the function and compute the sum

$$\sum_{i=1}^{n} f(a + i\Delta x)\Delta x,$$

where $\Delta x = (b - a)/n$, for $n = 10,\ 50$, and 100. Then make an estimate of the exact area under the curve and comment on its accuracy.

6. $f(x) = 1/x, \quad 1 \leq x \leq 2$
7. $f(x) = \sin(x^2/\pi), \quad 0 \leq x \leq \pi$

8. Jump ahead to Section 5.6, "Programming notes." Enter the program "**leftbox()**" given there and do Exercises 1–3 in that section.

5.3 The Definite Integral and the Fundamental Theorem of Calculus

Given a continuous function f, defined on an interval $[a, b]$, and the $n+1$ equally spaced points

$$x_0 = a,\ x_1 = a + \Delta x,\ x_2 = a + 2\Delta x,\ \ldots,\ x_n = b,$$

where $\Delta x = (b-a)/n$, we define *the definite integral of f from a to b* by

$$\int_a^b f(x)\,dx = \lim_{n\to\infty} \sum_{i=1}^{n} f(x_i^*)\Delta x,$$

where, for each i, x_i^* is any point chosen from the interval $[x_{i-1}, x_i]$. This is well-defined because it turns out that the defining limit has the same value regardless of how the x_i^*'s are chosen. Thus the definite integral can be computed as the limit of right-endpoint, left-endpoint, or midpoint approximations.

The definite integral $\int_a^b f(x)dx$ can be interpreted in terms of area. There are essentially three cases to consider. These are described in the following three examples.

• EXAMPLE 1. If $f(x) \geq 0$ on $[a, b]$, then the definite integral gives the area under the graph of f between a and b. Consider

$$\int_0^1 (2 - x^2)dx:$$

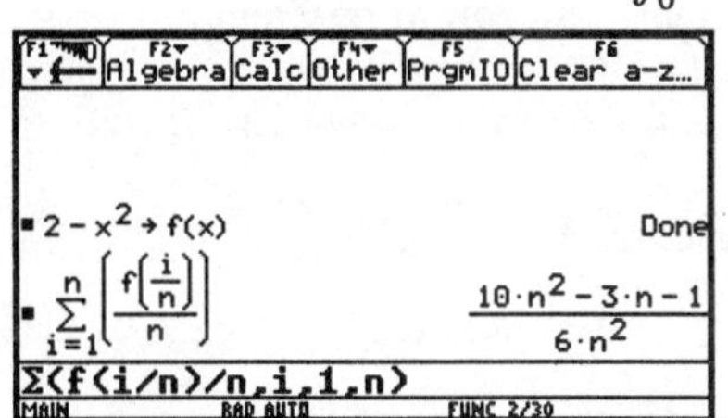

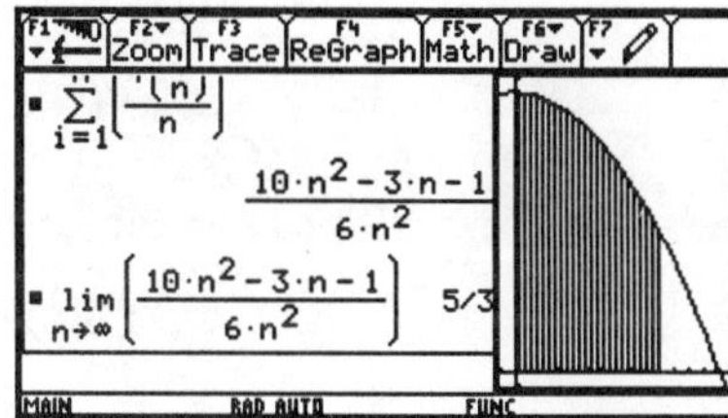

• EXAMPLE 2. If $f(x) \leq 0$ on $[a, b]$, then the definite integral gives the negative of the area "above" the graph of f between a and b. Consider

$$\int_0^1 (x^3 - 1)dx:$$

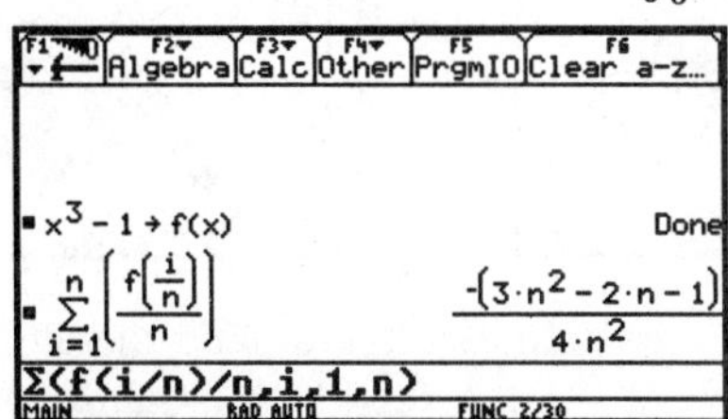

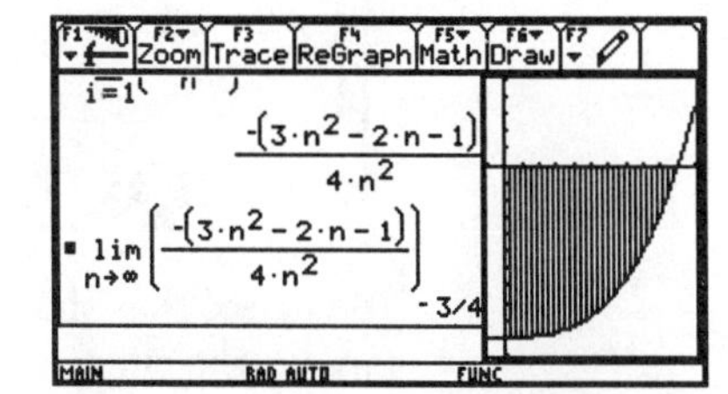

• EXAMPLE 3. If $f(x)$ changes sign on $[a, b]$, then the definite integral gives the "net area" under the graph of f between a and b. By "net area" we mean the area above the x-axis minus the area below. Consider

$$\int_0^2 (x^2 - 1)dx :$$

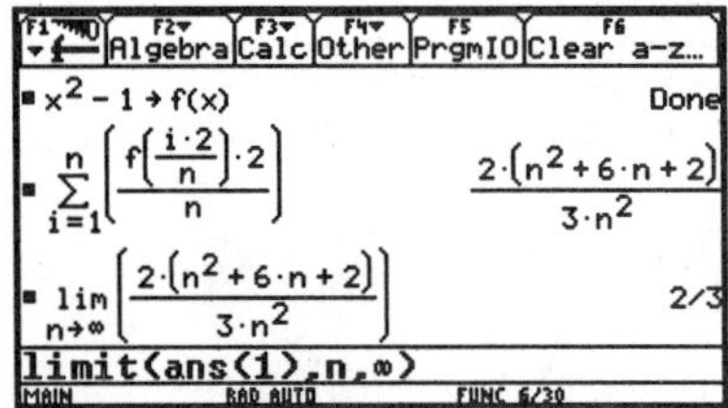

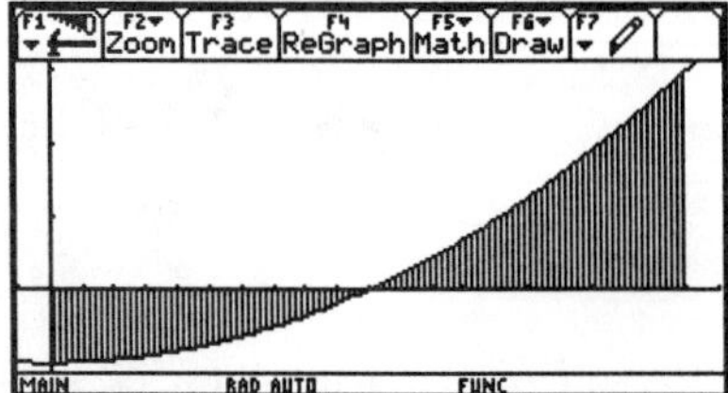

The Fundamental Theorem of Calculus. If $F'(x) = f(x)$ for all x in the interval $[a, b]$, then

$$\int_a^b f(x)\,dx \quad = \quad F(x)\Big|_a^b \quad = \quad F(b) - F(a).$$

This very powerful theorem allows us to calculate definite integrals very easily—when an antiderivative is available—without using the basic limit-definition of the definite integral.

• EXAMPLE 4. In Section 2 of this chapter we conjectured, based on numerical evidence, that $\int_0^\pi \sin x\,dx = 2$. An antiderivative of $\sin x$ is $-\cos x$; so according to the Fundamental Theorem of Calculus,

$$\int_0^\pi \sin x\,dx = -\cos x\Big|_0^\pi = (-\cos\pi) - (-\cos 0) = 2.$$

Thus our conjecture was correct.

• EXAMPLE 5. Let's use the Fundamental Theorem of Calculus to compute

$$\int_0^2 x^3 e^{-x^2}\,dx.$$

We will use the $\int()$ operator to find an antiderivative of $x^3e^{-x^2}$.

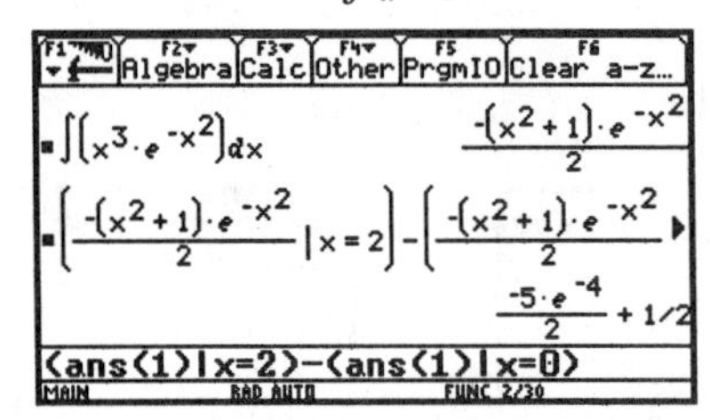

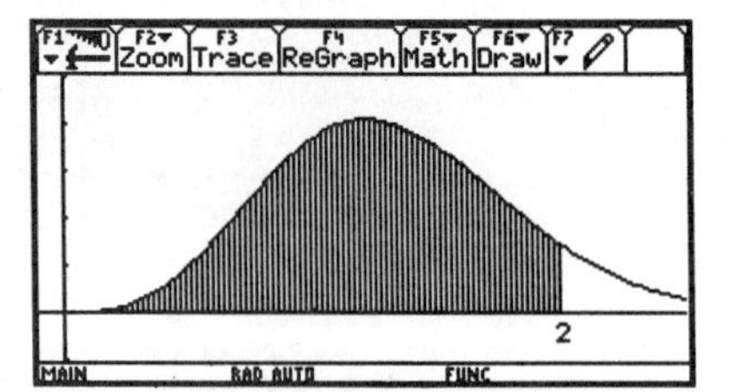

So we see that the value of the definite integral, and therefore the shaded area under the curve, is *exactly* $(1 - 5e^{-4})/2$, which is *approximately* 0.454211.

The **TI-89/92**'s **∫()** operator can also be used to compute definite integrals directly. The limits of integration are simply entered as third and fourth arguments: **∫(f(x),x,a,b)**. When the **Exact/Approx MODE** is set to **AUTO**, **∫()** returns an exact form (using an antiderivative and the Fundamental Theorem of Calculus) whenever possible and a numerical approximation when no antiderivative can be found.

```
F1 F2 Algebra F3 Calc F4 Other F5 PrgmIO F6 Clear a-z...
■ ∫_a^b (x^2)dx                    -a^3/3 + b^3/3
■ ∫_0^2 √(2-x) dx                  4·√2/3
■ ∫_0^(π/3) (x·cos(x))dx           π·√3/6 - 1/2
∫(x*cos(x),x,0,π/3)
MAIN        RAD AUTO        FUNC 3/30
```

```
F1 F2 Algebra F3 Calc F4 Other F5 PrgmIO F6 Clear a-z...
■ ∫_0^(π/3) (ln(x + 1)·cos(x))dx   .31634
■ ∫_1^2 sin(x^2)dx                 .494508
■ ∫_-1^1 (e^(-x^2))dx              1.49365
∫(e^(-x^2),x,-1,1)
MAIN        RAD AUTO        FUNC 6/30
```

Another equally important part of the Fundamental Theorem of Calculus (indeed the *first part*— see Section 5.3 of Stewart's *Calculus*) states that any continuous function f on an interval $[a, b]$ has an antiderivative on that interval, given by

$$F(x) = \int_a^x f(t)\, dt.$$

This is especially important when f does not have an antiderivative that can be expressed in terms of elementary functions.

• EXAMPLE 6. The function $f(x) = \sin x^2$ has an antiderivative, namely,

$$F(x) = \int_0^x \sin t^2\, dt.$$

We can define this function on the **TI-89/92**, compute values, find its derivative, and even graph it.

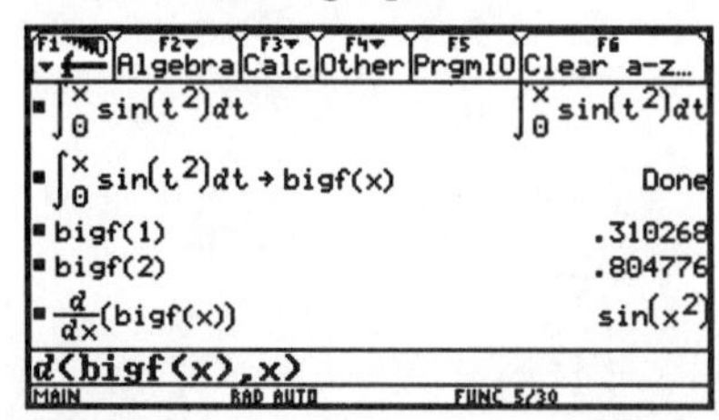

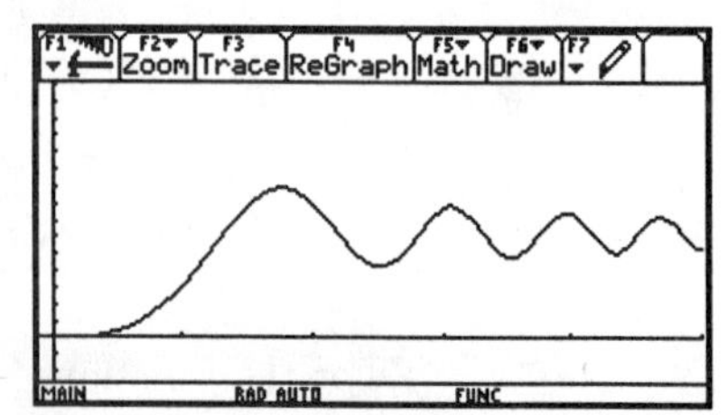

Warning: It takes quite some time for the **TI-89/92** to plot this graph. This is not surprising, since each evaluation of the function is done by a computationally intensive approximation procedure. It does speed things up a bit, however, if the window variable **xres** is increased to at least 4 or 5.

The function $F(x) = \int_0^x \sin t^2\, dt$ in Example 6 is related to (in particular a multiple of) the *Fresnel function*, or *Fresnel Sine Integral.*

Finally, let's look at a very nice item in the **Math** menu of the Graph screen. There you'll find **∫f(x)dx** (**F5-7**), a command that prompts you for lower and upper bounds and then computes the definite integral of the graphed function between those bounds—and simultaneously shades the region between the graph and the x-axis! The screens below show this done for the function $f(x) = \sin(x^2)$, plotted on the interval $0 \le x \le 4$, and integrated from $x = 0$ to $x = 3$.

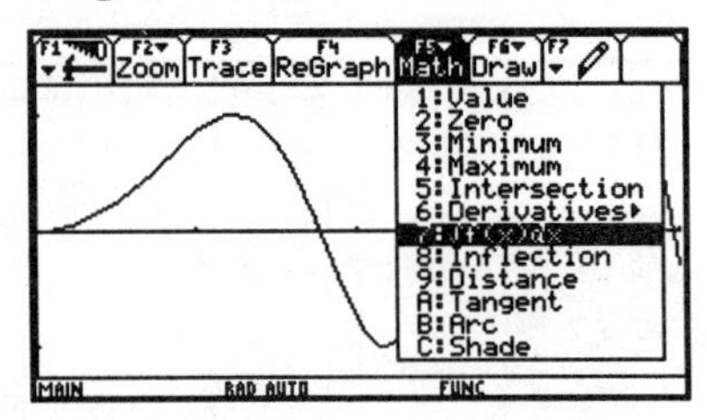

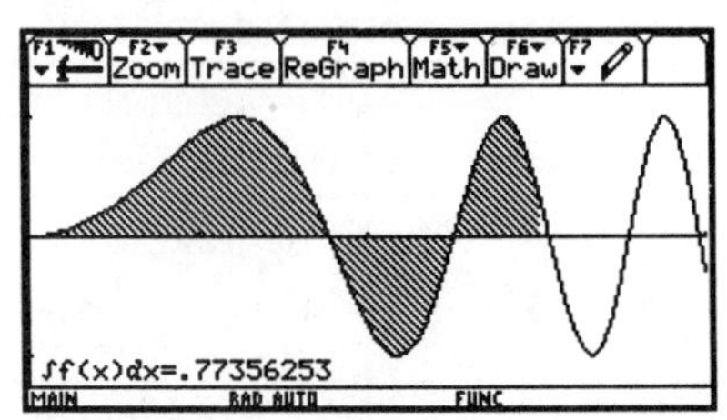

Exercises

In Exercises 1–5, evaluate $\int_a^b f(x)dx$ by first finding an antiderivative $F(x)$ and then evaluating $F(x)\Big|_a^b$. Check your answer by entering **∫(f(x),x,a,b)**.

1. $\displaystyle\int_0^1 \frac{dx}{x+1}$
2. $\displaystyle\int_{-1}^1 \frac{dx}{1+x^2}$
3. $\displaystyle\int_0^3 x\,e^{-x}\,dx$
4. $\displaystyle\int_{-1}^1 x\cos \pi x\,dx$
5. $\displaystyle\int_{-1}^1 x\sin \pi x\,dx$

In Exercises 6–10, compute the definite integral. Graph the function on the relevant interval and then interpret the value of the integral as an area, the negative of an area, or a "net" area.

6. $\displaystyle\int_0^1 20x^3(x-1)\,dx$
7. $\displaystyle\int_0^{2\pi} \sin x\,dx$
8. $\displaystyle\int_0^{2\pi} \sin^2 x\,dx$
9. $\displaystyle\int_{-1}^1 x\sin \pi x\,dx$
10. $\displaystyle\int_{-1}^1 x^2\sin \pi x\,dx$

For the functions in Exercises 11 and 12,

a) evaluate $f(x)$ for $x = -2, -1, 0, 1$, and 2;

b) compute $f'(x)$;

c) graph $f(x)$ together with $f'(x)$ for $0 \le x \le 2$. (Set the window variable **xres** to 5. Even with **xres** = 5, the plot will take a while.)

11. $\displaystyle f(x) = \int_0^x e^{-t^2}\,dt$
12. $\displaystyle f(x) = \int_0^x |\sin(3\pi t)|\,dt$

5.4 Approximate integration

Approximation procedures for definite integrals are of primary importance for at least two reasons. First, as we saw in the preceding section, many functions do not have antiderivatives that can be expressed in terms of other elementary functions. Also, many important integration problems arise in which the only thing known about the function to be integrated is a set of values at discrete points.

The Trapezoidal Rule. This procedure is motivated by the idea of using adjacent, non-overlapping, trapezoids (rather than rectangles) to approximate the area under a curve. The base of each trapezoid is on the x-axis, the sides are vertical, and each of the top corners touch the graph of the function. Such a trapezoid has an area of $\left(f(x_{i-1}) + f(x_i)\right)\frac{\Delta x}{2}$, where x_{i-1} and x_i are the endpoints of the base and $\Delta x = x_i - x_{i-1}$ is the width of the base. Summing the areas of n trapezoids produces the *Trapezoidal Rule*:

$$\int_a^b f(x)\,dx \approx \frac{\Delta x}{2}\left(f(a) + 2\sum_{i=1}^{n-1} f(x_i) + f(b)\right),$$

where $\Delta x = (b-a)/n$ and $x_i = a + i\,\Delta x$.

Let's enter this formula as a function **trap**(n):

(b–a)/(2n)∗(f(a)+2∑(f(a+i∗(b–a)/n), i, 1, n–1)+f(b)) →trap(n)

Then we'll define $f(x)$, a, and b. Here we use as an example the function $f(x) = e^{-x^3}$ on the interval $[0, 2]$.

```
F1 F2 Algebra F3 Calc F4 Other F5 PrgmIO F6 Clear a-z...
■ (b-a)/(2·n)·(f(a) + 2·Σ(i=1..n-1) f(a + i·(b-a)/n) + f(b))▸
                                                      Done
...a)/n),i,1,n-1)+f(b))→trap(n)
MAIN        RAD AUTO        FUNC 1/30
```

```
F1 F2 Algebra F3 Calc F4 Other F5 PrgmIO F6 Clear a-z...
■ ◂a) + 2·Σ(i=1..n-1) f(a + i·(b-a)/n) + f(b) → trap(n)
                                                      Done
■ e^(-x^3) → f(x)                                     Done
■ 0 → a                                                  0
■ 2 → b                                                  2
2→b
MAIN        RAD AUTO        FUNC 4/30
```

Now we can compute and compare trapezoidal rule approximations for increasing numbers of trapezoids. The shaded region in the graph indicates the area that is computed when $n = 4$.

```
F1 F2 Algebra F3 Calc F4 Other F5 PrgmIO F6 Clear a-z...
■ 0 → a                                                  0
■ 2 → b                                                  2
■ trap(2)                                          .868047
■ trap(4)                                          .892381
■ trap(8)                                          .892902
■ trap(16)                                         .892946
■ trap(32)                                         .892952
trap(32)
MAIN        RAD AUTO        FUNC 8/30
```

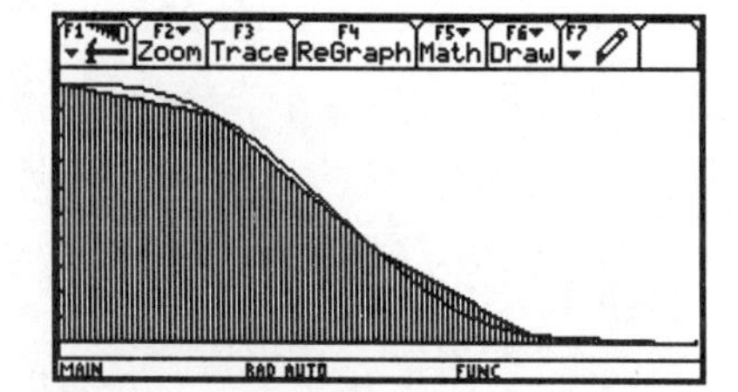

These computations suggest that the value of the integral, to three decimal places, is 0.893. Enter **∫(f(x),x,a,b)** yourself and see how the trapezoidal approximations compare.

Simpson's Rule. The geometric idea behind Simpson's Rule is to approximate the graph of f with a collection of connected parabolas, each of which is determined by three points on the curve. Then the integral of f is approximated by the integral of the resulting piecewise-quadratic approximation to f. All of this results in the formula

$$\int_a^b f(x)\,dx \approx \frac{h}{3}\big(f(a) + 4f(x_1) + 2f(x_2) + 4f(x_3) + \cdots + 2f(x_{n-2}) + 4f(x_{n-1}) + f(b)\big),$$

where $h = (b-a)/n$ and $x_i = a + i\,h$. Note that n must an even number. Also note that the alternating coefficients $4, 2, 4, 2, \ldots$ can be described by the formula $3 + (-1)^{i-1}$, $i = 1, 2, 3, \ldots$. Let's use this in entering the Simpson's rule formula as a function **simp**(n):

3+(-1)^(i–1) →c(i)

(b–a)/(3n)∗(f(a)+∑(c(i)∗f(a+i∗(b–a)/n), i, 1, n–1)+f(b)) →simp(n)

Then we'll apply Simpson's Rule to the integral $\int_0^2 e^{-x^3}\,dx$.

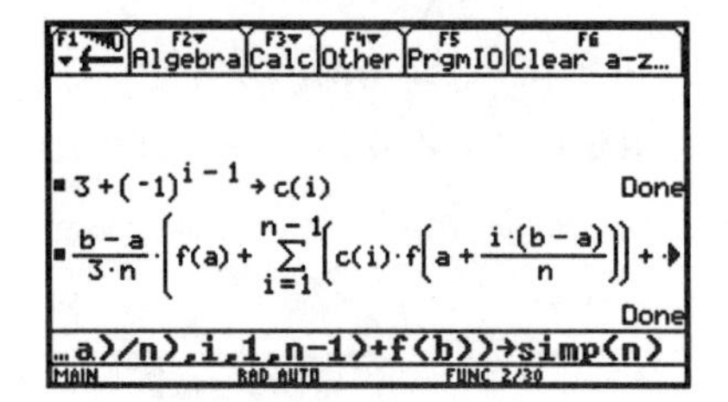

F1 F2 Algebra F3 Calc F4 Other F5 PrgmIO F6 Clear a-z...
e^{-x^3} → f(x) Done
0 → a 0
2 → b 2
simp(4) .900492
simp(8) .893076
simp(16) .892961
simp(32) .892954
simp(32)
MAIN RAD AUTO FUNC 3/30

Approximate integration is essential when function values are known only at discrete points. Suppose for example that we wish to determine the area of an irregularly shaped plot of land. The plot is bounded on two opposite sides by straight, parallel roads. It is bounded on the other two sides by meandering rivers. Eleven equally spaced width measurements, parallel to the two roads, are made by surveyors and shown in the figure below. From this data, we wish to approximate the area of the plot of land.

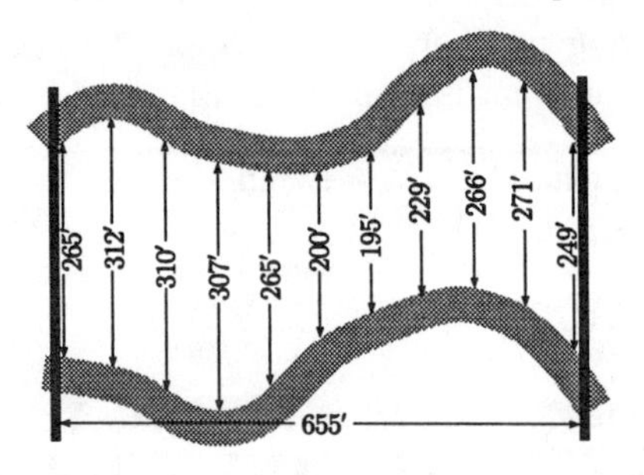

The area is the integral of the width function from 0 to 655. The first step toward approximating this integral is to enter the width data as a list. To do this, press the **APPS** key; then select the **Data/Matrix Editor** and **New...** (**6-3**). In the resulting dialog box, specify **Type: List** and **Variable: y**.

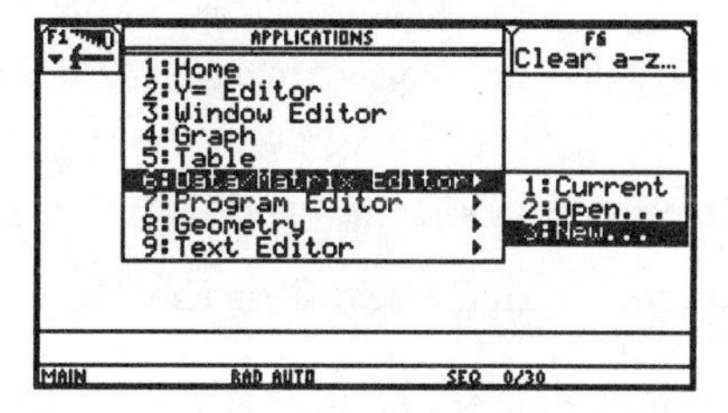

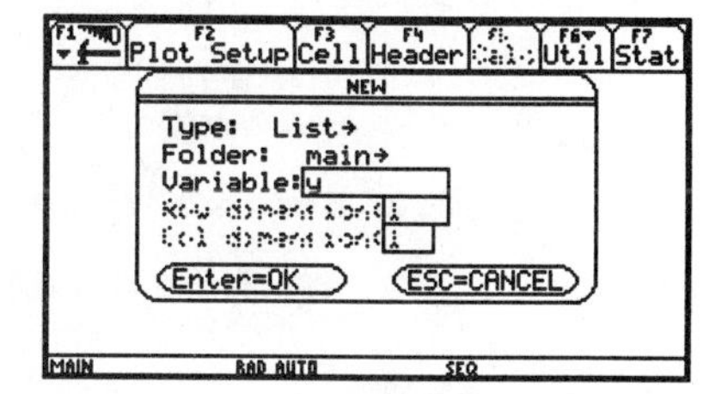

Then we enter the eleven width measurements into the first column and return to the Home screen to modify the formula for Simpson's Rule to handle list data. Note that since the list index begins at 1, the function values are $y_1, y_2, \ldots, y_{n+1}$ rather than $f(x_0), f(x_1), f(x_2), \ldots, f(x_n)$. So we'll redefine **simp(n)** by entering

(b–a)/(3n)∗(y[1] + ∑(c(i)∗y[i+1], i, 1, n–1) + y[n+1]) →simp(n)

LIST	c1	c2	c3	c4	c5
5	265				
6	200				
7	195				
8	229				
9	266				
10	271				
11	249				

r11c1=249

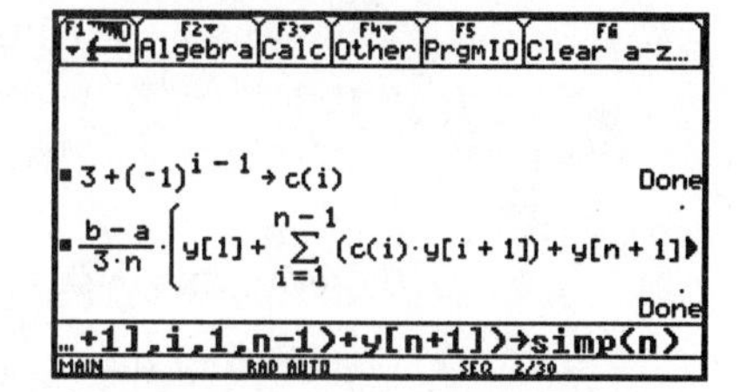

Now, with these chores done, we are ready to calculate the Simpson's Rule approximation to the area.

```
■ 3+(-1)^(i-1) → c(i)                              Done
■ (b-a)/(3·n)·(y[1]+ Σ_{i=1}^{n-1} (c(i)·y[i+1])+y[n+1]▸
                                                   Done
■ 0 → a                                               0
■ 655 → b                                           655
■ simp(10)                                      171654.
simp(10)
MAIN          RAD AUTO          SEQ 5/30
```

Thus the area of the plot of land is approximately 172,000 square feet.

For details on the derivation of the Trapezoidal Rule and Simpson's Rule, see Section 8.7 of Stewart's *Calculus*. Of particular importance are the error bounds discussed there.

Exercises

1. Approximate $\int_{-1}^{1} \sqrt{1+x^3}\,dx$ using the Trapezoidal Rule with $n = 4$, $n = 8$, $n = 16$, and $n = 32$. Compare each result to the value reported by the $\int()$ operator. To how many decimal places are each of the approximations accurate?

2. Rework Exercise 1 using Simpson's Rule.

3. Another method for approximating integrals is the *Midpoint Rule*:

$$\int_a^b f(x)dx \approx h\sum_{i=1}^{n} f((x_{i-1}+x_i)/2)$$

where $h = (b-a)/n$ and $x_k = a + k\,h$. (See Section 8.7 in Stewart's *Calculus* and Section 5.2 of this manual.) Rework Exercise 1 using the Midpoint Rule and compare the accuracy of the results to those obtained with the Trapezoidal Rule.

4. Let M_n, T_n, and S_n denote approximations to $\int_a^b f(x)dx$, using n subintervals, obtained from the Midpoint, Trapezoidal, and Simpson's Rules, respectively. It can be shown that

$$S_{2n} = \frac{1}{3}T_n + \frac{2}{3}M_n.$$

Verify this with the results of Exercises 1, 2, and 3.

5. A factory discharges effluent into a river. The rate of effluent discharge (in cubic meters per minute) is recorded hourly over a 24-hour period. The measurements are shown in the following table.

t	0	1	2	3	4	5	6	7	8	9	10	11	12	13	14	15	16	17	18	19	20	21	22	23	24
$r(t)$	3.2	2.4	4.3	5.5	5.8	6.3	4.7	4.0	5.9	7.2	8.3	8.0	7.3	6.6	5.1	3.8	3.5	3.1	2.2	4.1	5.5	5.1	4.8	4.1	3.7

The total effluent discharge over the 24-hour period is the integral of the rate of discharge. (See the "Total Change Theorem" in Section 5.4 of Stewart's *Calculus*.) Convert each of the rate measurements to cubic meters per hour and use Simpson's rule to approximate the factory's total effluent discharge over this 24-hour period.

6. Apply Simpson's Rule with $n = 4$, $n = 8$, and $n = 16$ to each of the integrals:

$$\int_0^1 x^2dx \qquad \int_0^1 x^3dx \qquad \int_0^1 x^4dx$$

Compare each approximation with the exact value of the integral. What do you observe? Is it surprising? Why?

5.5 Improper integrals

The definition of the definite integral $\int_a^b f(x)dx$ assumes two crucial things: 1) that the interval $[a,b]$ is bounded (i.e., a and b are each finite); and 2) that the integrand f is continuous on $[a,b]$. (Actually, this second assumption can be relaxed considerably, but it is essential that f be defined and bounded on $[a,b]$.) An *improper integral* involves either an interval that is not bounded (Type 1) or an integrand that has a vertical asymptote at one of the endpoints of the interval (Type 2). In all cases, an improper integral is, by definition, a limit of definite integrals.

As described in Section 8.8 of Stewart's *Calculus*, an improper integral of Type 1 (where the interval of integration is of the form $[a,\infty)$) is defined as

$$\int_a^\infty f(x)\,dx = \lim_{t\to\infty}\int_a^t f(x)\,dx.$$

A Type 1 improper integral of the form $\int_{-\infty}^b f(x)\,dx$ is defined similarly.

An improper integral of Type 2, where f is continuous on $(a,b]$ but has a vertical asymptote at $x=a$, is defined as

$$\int_a^b f(x)\,dx = \lim_{t\to a^+}\int_t^b f(x)\,dx.$$

An improper integral of Type 2, where f is continuous on $[a,b)$ but has a vertical asymptote at $x=b$, is defined as

$$\int_a^b f(x)\,dx = \lim_{t\to b^-}\int_a^t f(x)\,dx.$$

An improper integral of any type is said to be convergent if the defining limit exists and divergent if the defining limit does not exist.

The **TI-89/92**'s **∫()** operator handles improper integrals automatically. But in order to get a good understanding of how improper integrals are defined, we should compute a few from scratch (more or less).

• EXAMPLE 1. A typical improper integral of Type 1 is $\int_1^\infty x^{-3/2}dx$. By definition,

$$\int_1^\infty \frac{dx}{x^{3/2}} = \lim_{t\to\infty}\int_1^t \frac{dx}{x^{3/2}}.$$

So first we'll compute $\int_1^t x^{-3/2}dx$ in terms of t and then take the limit as $t\to\infty$. The result is shown below along with a picture of the area under the graph of $y=x^{-3/2}$ to the right of $x=1$.

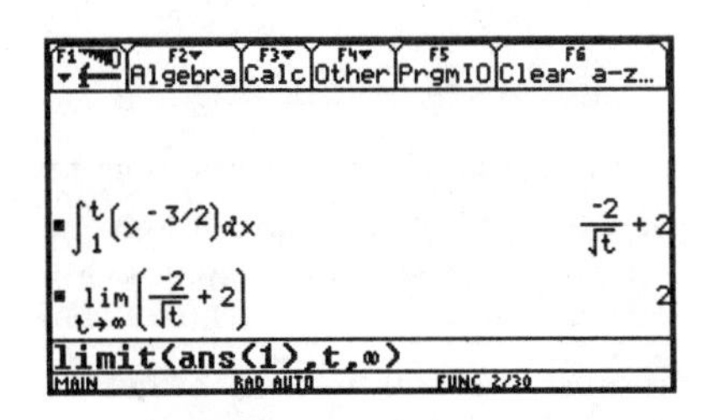

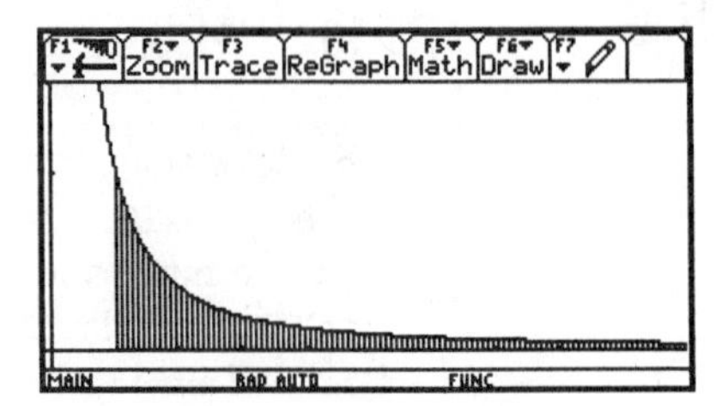

• EXAMPLE 2. A typical improper integral of Type 2 is $\int_0^2 1/\sqrt{e^x - 1}\,dx$. Note that the integrand has a vertical asymptote at $x = 0$. We'll compute $\int_t^2 1/\sqrt{e^x - 1}\,dx$ in terms of t and then take the limit as $t \to 0^+$.

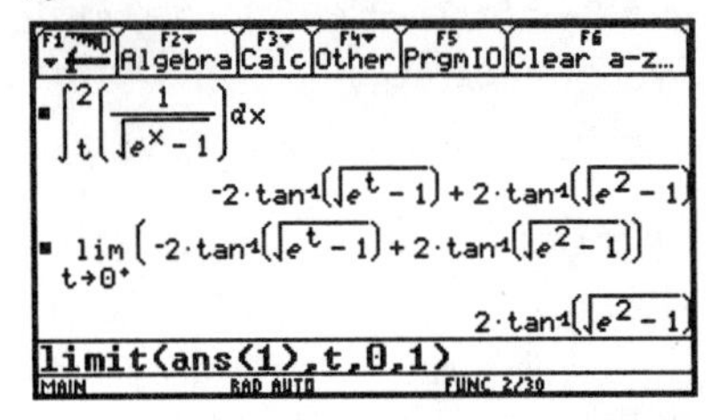

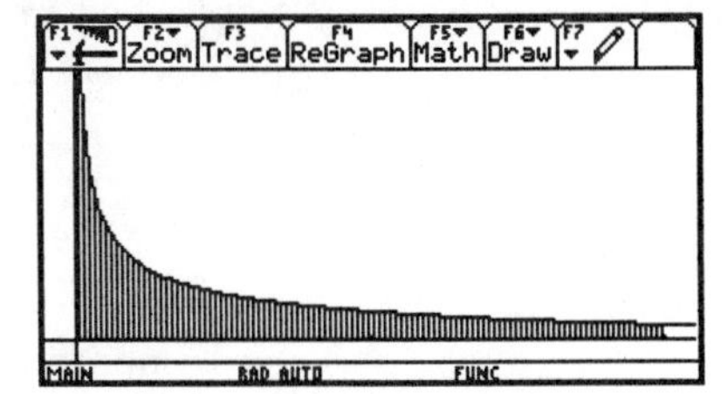

• EXAMPLE 3. Another improper integral of Type 2 is $\int_0^{\pi/2} \tan x\,dx$. Note that the integrand has a vertical asymptote at $x = \pi/2$. We'll compute $\int_0^t \tan x\,dx$ in terms of t and then take the limit as $t \to (\pi/2)^-$.

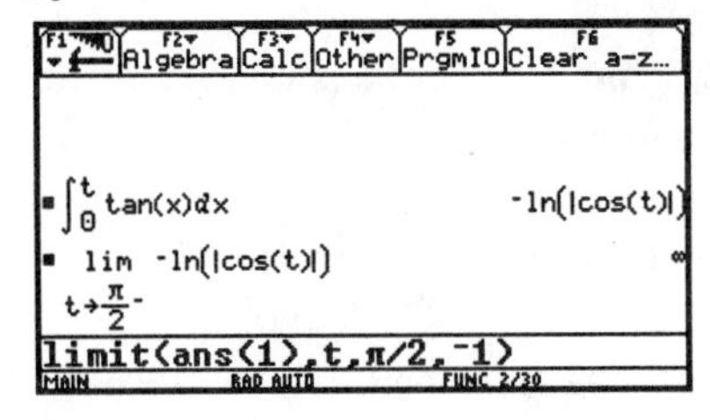

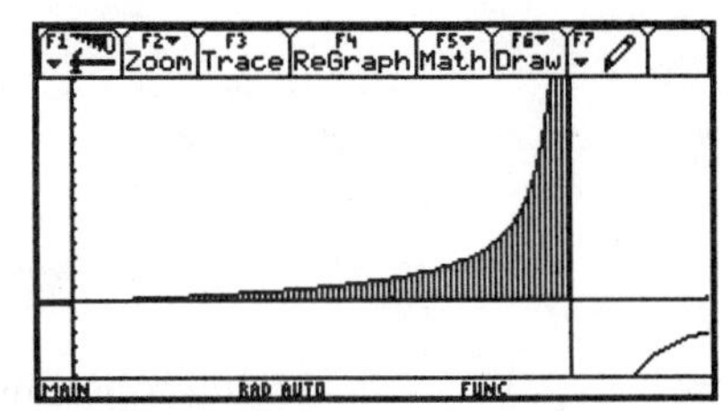

So it turns out that this improper integral is divergent.

As we mentioned earlier, the **TI-89/92**'s **∫()** operator handles improper integrals automatically. Let's recompute the improper integral in each of the preceding examples.

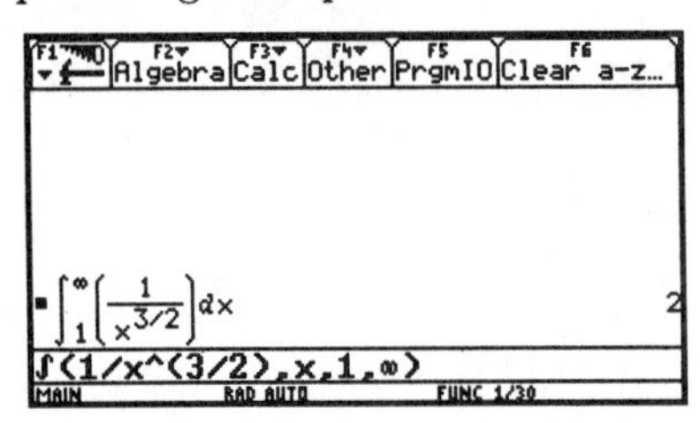

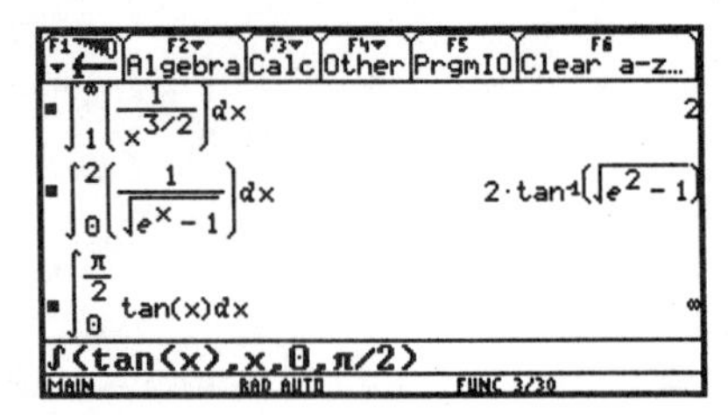

Exercises

For each of the improper integrals in Exercises 1–6, find the value or determine divergence by using the appropriate limit definition. Also graph the integrand and shade the relevant region.

1. $\displaystyle\int_0^2 \frac{dx}{4-x^2}$

2. $\displaystyle\int_0^2 \frac{dx}{x^3+3x^2}$

3. $\displaystyle\int_0^\infty e^{-3x}\,dx$

4. $\displaystyle\int_2^\infty \frac{dx}{x^3-1}$

5. $\displaystyle\int_1^\infty \frac{dx}{\sqrt{x}\,e^{\sqrt{x}}}$

6. $\displaystyle\int_0^1 \frac{dx}{\sqrt{x}\,e^{\sqrt{x}}}$

7. Considering the answers to Exercises 5 and 6, what should be the value of
$$\int_0^\infty \frac{dx}{\sqrt{x}\,e^{\sqrt{x}}}\ ?$$

8. Numerical approximation of an improper integral is a difficult problem because of the fact that either the function or the interval is unbounded. In situations where an approximation to an improper integral is needed, it is often better to compute a few definite integrals and estimate the limit than to try to compute the improper integral to high accuracy all at once.

   a) Estimate the value of $\displaystyle\int_0^\infty \frac{\sqrt{x}\,dx}{e^x+1}$ by computing $\displaystyle\int_0^b \frac{\sqrt{x}\,dx}{e^x+1}$ for $b =$ 10, 20, and 50 with the $\int()$ operator. Then use $\int()$ to evaluate the improper integral. What happens?

   b) Estimate the value of $\displaystyle\int_0^1 \frac{dx}{e^{\sqrt[3]{x}}-1}$ by computing $\displaystyle\int_a^1 \frac{dx}{e^{\sqrt[3]{x}}-1}$ for $a =$ .1, .001, and .00001 with the $\int()$ operator. Then use $\int()$ to evaluate the improper integral. What happens?

9. A function f defined on $(-\infty,\infty)$ is a *probability density function* if:
$$\text{(i) } f(x)\ge 0 \text{ for all } x \text{ in } (-\infty,\infty), \text{ and } \quad \text{(ii) } \int_{-\infty}^{\infty} f(x)dx = 1.$$
(See Section 9.5 in Stewart's *Calculus*.) Find the number k (at least approximately) so that each of the following is a probability density function on $(-\infty,\infty)$. Then graph f.

   a) $f(x) = \dfrac{k}{1+x^2}$

   b) $f(x) = \dfrac{k}{(1+x^2/2)^{3/2}}$

   c) $f(x) = ke^{-x^2/2}$

   d) $f(x) = \begin{cases} kx^2e^{-x/2}, & \text{if } x \ge 0 \\ 0, & \text{if } x < 0 \end{cases}$

5.6 Programming notes

This section is devoted to the creation of three programs named **leftbox()**, **rightbox()**, and **midbox()**. These will graphically depict left-endpoint, right-endpoint, and midpoint approximations, respectively, to the integral of a function. They will also compute and display the resulting approximation.

The first of these programs, **leftbox()**, takes arguments as in

leftbox(f, $xvar$, a, b, n**)**,

and produces the left-endpoint approximation to $\int_a^b f$ with respect to the variable $xvar$ using n rectangles. The code is as follows.

```
: leftbox(f, xvar, a, b, n)
: Prgm:Local i, j, h, xi, yi, s
: ClrGraph: ClrDraw: PlotsOff :FnOff
: string(f | xvar=xx) →ff
: Graph expr(ff), xx
: (b–a)/n →h
: 0. →s
: For i, 1, n
:     a+(i–1)*h →xi
:     f | xvar=xi →yi
:     s+yi*h →s
:     Line xi+h, 0, xi+h, yi
:     Line xi, yi, xi+h, yi
:     Line xi, 0, xi, yi
: EndFor
: PxlText string(s), 95, 0
: DelVar ff
: EndPrgm
```

With this program residing in your **myprogs** folder, entering

myprogs\leftbox(sin(x),x,0,π,20)

produces (in a $[0, \pi] \times [-.5, 1.5]$ window) the screen

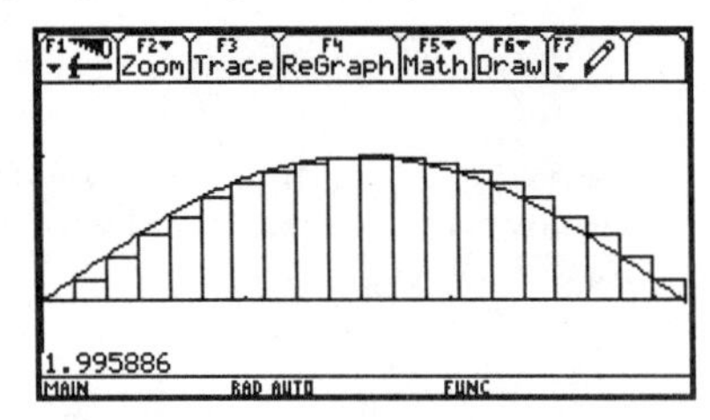

thus giving the approximation $\int_0^\pi \sin x \, dx \approx 1.996$.

Exercises

1. After entering the **leftbox()** program into your **myprogs** folder, use it to compute left-endpoint approximations to the integral

$$\int_1^3 \frac{1}{x}\,dx$$

with $n = 5$, 15, 30, and 60 rectangles.

2. Modify the **leftbox()** program to produce a program **rightbox()** that graphically depicts and computes a right-endpoint approximation to an integral. Then use it to compute right-endpoint approximations analogous to the left-endpoint approximations in Exercise 1.

3. Modify the **leftbox()** program to produce a program **midbox()** that graphically depicts and computes a midpoint approximation to the integral. Then use it to compute midpoint approximations analogous to the left-endpoint approximations in Exercise 1.

4. **Programming challenge.** Modify the **leftbox()** program to produce a program **trapez()** that graphically depicts and computes a trapezoidal approximation to an integral. Use it to compute trapezoidal approximations analogous to the left-endpoint approximations in Exercise 1.

5. **Programming challenge.** Create a function,

simpson($ylist$, h**)**,

that returns the Simpson's rule approximation to the integral of a function with values at equally spaced points specified in *ylist*.

6 Applications of the Integral

Applications of integration abound in mathematics, science, engineering, economics, and numerous other fields. To supplement Chapters 6 and 9 of Stewart's *Calculus*, we will look at a few standard geometric problems that we hope illustrate some of underlying philosophy behind these diverse applications.

6.1 Area

The problem of finding the area under the graph of a function was our original motivation in defining the definite integral. A rather straightforward extension of this idea allows us to calculate the area of a region bounded by two graphs. The principle involved is that *area is the integral of the length of a typical cross-section taken perpendicular to a coordinate axis.*

• EXAMPLE 1. *Find the area of the region bounded by the graphs of*

$$f(x) = 12x(1-x) + \sin(5\pi x) \quad \textit{and} \quad g(x) = \frac{1}{2}\sin(7\pi x).$$

The first step is to plot the graphs to get a picture of the region and find the points of intersection of the two graphs. The points of intersection can be found from the Graph screen with the **Intersection** command (**F5-5**). Then we'll shade the region bounded by the graphs (**F5-C**).

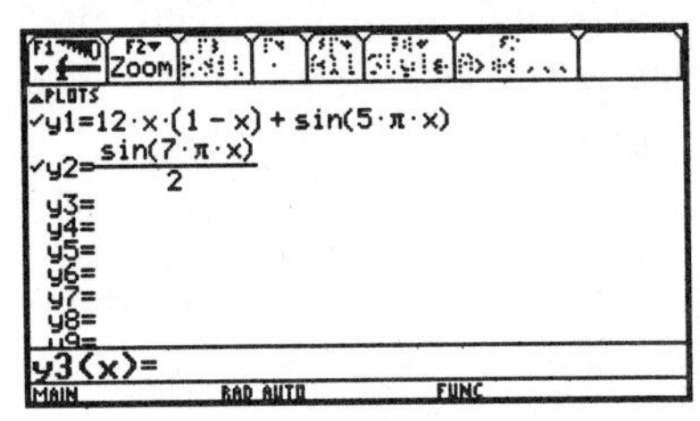

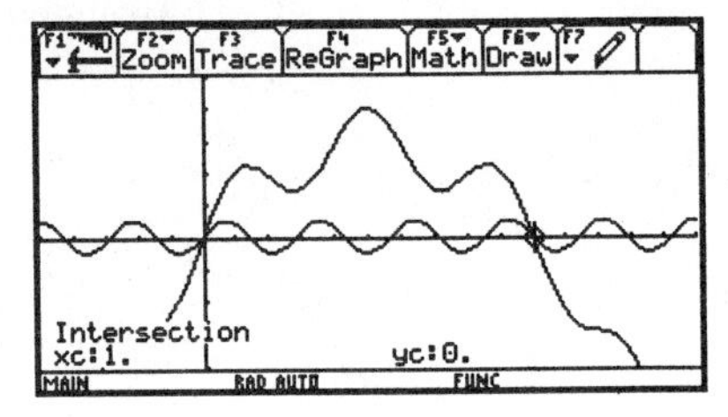

Vertical cross-sections have length given by

$$f(x) - g(x) = 12x(1-x) + \sin(5\pi x) - \frac{1}{2}\sin(7\pi x)$$

for $0 \le x \le 1$. Thus the area of the region is $\int_0^1 (f(x) - g(x))\,dx$.

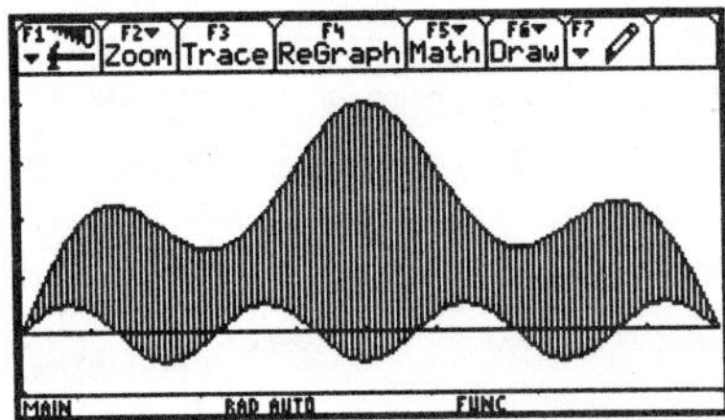

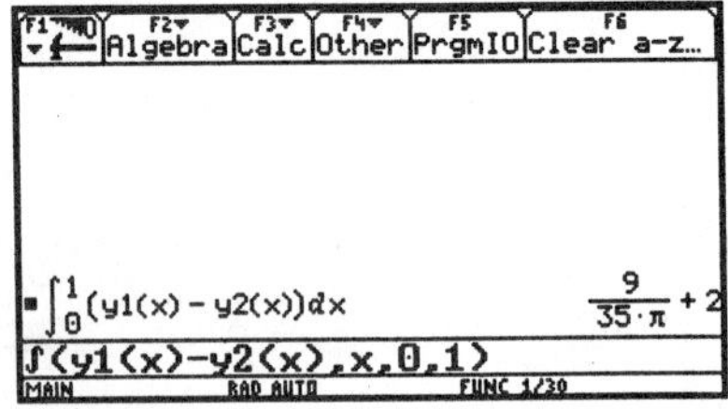

So the exact value of the area is $9/(35\pi) + 2$ square units.

Our second example is one in which it is more convenient to integrate with respect to y than with respect to x.

• EXAMPLE 2. *Find the area of the region bounded by the graphs of $xy = 4$ and $x = y(5 - y)/2$.*

Let's first graph each equation to get a picture of the region. We'll first enter $y = 4/x$ in the **Y=** Editor and then use the **DrawInv** command from the Home screen to plot the inverse relation of $y = x(5 - x)/2$.

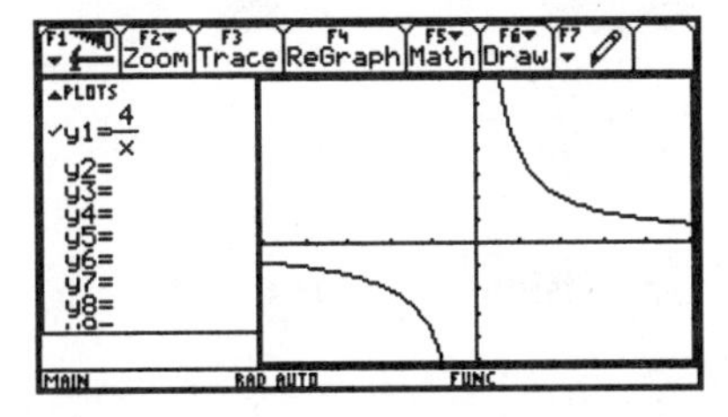

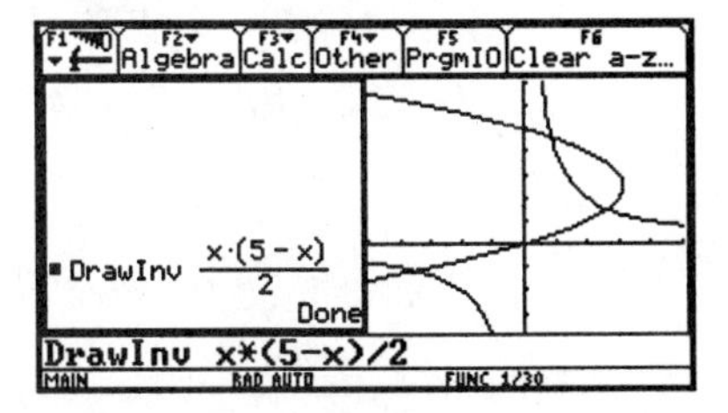

Because of the particular geometry of this region, it is simpler to describe horizontal cross-sections than it is to describe vertical cross-sections. Horizontal cross-sections have length $y(5 - y)/2 - 4/y$. To determine the limits of integration, we need to find the positive solutions of the equation $y(5 - y)/2 = 4/y$.

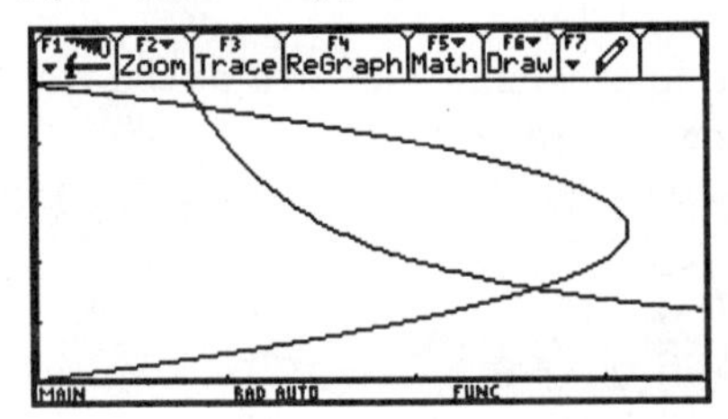

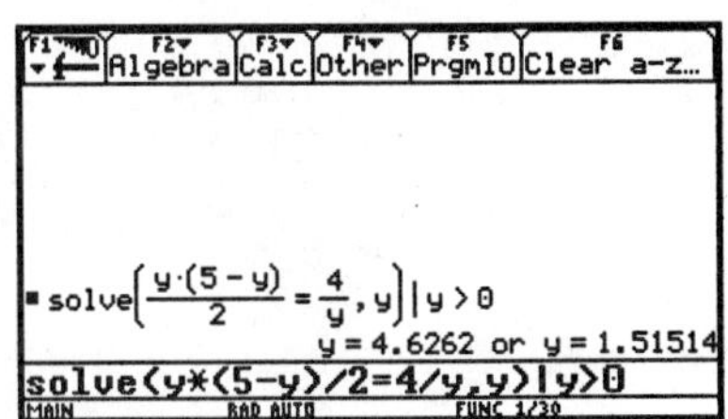

So the area of the region is given by the integral

$$\int_{1.51514}^{4.6262} \left(y(5 - y)/2 - 4/y\right) dy.$$

Computation of this integral shows that the area is 3.49595 square units.

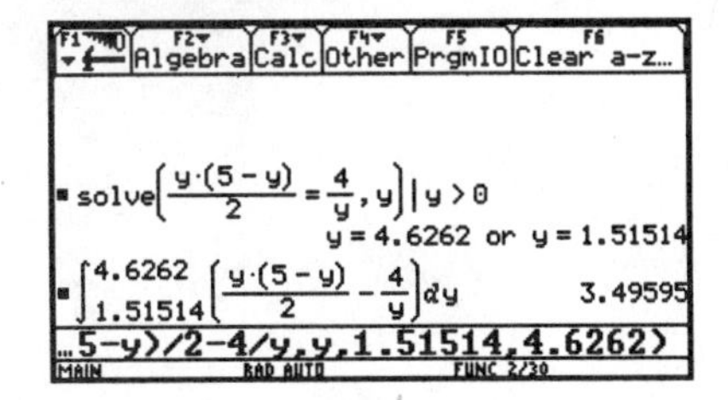

Exercises

In Exercises 1–6, find the area of the planar region bounded by the graphs of the given equations.

1. $y = x^3$ and $y = 2x^2$
2. $y = x(2 - x)$, $y = 1 - x^2$, and $y = -(x - 1)(x - 2)$
3. $y = x/2 + \sin x$ and $y = 3x^{0.3}$
4. $y = 3 \sin x^2$ and $y = x^2$
5. $xy = 1$ and $x^2 + y = 3$
6. $x = y^2$ and $y = x(x - 4)/5$

7. Find the area of the region inside the circle $(x - 1)^2 + y^2 = 4$ and outside the circle $x^2 + y^2 = 4$.

8. Find, in terms of a and m, the area bounded by the parabola $y = ax^2$ and the line $y = mx$.

9. Find, in terms of R and b, the area of the region inside the circle $x^2 + y^2 = R^2$ and above the line $y = b$, where $0 \leq b \leq R$.

10. Use Simpson's rule to approximate the area of the region in the figure. The distance between measurements is 49 feet.

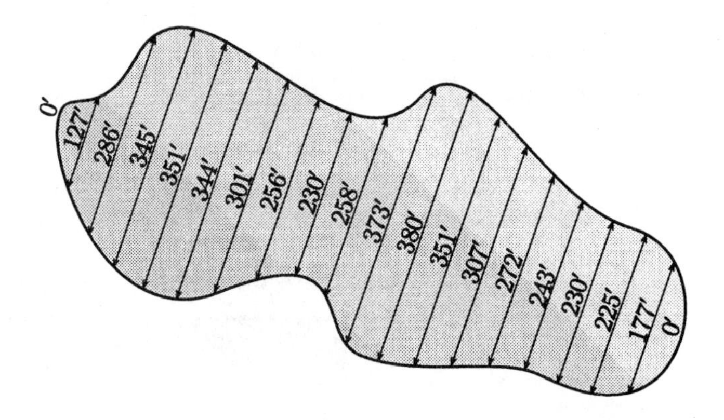

6.2 Volume

In this section we will look at two examples of volume calculations for solids of revolution. The first of these examples illustrates the technique of integrating cross-sectional area to compute volume.

• EXAMPLE 1. *The inside of a vase, ten inches tall, can be described as the "solid" obtained by revolving the region under the graph of*

$$y = 2\sin^2\left(\frac{9\pi}{200}(2x+5)\right) + 1, \quad 0 \le x \le 10,$$

about the x-axis. Find the volume of water that the vase will hold.

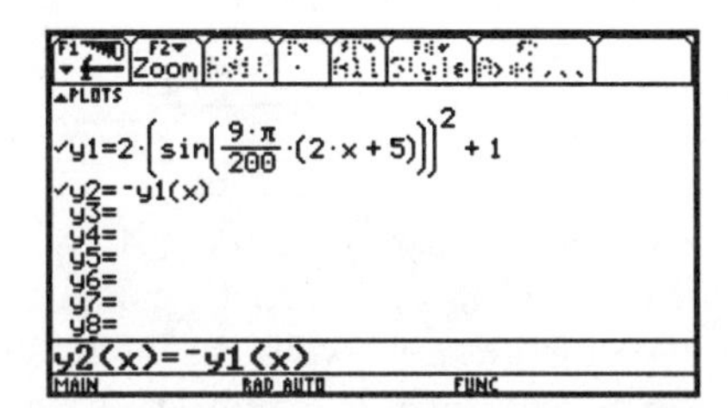

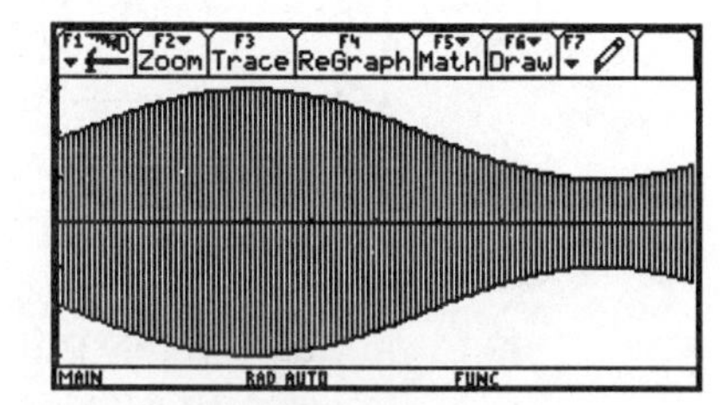

The picture on the right above shows the central cross-section of the vase, parallel to the x-axis. A typical cross-section perpendicular to the x-axis will be a disk with radius $r = y$ and therefore area given by

$$A(x) = \pi y^2 = 4\pi\left(\sin^2\left(\frac{9\pi}{200}(2x+5)\right) + 1\right)^2.$$

The volume inside the vase can be found by integrating this cross-sectional area from $x = 0$ to $x = 10$. Since the radius is already entered as **y1(x)**, we need only enter **∫(y1(x),x,0,10)** to obtain the volume. Seeing that the exact value of this integral is very complex and of little practical use, we'll re-do the calculation with the numerical integration operator, **nInt()**. (We could have just as well ◇-entered **ans(1)** after computing the exact value.)

```
∫(π*(y1(x))^2,x,0,10)
```

```
nInt(π·(y1(x))^2,x,0,10)        148.567
```

Thus the volume of the vase is approximately 148.57 cubic inches.

The second of our examples illustrates the cylindrical shells technique for computing volume.

• EXAMPLE 2. *Let A denote the region in the first quadrant bounded by the graphs of $y = 2\sin 2x$ and $y = x^2$. Find the volume of the solid generated by revolving A about the y-axis.*

The first step is to get a picture of the region A and find the point of intersection of the two graphs (**F5-5**). Then by plotting the reflection of A about the y-axis, we can see the central vertical cross-section of the solid. (The ellipses have been added with a graphics editor to suggest a 3-dimensional view of the solid.)

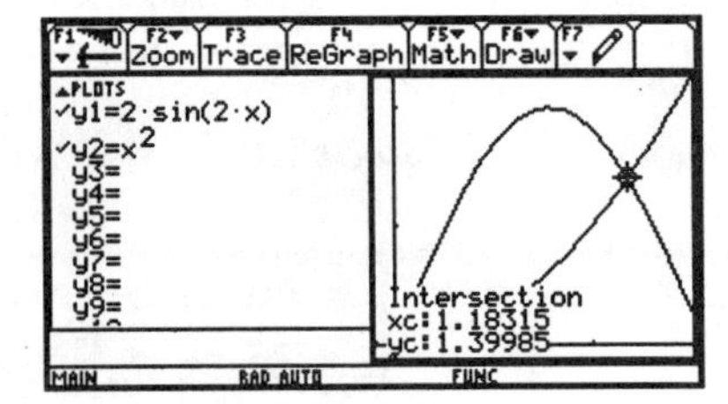

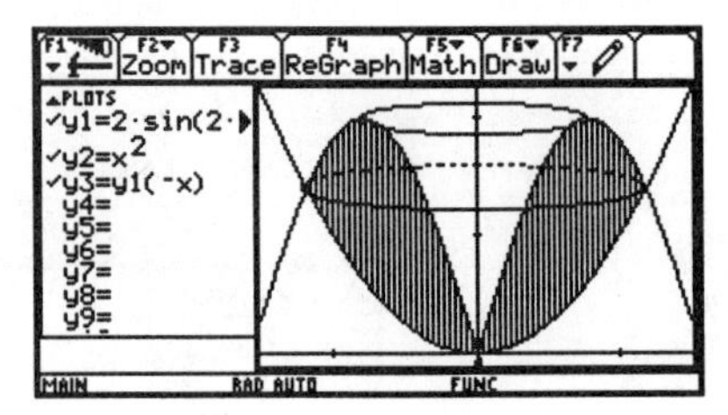

Notice that vertical cross-sections of A are far simpler to describe than horizontal cross-sections. So the simplest approach to calculating the volume in this example is to sum—by integration—volumes of cylindrical shells obtained by revolving vertical "slices" of the region A—each with thickness dx—about the y-axis. The volume dV of a typical cylindrical shell is

$$dV = 2\pi x(2\sin 2x - x^2)dx.$$

Since $0 \le x \le 1.18315$ in the region A, the volume is

$$V = \int_0^{1.18315} 2\pi x(2\sin 2x - x^2)dx.$$

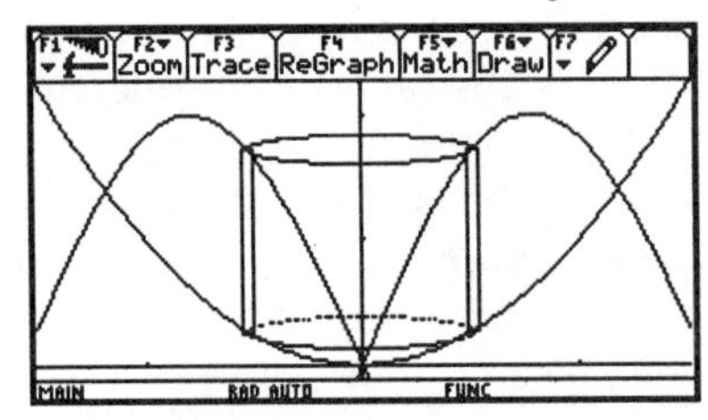

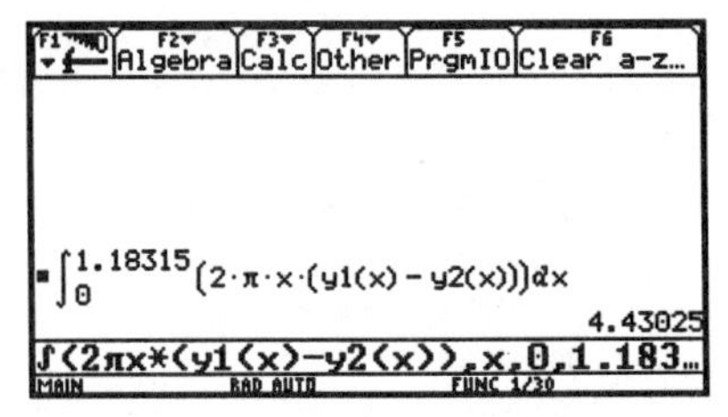

So the volume of the solid is 4.43025 cubic units.

Exercises

In Exercises 1–4, find the volume of the solid generated by revolving about the x-axis the region bounded by the given graphs.

1. $y = x^3$ and $y = 2x^2$
2. $y = x(2 - x)$, $y = 1 - x^2$, and $y = -(x - 1)(x - 2)$

3. $y = x/2 + \sin x, \quad x = 2\pi$, and $y = 0$

4. $xy = 1$ and $x^2 + y = 3$

In Exercises 5–8, find the volume of the solid generated by revolving about the y-axis the region bounded by the given graphs.

5. $y = x^3$ and $y = 2x^2$

6. $y = x(2-x), \quad y = 1 - x^2$, and $y = -(x-1)(x-2)$

7. $x = y^2$ and $y = x(x-4)/5$

8. $xy = 1$ and $x^2 + y = 3$

9. Find, in terms of R and h, the volume of the solid generated by revolving the region inside the circle $x^2 + y^2 = R^2$, to the right of the y-axis and below the line $y = h - R$, where $0 \le h \le 2R$.

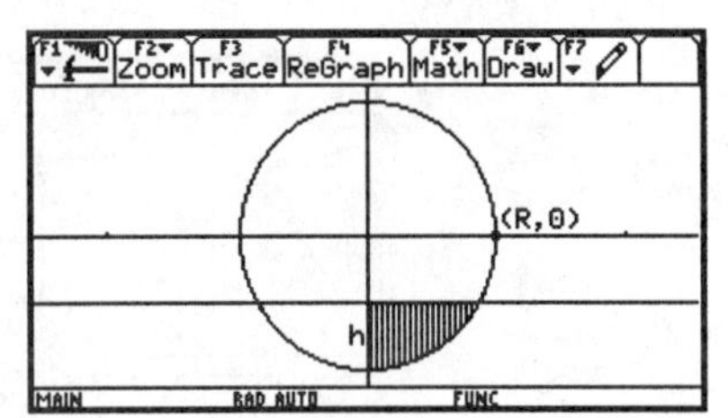

10. Find the volume of the torus generated by revolving each of the following disks about the y-axis.

    a) $(x-1)^2 + y^2 \le 1$

    b) $(x-3)^2 + y^2 \le 4$

    b) $(x-R)^2 + y^2 \le r^2$ where $0 < r \le R$

6.3 Arc length and surface area

Arc length. The length of a curve described by $y = f(x)$, $a \le x \le b$, is given by the integral

$$\ell = \int_a^b \sqrt{1 + (f'(x))^2}\, dx.$$

The capabilities of the **TI-89/92** are especially useful for this type of problem, since antiderivatives for the integrands in integrals such as this are typically very difficult—if not impossible—to find (in terms of elementary functions) even when the function f is quite simple.

• EXAMPLE 1. *Find the length of the curve* $y = x^3$, $0 \le x \le 1$.

The derivative here is $f'(x) = 3x^2$, and so the arc length is given by the integral $\int_0^1 \sqrt{1 + 9x^4}\, dx$.

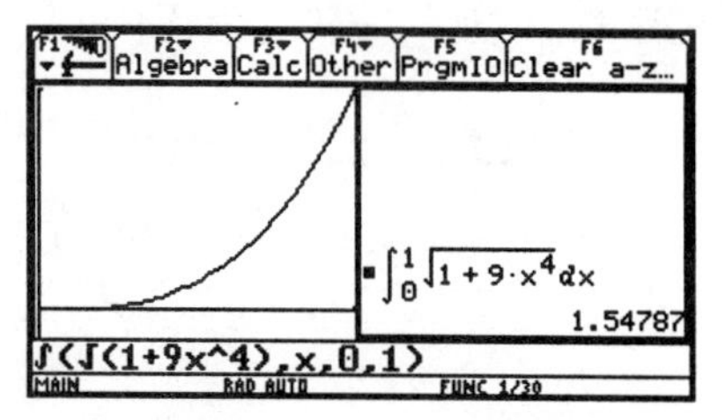

• EXAMPLE 2. *Find the length of the curve* $y = \sin x$, $0 \le x \le \pi$.

The derivative here is $f'(x) = \cos x$, and so the arc length is given by the integral $\int_0^\pi \sqrt{1 + \cos^2 x}\, dx$.

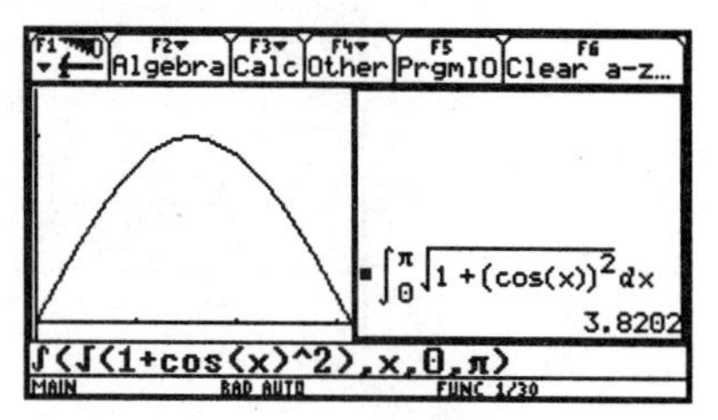

Surface area. The areas of the surfaces generated by revolving a curve $y = f(x)$, $a \le x \le b$, about the x- and y-axes, respectively, are

$$\int_a^b 2\pi f(x)\sqrt{1 + (f'(x))^2}\, dx \quad \text{and} \quad \int_a^b 2\pi x\sqrt{1 + (f'(x))^2}\, dx.$$

Likewise, the surface areas generated by revolving a curve $x = f(y)$, $c \le y \le d$, about the x-axis and y-axis, respectively, are

$$\int_c^d 2\pi y\sqrt{1 + (f'(y))^2}\, dy \quad \text{and} \quad \int_c^d 2\pi f(y)\sqrt{1 + (f'(y))^2}\, dy.$$

• EXAMPLE 3. *Find the surface area generated by revolving the curve* $y = \sin x$, $0 \le x \le \pi$, *about the* y*-axis.*

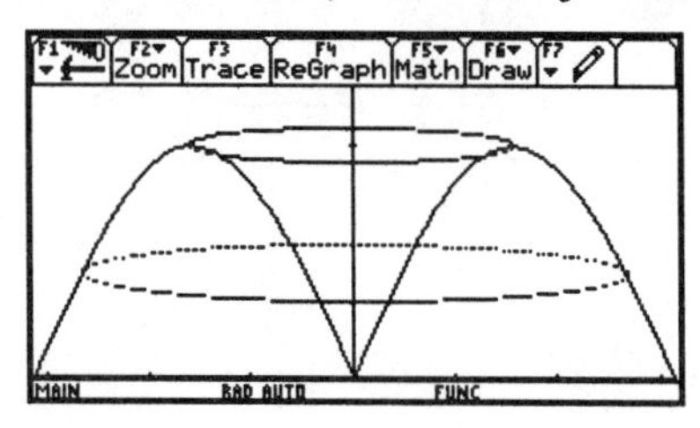

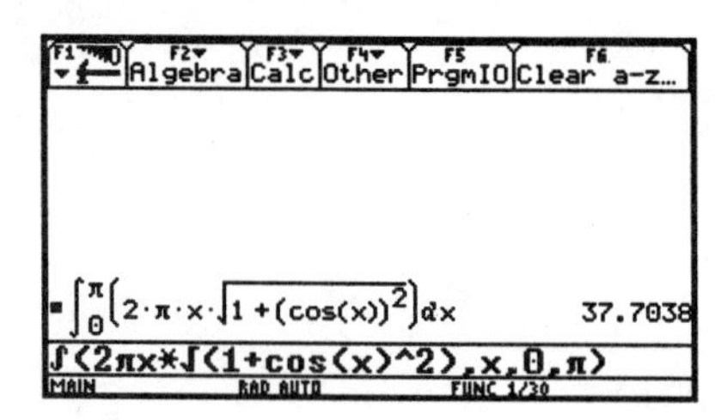

• EXAMPLE 4. *Find the surface area generated by revolving the curve* $y = \sin x$, $0 \le x \le \pi$, *about the x-axis.*

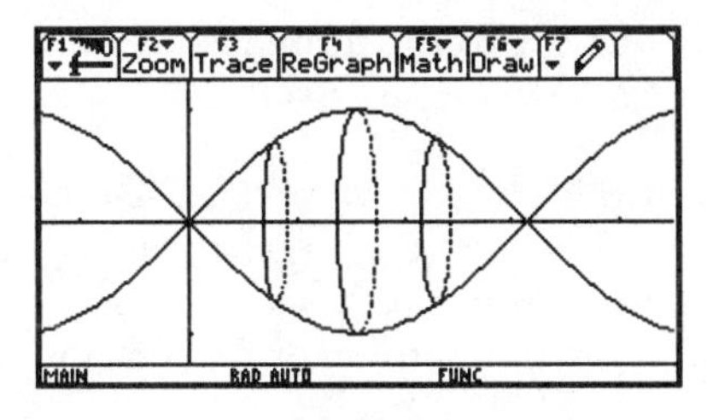

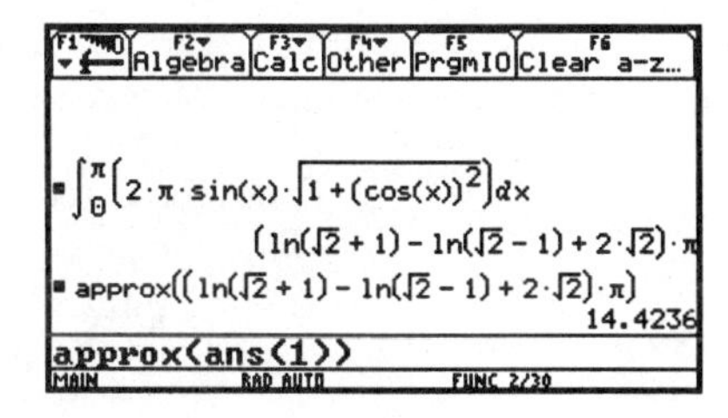

Exercises

In Exercises 1–4, find:

a) the length of the given arc;

b) the area of the surface generated by revolving the arc about the x-axis;

c) the area of the surface generated by revolving the arc about the y-axis.

1. $y = 1/x, \ 1 \le x \le 2$

2. $y = (e^x + e^{-x})/2, \ 0 \le x \le 1$

3. $y = \tan x, \ 0 \le x \le \pi/3$

4. $y = \sqrt{x}, \ 0 \le x \le 1$

5. Find, in terms of R and h, the area of the surface generated by revolving about the y-axis the portion of the circle $x^2 + y^2 = R^2$, to the right of the y-axis and below the line $y = h - R$, where $0 \le h \le 2R$.

6. Using Simpson's rule and the following table of values, approximate:

a) the length of the arc $y = f(x)$, $0 \le x \le 5$;

b) the area of the surface generated by revolving the arc about the x-axis;

c) the area of the surface generated by revolving the arc about the y-axis.

x	0	0.5	1.0	1.5	2.0	2.5	3.0	3.5	4.0	4.5	5.0
$f(x)$	1.2	3.1	3.9	4.5	5.3	4.7	4.3	3.3	2.5	1.3	0.7
$f'(x)$	3.5	2.3	1.6	1.3	-0.2	-1.1	-1.5	-2.1	-1.9	-1.2	-0.2

6.4 Moments and centers of mass

Note: For a detailed development of the formulas below, see Section 9.3 in Stewart's *Calculus*.

Consider a region $\mathcal{R}$ in the plane, with area A, bounded by $y = f(x)$, $y = g(x)$, $x = a$, and $x = b$, where $f(x) > g(x)$ for $a \le x \le b$. If a flat plate with uniform thickness and constant mass density occupies $\mathcal{R}$, then the plate's *center of gravity*, or the *centroid* of $\mathcal{R}$, is the point

$$(\bar{x}, \bar{y}) = \left(\frac{M_y}{A}, \frac{M_x}{A}\right)$$

where M_x and M_y are its *moments* about the x- and y-axes given by

$$M_x = \frac{1}{2}\int_a^b \left(\left(f(x)\right)^2 - \left(g(x)\right)^2\right) dx,$$

$$M_y = \int_a^b x\left(f(x) - g(x)\right) dx.$$

• EXAMPLE 1. *Find the centroid of the region in the first quadrant bounded by $y = e^{x-1}$ and $y = x^2$.*

The region is shown in the screen on the left below. It is easy to observe that the point of intersection of the two graphs is $(1, 1)$. The area calculation is shown on the right.

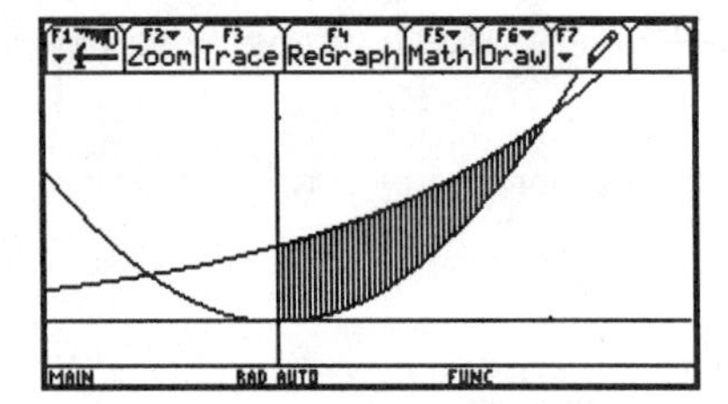

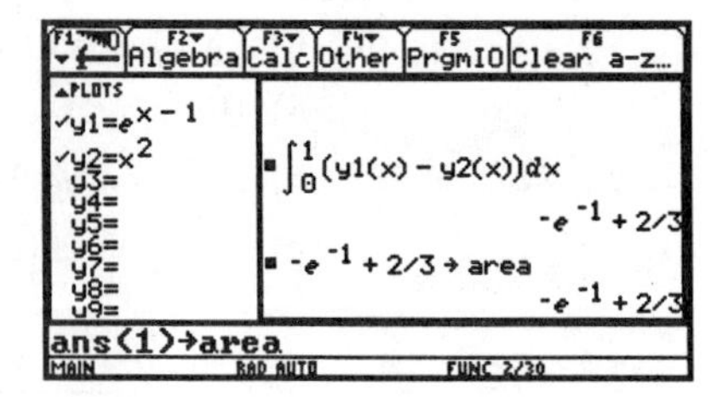

Next we'll calculate the moments M_y and M_x.

Finally, we complete the calculation of $(\bar{x}, \bar{y})$ and show its approximate location.

```
F1 F2 Algebra F3 Calc F4 Other F5 PrgmIO F6 Clear a-z...
                                         20
(3·e^2 - 5)·e^-2                (3·e^2 - 5)·e^-2
---------------- → mx           ----------------
       20                              20
approx(my/area)                     .39452638
approx(mx/area)                     .38879232
approx(mx/area)
MAIN        RAD AUTO        FUNC 8/30
```

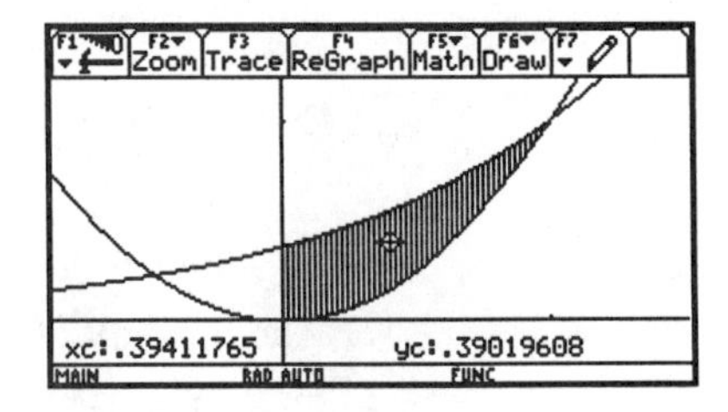

• EXAMPLE 2. *A boomerang with uniform thickness and constant mass has the shape of the region bounded by the graphs of $y = \sin x$ and $y = \frac{9}{10}\sin^4 x$ between $x = 0$ and $x = \pi$. Find its center of mass.*

The region is shown in the first screen below. It is obvious that $\bar{x} = \pi/2$ because of symmetry. So we need to find $\bar{y}$. The screen on the right shows the area calculation.

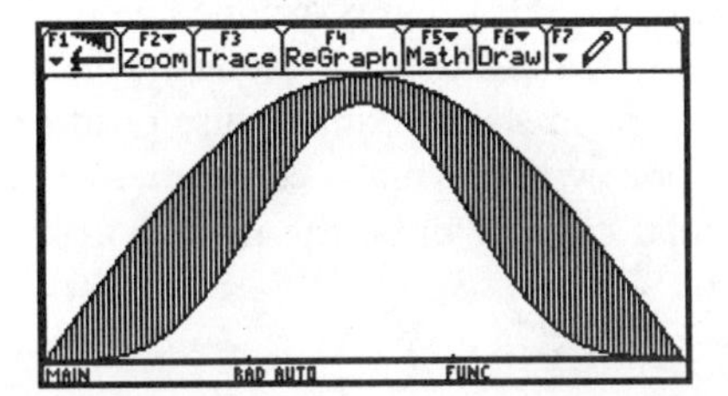

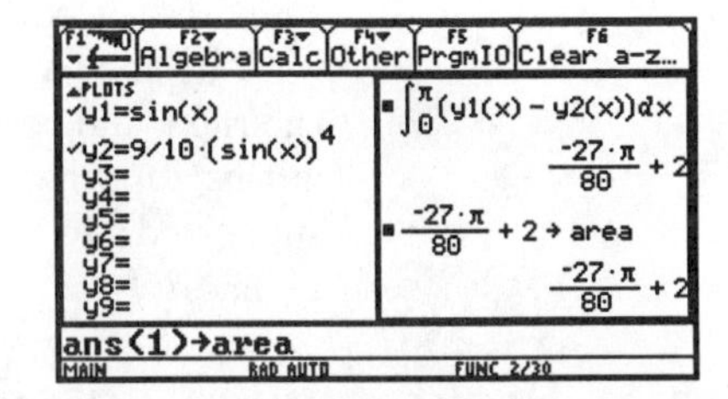

Now we need to compute M_x. This calculation is shown on the left below. On the right we see the (approximate) location of the centroid.

```
F1 F2 Algebra F3 Calc F4 Other F5 PrgmIO F6 Clear a-z...
∫[0,π](y1(x) - y2(x))dx                 -27·π/80 + 2
-27·π/80 + 2 → area                     -27·π/80 + 2
1/2·∫[0,π]((y1(x))^2 - (y2(x))^2)dx      713·π/5120
1/2 ∫(y1(x)^2-y2(x)^2,x,0,π)
MAIN        RAD AUTO        FUNC 3/30
```

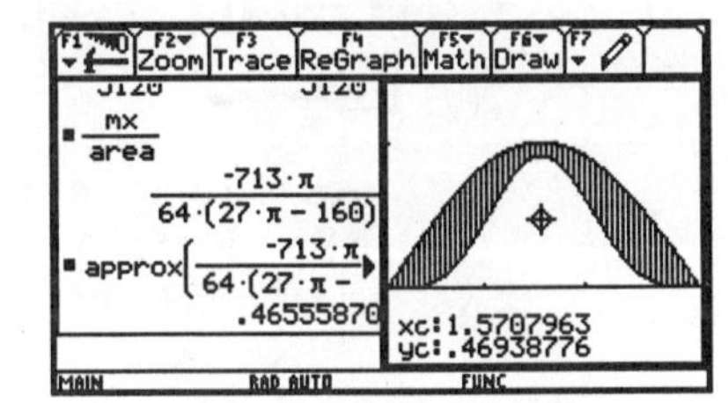

Exercises

In Exercises 1–5, find the centroid of the region bounded by the graphs of the given equations. Use symmetry whenever possible. Plot and shade each region.

1. $y = x^2$, $y = 4$
2. $y = x^2$, $y = x$
3. $y = e^{x-1}$, $y = x^2$
4. $y = \cos x$, $y = 0$, $x = \pm\pi/2$

5. $y = \sin x$, $y = \cos x$, $x = 0$, $x = \pi/4$

In Exercises 6–10, find the centroid of the described region. Use symmetry whenever possible.

6. The quarter of the unit disk $x^2 + y^2 \leq 1$ in the first quadrant
7. The top half of the unit disk $x^2 + y^2 \leq 1$
8. The triangle in the first quadrant under the line $x + y = 1$
9. The square $-1 \leq x \leq 1,\ 0 \leq y \leq 2$, surmounted by a half-disk
10. The portion of the unit disk above the line $y = 1/2$

11. Consider the region in Example 1 to be a flat plate with uniform thickness and constant mass density. Imagine also that the x-axis is a flat surface upon which the plate is balanced on its edge. Clearly the plate must be supported in some way in order to be positioned as it is, and if the plate were no longer supported, it would roll along its bottom edge toward the right and come to rest in some "equilibrium position." The question is this: In what position will the plate come to rest if it is no longer supported? In particular, what will be the new coordinates of its center of mass?

6.5 Programming notes

The first exercise below has you modify the **midbox()** program from Section 5.6 so that it will illustrate midpoint approximations to the area between two curves.

In Exercise 2 you are asked to create two functions for computing the area of a surface of revolution. Note that the **TI-89/92** already has two built-in functions for computing arc length. These are **arcLen()**, found in the Home screen's **Calc** menu, and **Arc**, found in the Graph screen's **Math** menu.

Exercise 3 asks you to create a function that will return the centroid of a region bounded by two graphs.

Exercises

1. Modify the program **midbox()** created in Exercise 3 of Section 5.6 so that it takes two functions, f and g, as arguments:

   midbox(f, g, $xvar$, a, b, n**)**,

   and graphically depicts and computes a midpoint approximation to the area between the graphs of f and g over the interval $a \leq x \leq b$ using n rectangles, assuming that $f(x) \geq g(x)$ on the interval. Test the program on the example

$$f(x) = \sin x, \quad g(x) = \sin^2 x, \quad 0 \leq x \leq \pi/2.$$

(Note that **midbox()** should behave exactly as before if the second argument g is zero.)

2. Create a function, **surfAx(**f, $xvar$, a, b**)**, that returns the area of the surface generated by revolving the graph of

$$y = f(xvar), \quad a \leq xvar \leq b,$$

about the $xvar$-axis. Test your function on the problem in Example 4 of Section 6.3. Also create an analogous function, **surfAy(**f, $xvar$, a, b**)**, that computes the area of the surface generated by revolving the same graph about the y-axis. Test this function on the problem in Example 3 of Section 6.3.

3. Create a function, **centroid(**f, ,g, $xvar$, a, b**)**, that returns the centroid of the region bounded by $y = f(xvar)$ and $y = g(xvar)$ between $xvar = a$ and $xvar = b$. Test the function on Examples 1 and 2 of Section 6.4. Then use it to rework Exercises 1–5 in Section 6.4.

7 Differential Equations

As described in Section 10.1 of Stewart's *Calculus*, many real-world phenomena can be described in terms of relationships between quantities and their rates of change, and such relationships give rise to differential equations. In this section we will explore some of the ways that the **TI-89/92** can be used to study differential equations.

7.1 Equations and solutions

The differential equations we will consider here can be written in the form

$$\frac{dy}{dt} = f(t, y),$$

where t is the *independent variable*, y is the *dependent variable*, and f is a given continuous function of two variables. Such an equation is called a *first-order* differential equation. An equation such as this is said to be a *first-order linear* differential equation if it can be written in the form

$$\frac{dy}{dt} + p(t)y = q(t),$$

where p and q are given, possibly nonlinear, functions of one variable. A function $y(t)$ is a *solution* of $\frac{dy}{dt} = f(t, y)$ on an interval I if $\frac{dy}{dt} = f(t, y(t))$ holds for all t in I.

• EXAMPLE 1. *Verify that $y = (1 - t^2)^{-1}$ is a solution of $\frac{dy}{dt} = 2t\,y^2$ on any interval that does not contain $t = \pm 1$. Also plot the solution.*

The most efficient way to verify a solution of this equation is to compute $\frac{dy}{dt} - 2t\,y^2$, hopefully obtaining 0. So we'll use the **d()** operator and enter

d(y(t),t)−2*t*y(t)^2 ,

after which we'll enter the function in the **Y=** Editor and plot the graph in a $[-3, 3] \times [-2, 3]$ window.

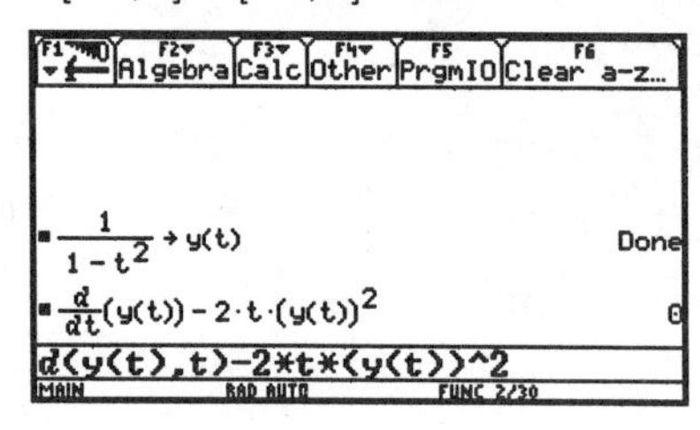

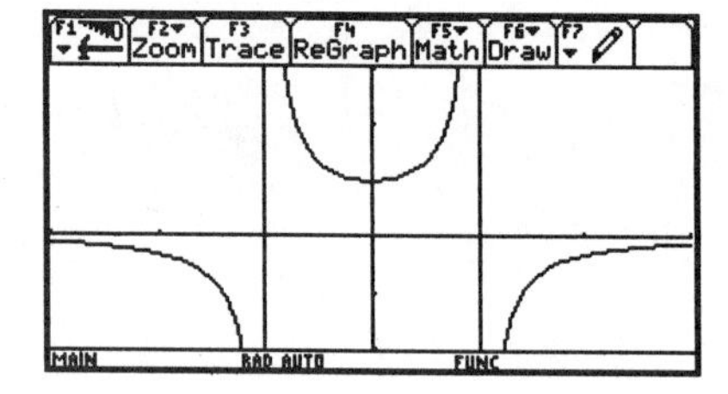

From the calculation and the graph, it is easy to see that the solution is undefined at $t = \pm 1$, but that the differential equation is satisfied at all other values of t.

• EXAMPLE 2. *Verify that* $y = (t^2+1)^{-2} + t^2 + 1$ *is a solution of the linear differential equation* $\frac{dy}{dt} + \frac{4t}{t^2+1}y = 6t$ *on any interval. Also plot the solution.*

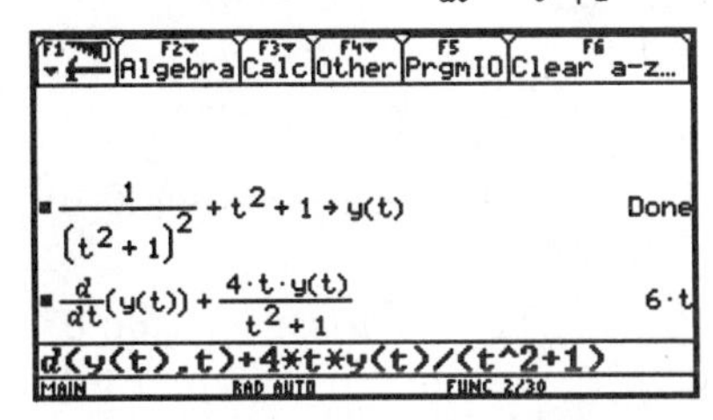

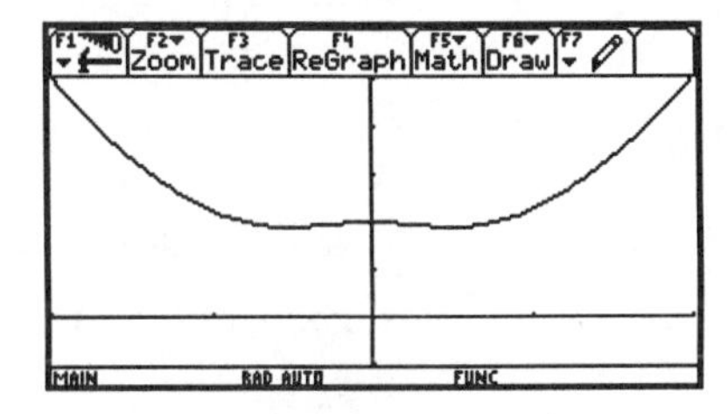

It is clear in this case that $y(t)$ satisfies the differential equation for all t and thus is a solution on any interval.

Differential equations typically have many solutions. First-order equations typically have a *one-parameter family* of solutions. This one-parameter family is called the *general solution* of the differential equation. In each of the following examples, the constant C represents the parameter.

• EXAMPLE 3. *Verify that* $y = (C - t^2)^{-1}$ *is a solution of* $\frac{dy}{dt} = 2t\,y^2$ *on any interval on which* $C - t^2 \neq 0$. *Plot solutions corresponding to* $C = -1, -.5, 0, 1, 2, 4$.

The verification is done exactly as in Example 1.

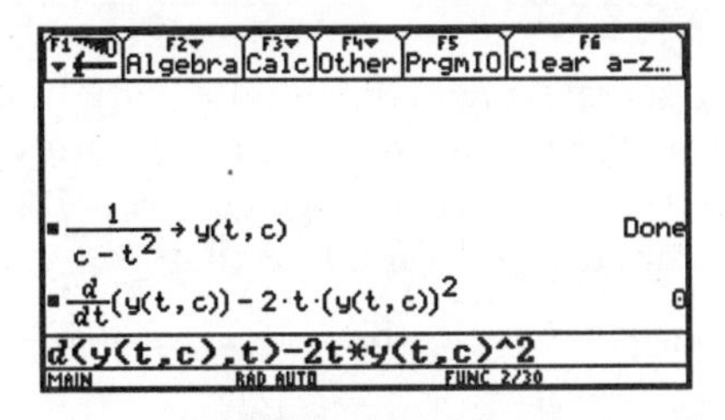

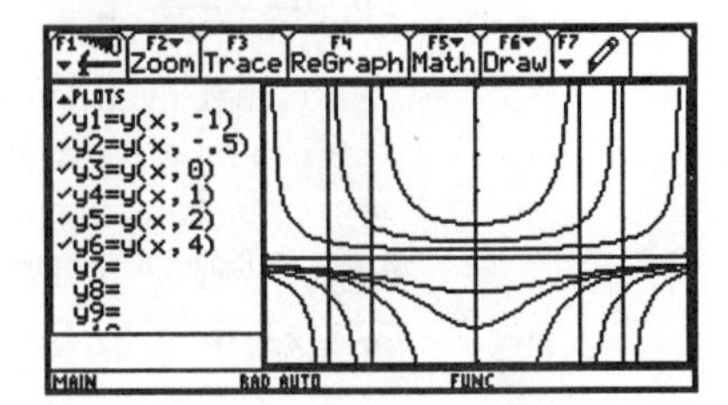

Notice that we defined the solution as a function of both t and C in order to simplify the task of entering the functions in the **Y=** Editor.

• EXAMPLE 4. *Verify that* $y = t^2 + 1 + C/(t^2+1)^2$ *is a solution of*

$$\frac{dy}{dt} + \frac{4t}{t^2+1}y = 6t$$

on any interval. Plot solutions corresponding to $C = -1, 0, 1, 2, 3, 4$.

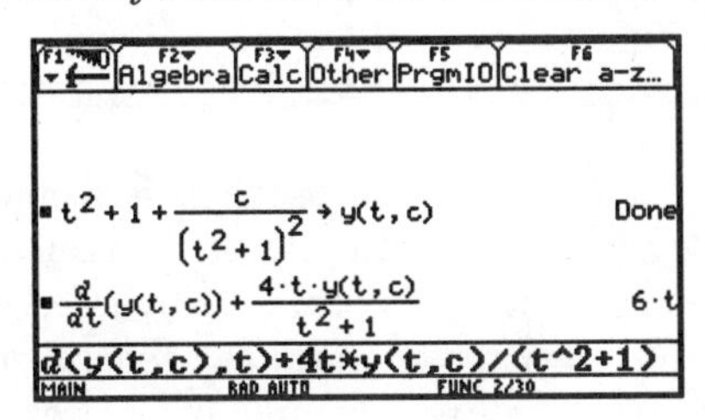

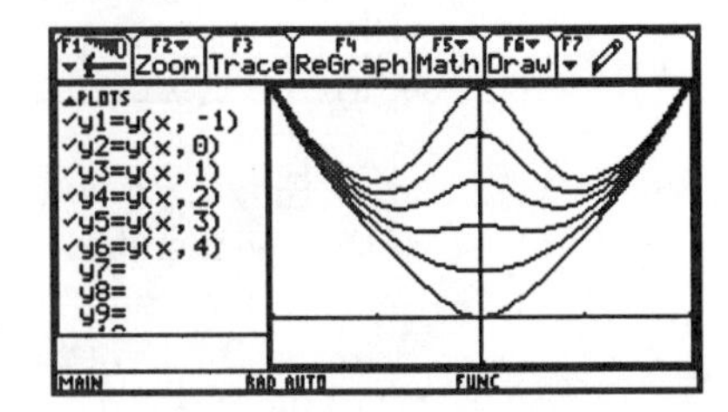

Initial value problems. A first-order *initial value problem* consists of a first-order differential equation and an *initial value*:

$$\frac{dy}{dt} = f(t, y), \quad y(t_0) = y_0.$$

Note that the initial value simply requires that the graph of the solution pass through the point (t_0, y_0).

• EXAMPLE 5. *Verify that* $y = 4/(1 - 4t^2)$ *is a solution of*

$$\frac{dy}{dt} = 2t\,y^2, \quad y(0) = 4$$

on the interval $-.5 < t < .5$. *Also plot the solution.*

Here we'll do essentially the same thing as in Example 1 to verify that y satisfies the differential equation (for all $t \neq \pm.5$). Then we'll do a simple evaluation to check the initial condition.

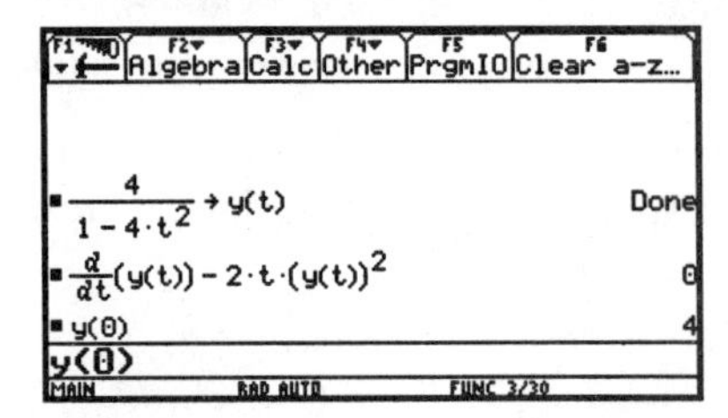

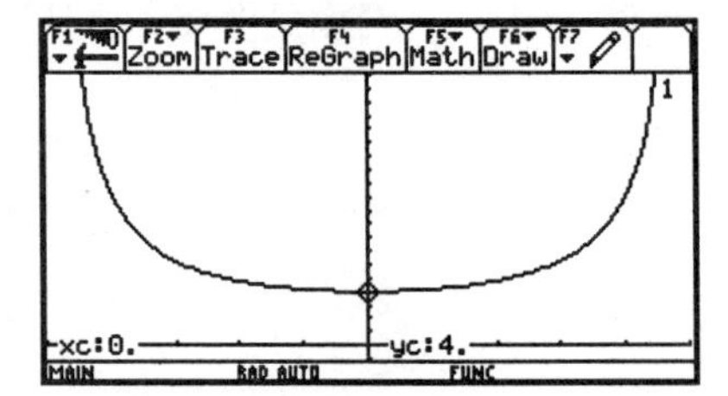

Note that even though y satisfies the differential equation at all $t \neq \pm.5$, it cannot be a solution of the differential equation on an interval that contains $t = \pm.5$. Therefore, since the initial value is given at $t = 0$, y cannot satisfy the initial value problem on any interval larger than $-.5 < t < .5$.

• EXAMPLE 6. *Verify that* $y = t^2 + 1 + 8/(t^2 + 1)^2$ *is a solution of*

$$\frac{dy}{dt} + \frac{4t}{t^2 + 1}y = 6t, \quad y(1) = 4$$

on any interval. Also plot the solution.

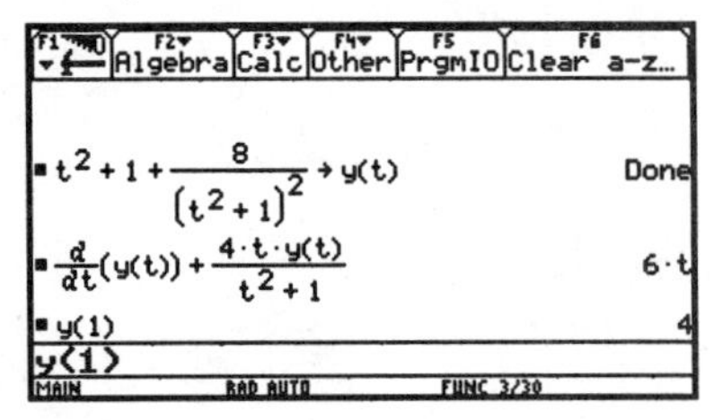

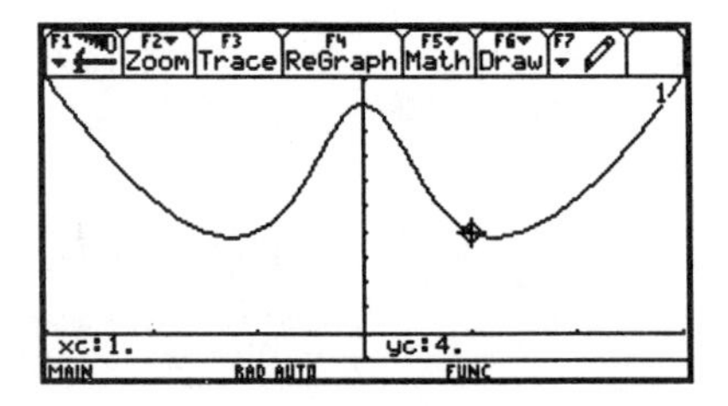

Finding a solution of an initial value problem is a matter of selecting the one member of the general solution family that satisfies the given initial condition. Such a solution can often be obtained by substituting the given initial values of t and y into the expression for the general solution and then solving for the parameter C. The next example illustrates this.

• EXAMPLE 7. *Check that $y = 10/(1+Ce^{-t/2})$ satisfies $\frac{dy}{dt} = .05y(10-y)$ for any constant C. Then find and plot the solution of the initial value problem*

$$\frac{dy}{dt} = .05y(10-y), \quad y(0) = 2.$$

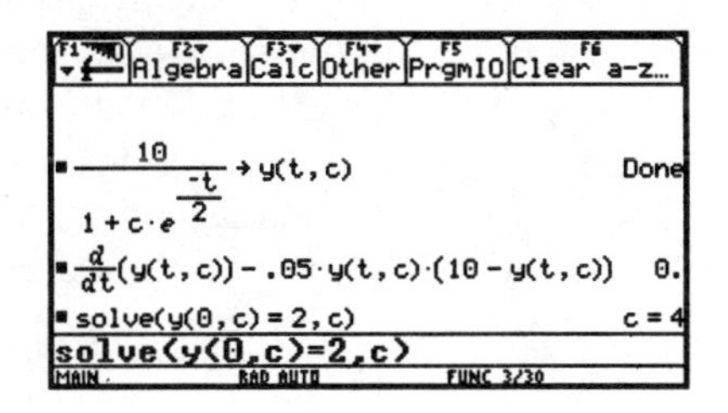

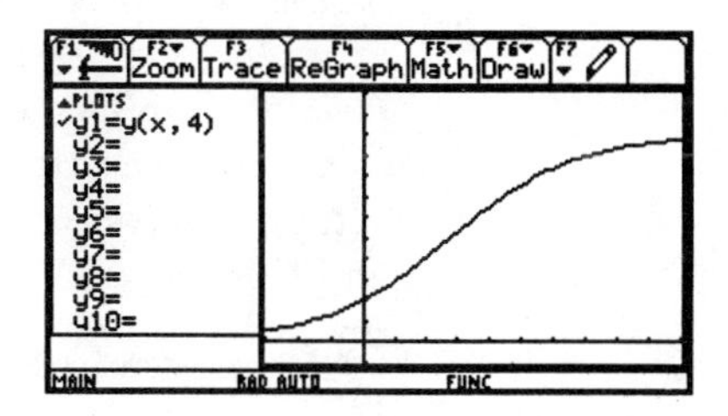

Exercises

In each of Exercises 1–5, check that the family of functions described in the first column satisfies the differential equation in the second column. Then plot the solution for each value of C in the third column on the interval given in the fourth column. *Suggestions: Set* **Graph Order** *to* **SIMUL** *(***F1-9***). For the plots in Exercise 4, use* **ymin** $= -75$ *and* **ymax** $= 100$.

1. $y = Ce^{-3t}$ $\quad \dfrac{dy}{dt} + 3y = 0$ $\quad C = \pm 1, \pm 2$ $\quad [-.25, 1]$

2. $y = \dfrac{2}{1+Ce^{-2t}}$ $\quad \dfrac{dy}{dt} - 2y = -y^2$ $\quad C = -9, -.5, 0, 3$ $\quad [0, 2]$

3. $y = \dfrac{(1+3t)^{4/3} + C}{\sqrt[3]{1+3t}}$ $\quad \dfrac{dy}{dt} + \dfrac{y}{1+3t} = 4$ $\quad C = 0, 25, 50$ $\quad [0, 10]$

4. $y = t^2(C + \ln t)$ $\quad \dfrac{dy}{dt} - 2y/t = t$ $\quad C = 1, -2, -3$ $\quad [0, 25]$

5. $y = \sin^{-1}(Ce^{-t})$ $\quad \dfrac{dy}{dt} + \tan y = 0$ $\quad C = \pm 1, \pm \frac{1}{2}$ $\quad [-1, 2]$

In Exercises 6–10, use the general solution from the corresponding Exercise 1–5 to solve the given initial value problem. Plot the solution.

6. $\dfrac{dy}{dt} + 3y = 0, \quad y(0) = 5$

7. $\dfrac{dy}{dt} + \dfrac{y}{1+3t} = 4, \quad y(21) = 0$

8. $\dfrac{dy}{dt} + \tan y = 0, \quad y(0) = \pi/6$

9. $\dfrac{dy}{dt} - 2y = -y^2, \quad y(0) = 5$

10. $\dfrac{dy}{dt} - 2y/t = t, \quad y(1) = -1/2$

7.2 Direction fields

If the graph of a solution of the differential equation

$$y'(t) = f(t, y)$$

passes through a point (t_1, y_1), then the slope of the graph at that point is $m = f(t_1, y_1)$. Thus, we can think of the function f as specifying a *slope field*, or *direction field*, in the ty-plane. This direction field can be visualized as an array of arrows at points (t, y), where each arrow has slope given by $f(t, y)$. The two screens below show the direction field given by $f(t, y) = t - y$ and then the direction field together with a few solution curves.

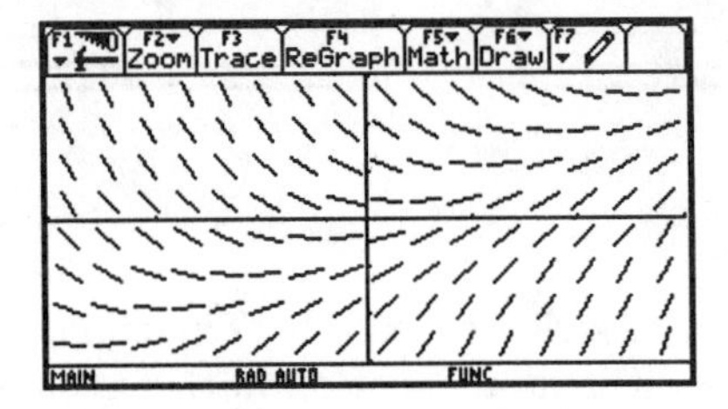

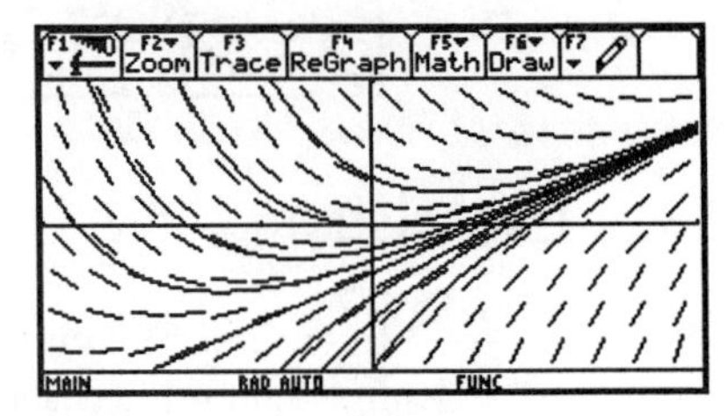

Creating direction field plots with the **TI-92** requires a little programming. The **TI-89** and **TI-92 Plus** have built-in capability for plotting direction fields, but for extra versatility—and for the experience of doing the programming yourself—this independent program will still be useful.

We will create a program called **slopefld()**. To create the program, first press the **APPS** key and select **Program Editor** and **New...** as usual. Then specify **Type: Program, Folder: main**, and **Variable: slopefld**. This takes us to the Program Editor, ready to begin entering the following program.

```
: slopefld(f, xvar, yvar, varrows)
: Prgm: Local i, j, hh, kk, xi, yj, dx, fij, mscr, len
: ClrGraph: ClrDraw: PlotsOff
: (xmax–xmin)/varrows/2 → hh
: (ymax–ymin)/varrows → kk
: kk/(2*hh) → mscr
: For i, 1, 2*varrows
:    xmin+(i–0.5)*hh→ xi
:    For j, 1, varrows
:       ymin+(j–0.5)*kk → yj
:       (f | xvar=xi and yvar=yj) → fij
:       e^(-abs(fij)/mscr) → wt
:       .8*wt*hh+(1–wt)*kk →len
:       .33*len/√(1+fij^2) →dx
:       Line xi–dx, yj–fij*dx, xi+dx, yj+fij*dx
:    EndFor
: EndFor
: EndPrgm
```

Once this program is entered, return to the Home screen. To use the program, enter **slopefld(f(t,y), t, y, 8)**. This will plot the direction field for $\frac{dy}{dt} = f(t, y)$ with 8 arrows in the vertical direction and 16 in the horizontal. (Change the 8 if you want more or fewer arrows.) Here's what you should see in a $[-3, 3] \times [-3.5, 3.5]$ window with $f(t, y) = -t\,y$.

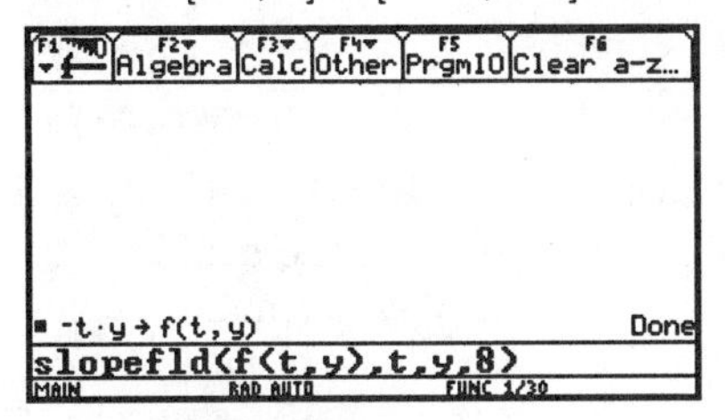

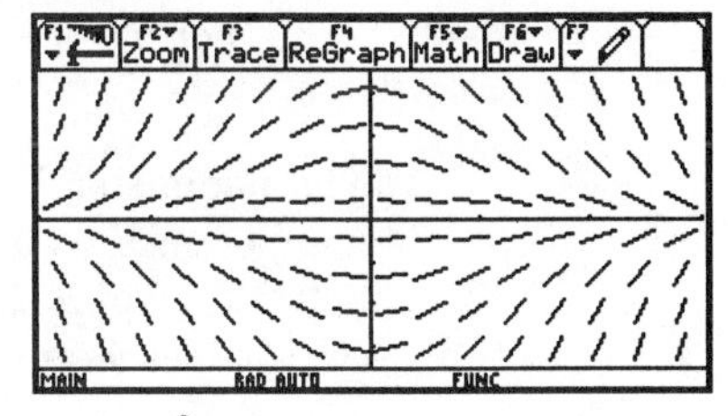

The general solution of this equation is $y = Ce^{-t^2/2}$. (Verify it.) So let's plot a few solutions on top of the direction field. (See the note below.) For each of the curves, we've set **Style** to **Thick** by pressing **F6-4** in the **Y=** Editor for each of the functions.

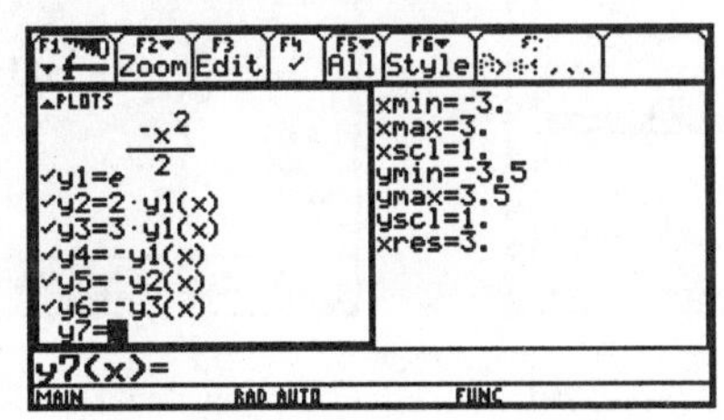

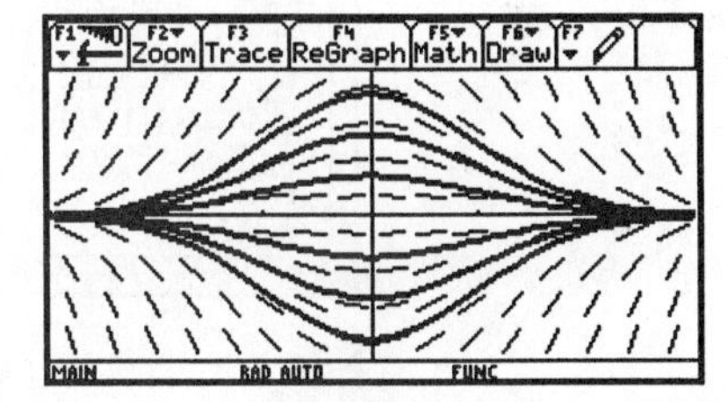

Note: slopefld() automatically clears all previous graphs and turns off (unchecks) all functions in the **Y=** Editor before plotting the direction field. So if you define solutions in the **Y=** Editor before plotting the direction field, you must reactivate them (press **F4** in the Y= Editor) in order to include them in the plot.

Exercises

For each of Exercises 1–5 in Section 7.1, replot all of the requested curves (with **Thick** Style) on top of the direction field for the differential equation.

7.3 Euler's method

There are many techniques for finding exact solutions of differential equations. Different types of equations require different techniques, two of which we will see in the next section. But situations in which exact solutions are either extremely difficult—or impossible—to find are quite common. So numerical methods for obtaining approximate solutions are extremely useful.

Typically, a numerical method for approximating the solution of

$$\frac{dy}{dt} = f(t, y), \quad y(t_0) = y_0$$

consists of a recurrence formula for computing an approximation at some time $t + h$ from an approximation at time t. This allows the computation of approximations y_1, y_2, y_3, ... to values of the solution at a sequence of times t_1, t_2, t_3, ..., where $t_k = t_0 + k\,h$.

The simplest such method is known as *Euler's method:*

For $n = 1, 2, 3, \ldots$, *compute*

$$y_n = y_{n-1} + h\,f(t_{n-1}, y_{n-1}), \quad \text{where } t_n = t_0 + n\,h.$$

The parameter h is called the *stepsize.* It is typically some small, positive number such as 0.05. Roughly speaking, the smaller the stepsize h is, the better the approximation will be. However, smaller values of h require more steps to reach any given value of t.

Euler's method is very simple to implement on the **TI-89/92**. First we set the Graph **MODE** to **SEQUENCE**. Then in the **Y=** Editor, we'll define sequences **u1(n)** and **u2(n)** to represent t_n and y_n, respectively.

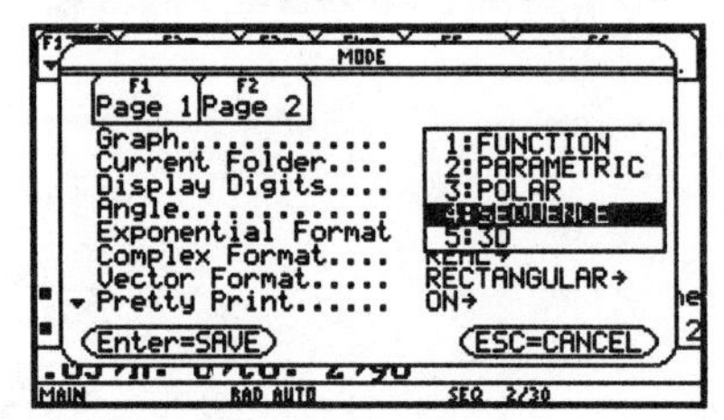

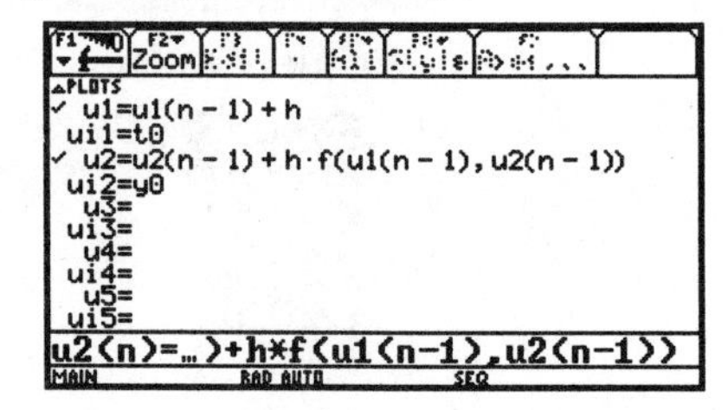

Now we need to press **F7** and specify the type of axes. We'll set **Axes: CUSTOM**, **X-Axis: u1** and **Y-axis: u2**. To complete the set-up, highlight **u2** and select the **Dot Style** (**F6-2**) for the graph.

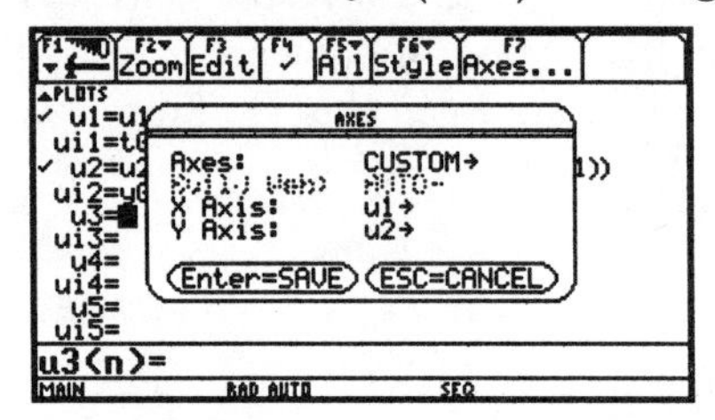

With this set-up in place, we're ready to compute some approximate solutions.

• EXAMPLE 1. *Plot an approximate solution to*

$$\frac{dy}{dt} = \sin(t\,y), \quad y(0) = 2,$$

on the interval $0 \le t \le 5$, *using Euler's method with stepsize* $h = 0.1$.

We need only enter a definition for the function **f(t,y)**, the values of **t0** and **y0**, and appropriate window variables. Note that with $h = 0.1$, we need to compute 50 steps in order to reach $t = 5$. Once all this is done, just press **◇GRAPH**.

```
F1 F2 Algebra F3 Calc F4 Other F5 PrgmIO F6 Clear a-z...
■ sin(t·y) → f(t, y)     Done
■ .1 → h                 .1
■ 0 → t0                 0
■ 2 → y0                 2
2→y0
nmin=0.
nmax=50.
plotstrt=1.
plotstep=1.
xmin=0.
xmax=5.
xscl=1.
ymin=-.5
ymax=3.
yscl=1.
MAIN   RAD AUTO   SEQ 4/30
```

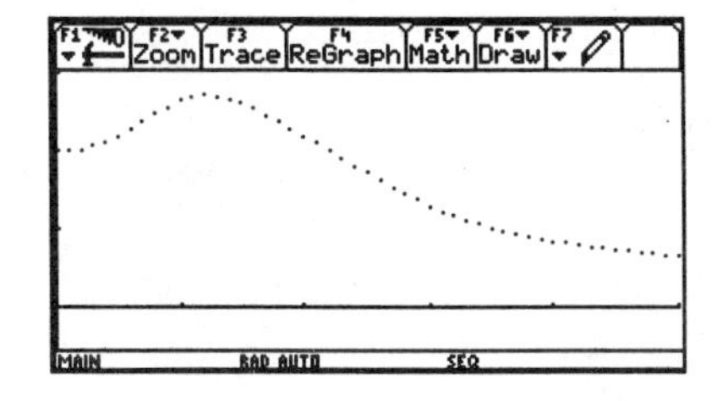

The purpose of selecting **Dot Style** (**F6-2**) for the graph was simply to emphasize that Euler's method really just computes approximations to values of the solution at discrete points. Changing that setting to **Line** or **Thick Style** causes a continuous graph to be drawn through the computed points.

• EXAMPLE 2. *Plot an approximate solution to*

$$\frac{dy}{dt} = \cos y - \sin t, \quad y(0) = 1,$$

on the interval $0 \le t \le 10$, *along with the direction field for the differential equation. Use Euler's method with stepsize* $h = 0.1$.

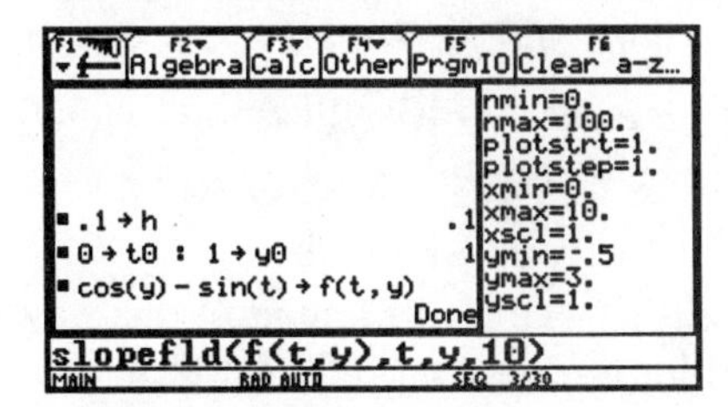

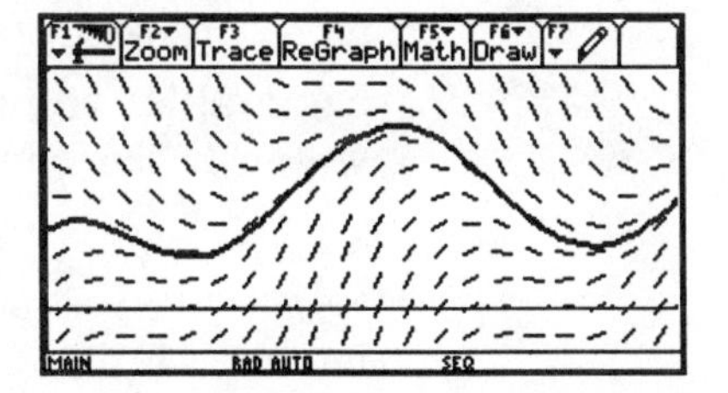

Note that (because of the **Custom** Axes setting) the approximate solution curve is automatically plotted each time a **slopefld** command is entered.

Exercises

1. Using the set-up outlined above, the viewing window $[0,3] \times [0,3]$, and **Line Style** for graphs, plot the following Euler's method approximations to the solution of

$$\frac{dy}{dt} = t^2 - 2y, \quad y(0) = 2, \quad \text{for } 0 \le t \le 3.$$

a) 3 steps with stepsize $h = 1$

b) 6 steps with stepsize $h = .5$

c) 12 steps with stepsize $h = .25$

d) 30 steps with stepsize $h = .1$

e) 60 steps with stepsize $h = .05$

In Exercises 2–5, use Euler's method with the suggested stepsize to plot an approximate solution on the indicated interval. If you have a **TI-89** or **TI-92 Plus**, also plot the solution with your calculator's built-in capability and compare.

2. $\dfrac{dy}{dt} = (e^{2-y} - 1)y, \quad y(0) = .1; \quad h = .1, \ 0 \le t \le 2$

3. $\dfrac{dy}{dt} = \sin(t^2 + y), \quad y(0) = 0; \quad h = .05, \ 0 \le t \le 5$

4. $\dfrac{dy}{dt} = \sin(t^2), \quad y(0) = 0; \quad h = .05, \ 0 \le t \le 5$

5. $\dfrac{dy}{dt} = .5y(3 - 2\sin 2\pi t - y), \quad y(0) = 1; \quad h = .05, \ 0 \le t \le 5$

7.4 Exact solutions

There are several types of differential equations for which there are standard methods for finding exact solutions. In this section we will look at methods for solving *separable* equations and *first-order linear* equations.

Separable equations. A first-order differential equation that can be written in the form

$$f(y)dy = g(t)dt$$

is said to be separable. With variables separated in this way, the differentials on each side of the equation may be antidifferentiated independently:

$$\int f(y)dy = \int g(t)dt$$

In some cases, the resulting equation can be solved algebraically for y in terms of t.

• EXAMPLE 1. *Find the general solution of* $\dfrac{dy}{dt} = y^{4/3}\sin t.$

The separated form of this equation is

$$y^{-4/3}dy = \sin t\, dt.$$

We'll apply the $\int()$ operator to each side of this equation, asking for the indefinite integral (involving a constant C) only on the right-hand side. The **solve()** function (**F2-1**) then finds y explicitly in terms of t.

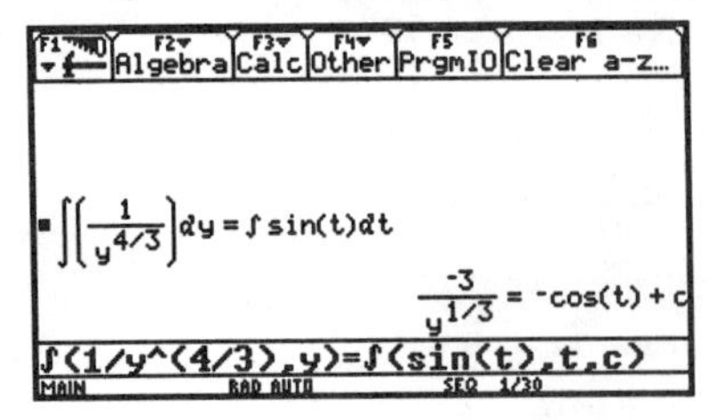

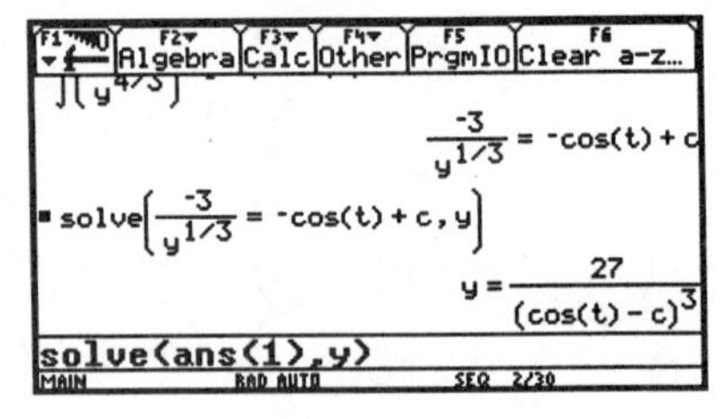

• EXAMPLE 2. *Find and graph the solution of the initial value problem*

$$\frac{dy}{dt} = -t\,y^2, \quad y(0) = 3.$$

The separated form of this equation is

$$y^{-2}dy = -t\,dt.$$

First let's apply the $\int()$ operator to each side of this equation just as in the previous example. Then we'll solve for the constant of integration C, using the "with" operator to substitute the initial values of t and y.

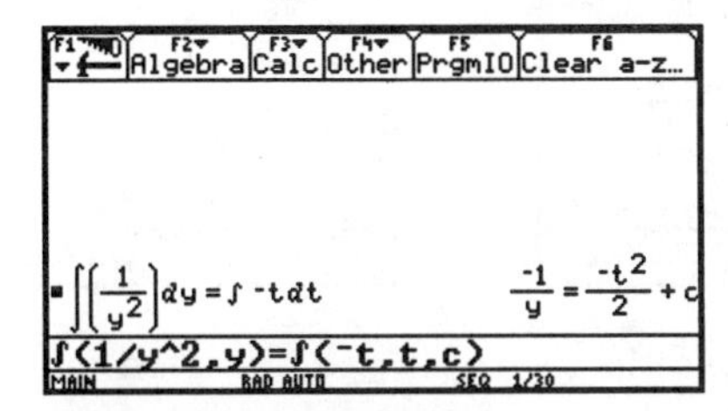

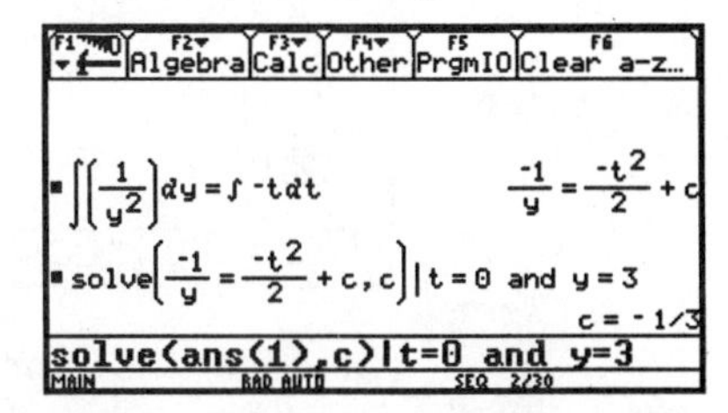

Then we substitute the value of C into the equation, solve for y in terms of t, and plot the solution.

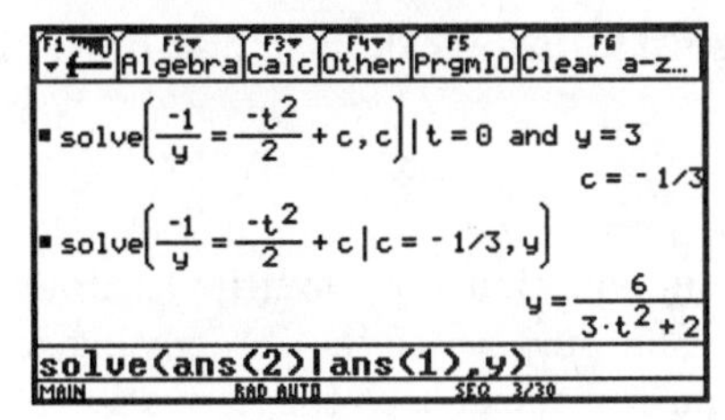

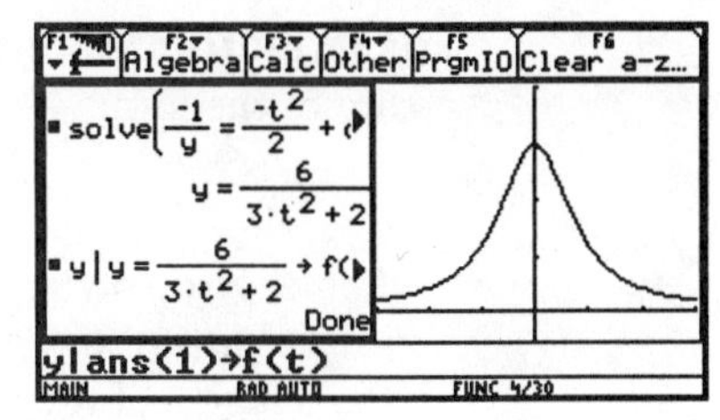

Integration between limits. A useful technique for solving initial value problems is definite—rather than indefinite—integration, i.e., *integration between limits.* Given a separable differential equation together with an initial value,

$$f(y)dy = g(t)dt, \quad y(t_0) = y_0,$$

we can find the solution by introducing dummy variables of integration and then integrating the left side of the equation from y_0 to y and the right side from t_0 to t:

$$\int_{y_0}^{y} f(u)du = \int_{t_0}^{t} g(s)ds.$$

• EXAMPLE 3. *Find the solution of the initial value problem*

$$\frac{dy}{dt} = -\cos(t)\sqrt{y}, \quad y(0) = 1.$$

First we separate variables and rewrite the problem as

$$\int_1^y \frac{du}{\sqrt{u}} = -\int_0^t \cos(s)ds.$$

Then we let the **TI-89/92** do the rest of the work by entering

∫(1/√(u),u,0,y)=∫(-cos(s),s,0,t)

and then using **solve()** to solve for y.

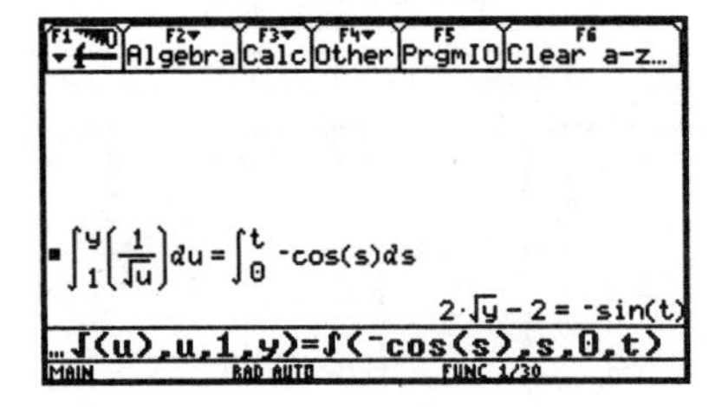

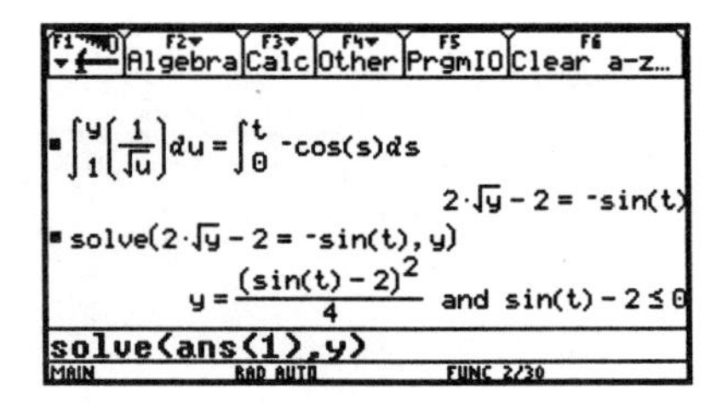

(Note that the second condition, $\sin t - 2 \leq 0$, reported by the calculator is superfluous in this problem because it's true for all t. Can you think of the reason why such a condition is reported?)

First-order linear equations. Recall that first-order linear equations are of the form

$$\frac{dy}{dt} + p(t)y = q(t).$$

The key to solving such an equation is to multiply through by an appropriate *integrating factor.* The purpose of the integrating factor is to make the left side of the equation recognizable as the derivative of a product. The integrating factor that serves this purpose is

$$e^{\int p},$$

where $\int p$ means any antiderivative of $p(t)$. Multiplying each side of the equation by $e^{\int p}$ gives

$$e^{\int p} \cdot \frac{dy}{dt} \;+\; p(t)e^{\int p} \cdot y \;=\; e^{\int p} q(t),$$

which can be rewritten as

$$\frac{d}{dt}\left(e^{\int p} y\right) = e^{\int p} q(t).$$

With this done, we can now find y by antidifferentiating each side of the equation and then multiplying through by $e^{-\int p}$.

• EXAMPLE 4. *Find the general solution of*

$$\frac{dy}{dt} + y = \sin t.$$

Here we have $p(t) = 1$ and $q(t) = \sin t$. The integrating factor is

$$e^{\int 1} = e^t.$$

Multiplying through the equation by e^t and recognizing the left side as the derivative of a product gives us

$$\frac{d}{dt}\left(e^t y\right) = e^t \sin t,$$

which implies that

$$e^t y = \int e^t \sin t \, dt.$$

We'll let the **TI-89/92** take it from here.

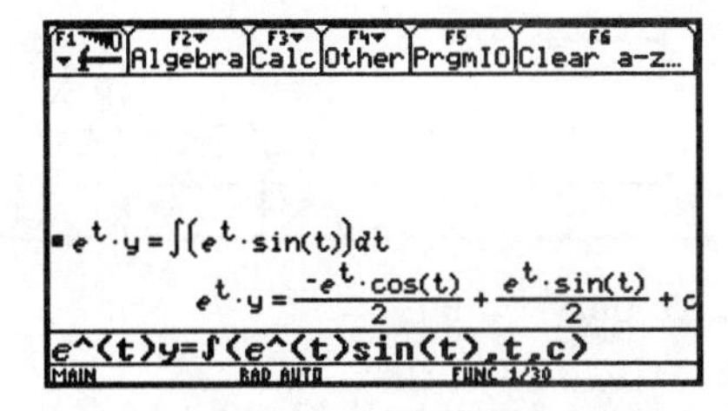

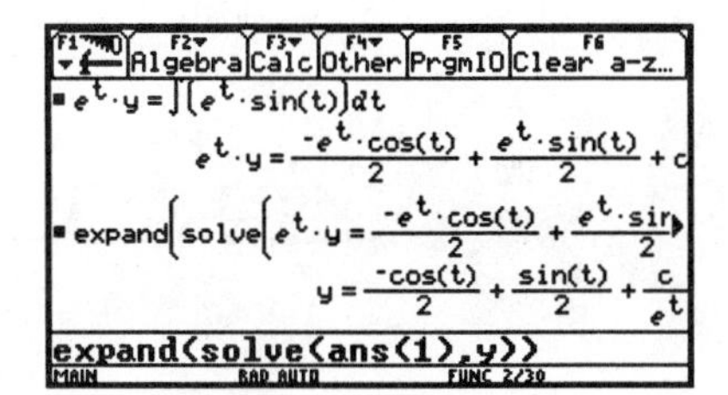

Thus the general solution is $y = (\sin t - \cos t)/2 + C\,e^{-t}$.

An integration-between-limits technique can be used to solve initial value problems that involve first-order linear equations. Suppose we have an initial value problem

$$\frac{dy}{dt} + p(t)y = q(t), \quad y(t_0) = y_0.$$

The integrating factor technique described above gives

$$\frac{d}{dt}\left(e^{\int p} y\right) = e^{\int p} q(t).$$

To simplify notation, let $\phi(t) = e^{\int p}$. Then after multiplying through by dt, we have

$$d\left(\phi(t)y(t)\right) = \phi(t)q(t)dt.$$

Now we replace t with a different dummy variable and then integrate each side of this equation from t_0 to t to produce

$$\phi(t)y(t) - \phi(t_0)y_0 = \int_{t_0}^{t} \phi(s)q(s)ds.$$

Thus we arrive at the solution

$$y(t) = \phi(t)^{-1}\left(\phi(t_0)y_0 + \int_{t_0}^{t} \phi(s)q(s)ds\right).$$

• EXAMPLE 5. *Find, for $t < 5$, the solution of*

$$\frac{dy}{dt} + \frac{3y}{5-t} = t, \quad y(0) = y_0.$$

Plot the graph of the solution for $0 \le t \le 5$ if $y_0 = 10$.

First we compute the integrating factor $\phi(t)$ for $t < 5$. Then we'll enter the formula derived above as

$$\phi\text{(t)}^{-1}(\phi\text{(0)*y0} + \textstyle\int(\phi\text{(s)*s,s,0,t)}) .$$

(Recall that the letter "ϕ" (phi) can be typed by pressing **[2nd]-G-F** on the **TI-92**, or **[2nd]-[( ]-F** on the **TI-89**.)

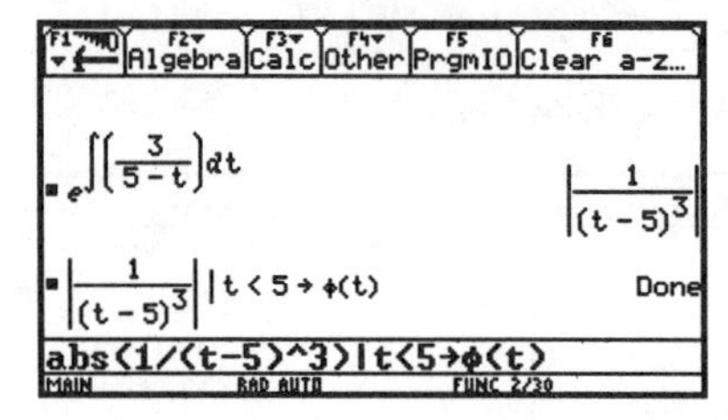

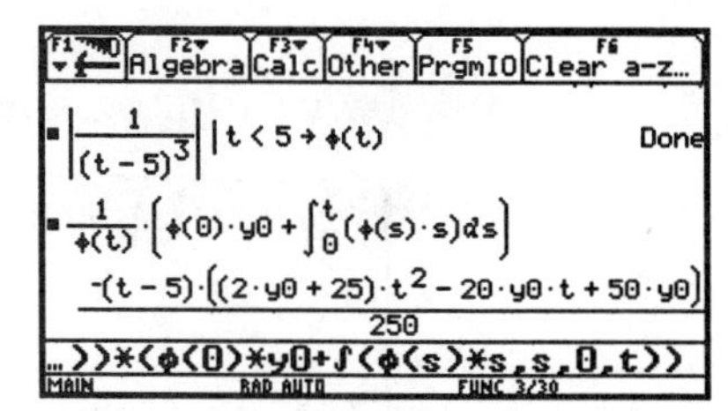

So, the solution is the cubic polynomial

$$y = \frac{y_0(5-t)}{250}\left(\left(\frac{2y_0+25}{y_0}\right)t^2 - 20t + 50\right).$$

Now let's use the "with" operator to substitute the initial value $y_0 = 10$ into the previous **ans()**wer and then plot the graph.

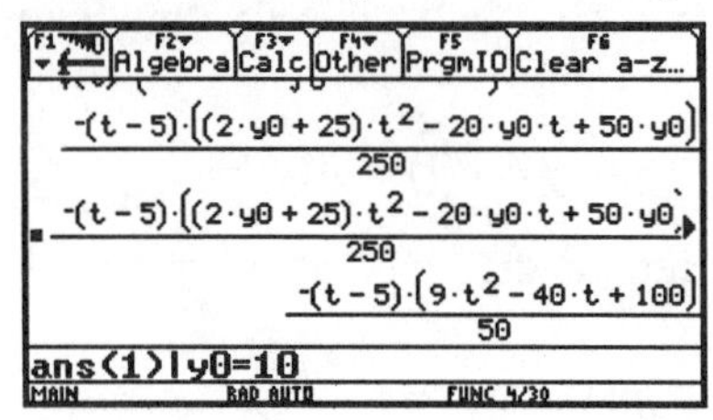

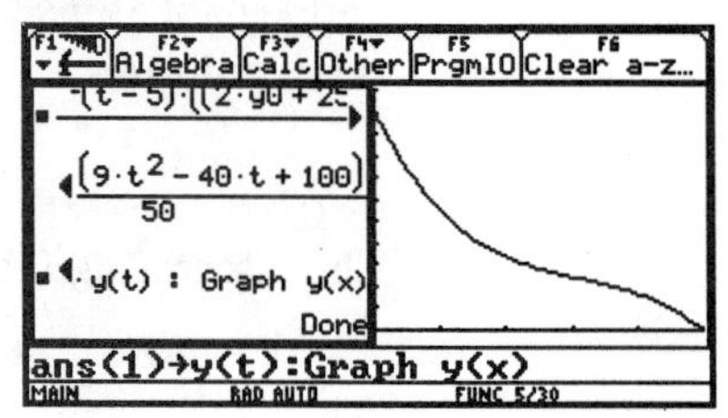

• EXAMPLE 6. *The velocity of a free-falling object, under the influence of constant gravitational force and air resistance, can be modeled by*

$$\frac{dv}{dt} + \frac{k}{m}v = -g, \quad v(0) = v_0.$$

We assume that k, m, and g are constants. Find the solution in terms of k, m, g, and v_0. Then plot the solution for $k = m = 1$, $g = 32$, and $v_0 = 0$.

The integrating factor here is

$$\phi(t) = e^{\int k/m} = e^{kt/m}.$$

So let's enter this and then the formula for the solution. Then we'll substitute the given values for k, m, g, and v_0.

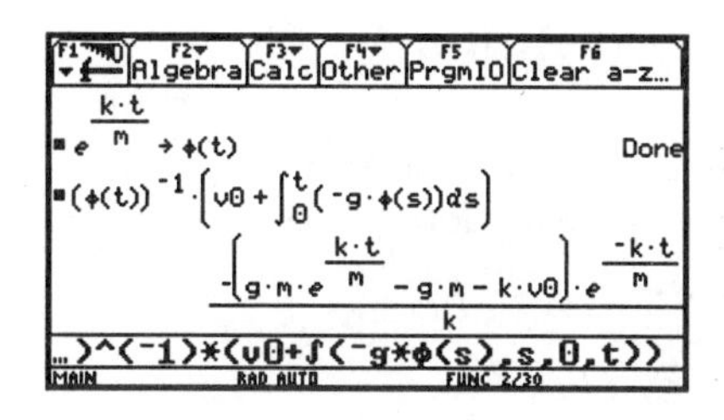

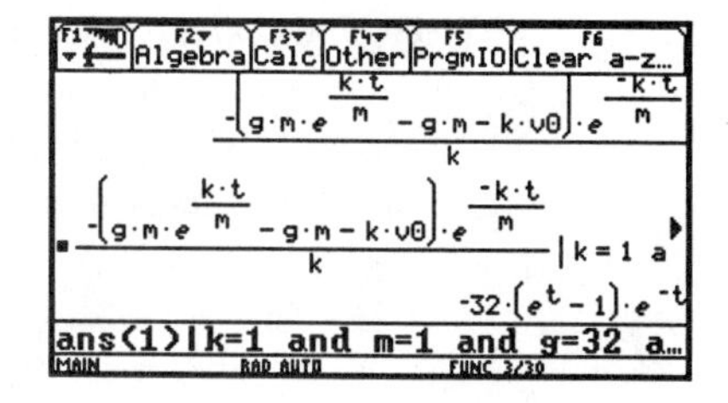

Only plotting the graph remains to be done.

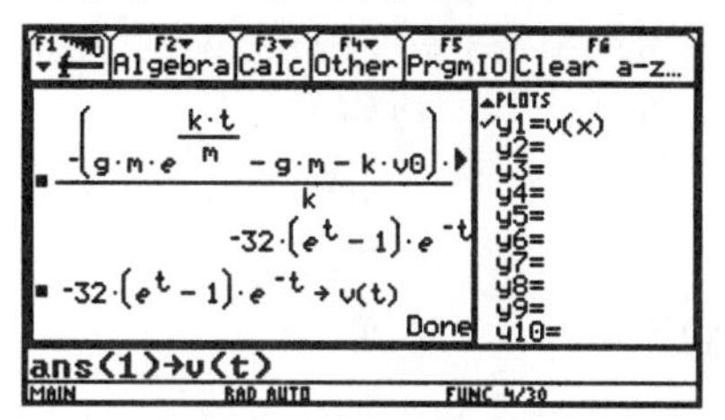

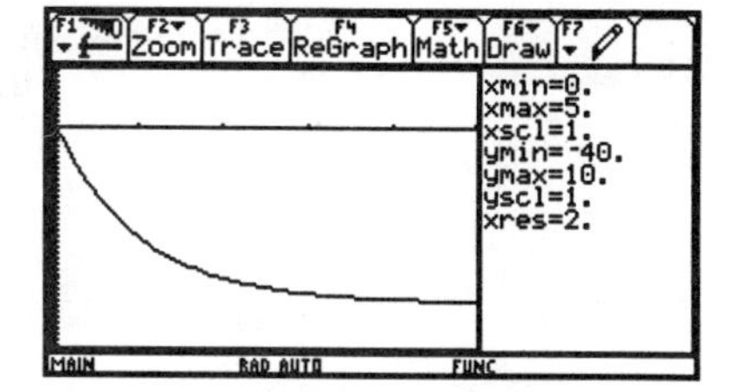

Note that because air resistance is included in the model, the velocity has a limit as $t \to \infty$. This limit is called the *terminal velocity.*

Exercises

In Exercises 1–6, find the general solution of the differential equation.

1. $\dfrac{dy}{dt} = -\sqrt{y}$

2. $\dfrac{dy}{dt} + y\cos t = \cos t$

3. $\dfrac{dy}{dt} + \dfrac{\cos t}{2+\sin t}y = \sin t$

4. $\dfrac{dy}{dt} = y(\ln y)^2$

5. $\dfrac{dy}{dt} - y = \sin t$

6. $\dfrac{dy}{dt} = (1+y)^2(1-t^2)$

7.–12. For each of the equations in Exercises 1–6, use the integration between limits technique (as in Examples 3 and 5) to find the solution satisfying the initial condition $y(0) = y_0$. For the equation in Exercise 4, assume that $y_0 > 0$.

13. Store the velocity function derived in Example 6 in terms of k, m, g, and v_0 as **v(t)**. Then compute the height function

$$y(t) = y_0 + \int_0^t v(s)ds.$$

7.5 Systems of differential equations

We consider here initial value problems involving pairs of coupled, *autonomous*, first-order differential equations. These are of the form

$$\frac{dx}{dt} = f(x, y), \quad x(0) = x_0;$$
$$\frac{dy}{dt} = g(x, y), \quad y(0) = y_0.$$

(The differential equations are called autonomous because f and g do not depend upon t.) Think of the solution of such a problem as a pair of parametric equations

$$x = x(t), \quad y = y(t),$$

whose graph is (naturally) a parametric curve.

• EXAMPLE 1. *Show that $x = 2\sin t + \cos t$ and $y = \sin t$ satisfy the system*

$$\frac{dx}{dt} = 2x - 5y, \quad x(0) = 1;$$
$$\frac{dy}{dt} = x - 2y, \quad y(0) = 0,$$

and plot the parametric curve (using Graph **Style: Path***).*

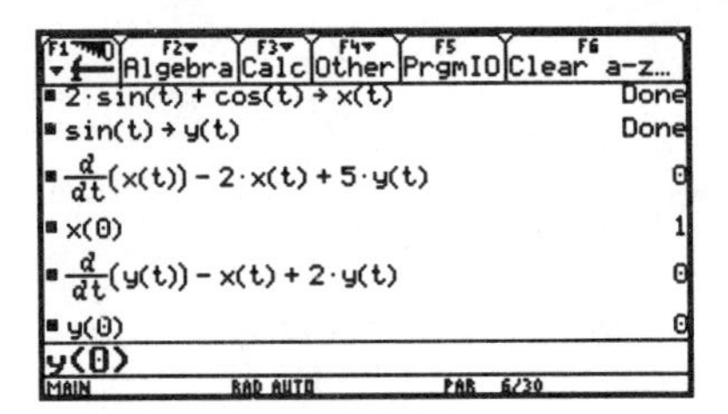

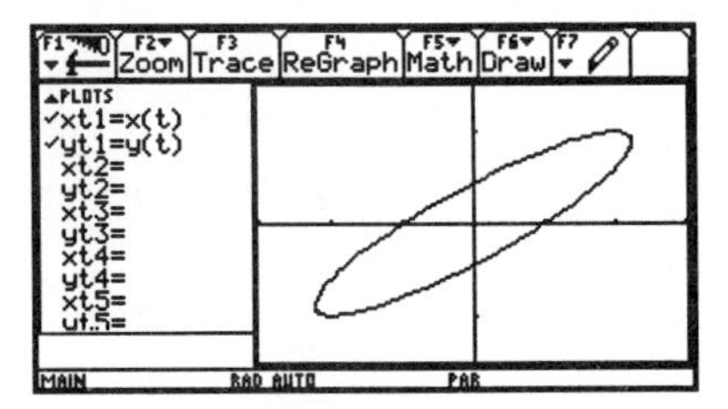

Direction fields. Let $x(t)$ and $y(t)$ satisfy the general autonomous system above. Because of the chain rule, the slope at any point on the parametric curve described by $x = x(t)$ and $y = y(t)$ is given by

$$\frac{dy}{dx} = \frac{dy/dt}{dx/dt} = \frac{g(x, y)}{f(x, y)}.$$

Therefore it makes sense to define the direction field of such a system to be the direction field for the single equation

$$\frac{dy}{dx} = \varphi(x, y) \quad \text{where} \quad \varphi(x, y) = g(x, y)/f(x, y).$$

As a result, **we can plot the direction field with our slopefld() program from Section 7.2.** (See page 113.)

• EXAMPLE 2. *In a* $[-3, 3] \times [-2, 2]$ *window, plot the direction field for the system*

$$\frac{dx}{dt} = -x - y, \quad \frac{dy}{dt} = 2x - y.$$

Determine from the direction field the behavior of solutions of the system.

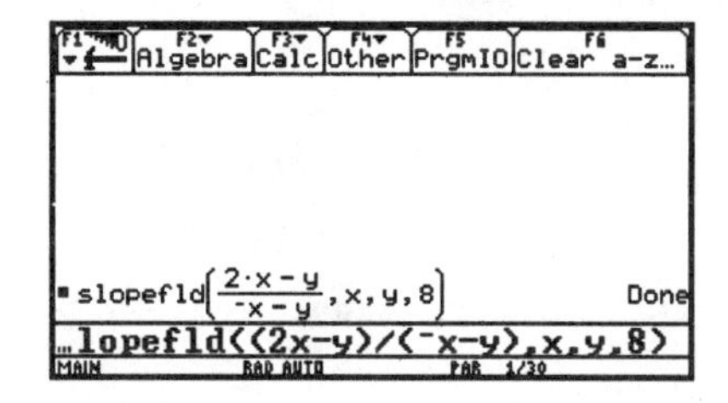

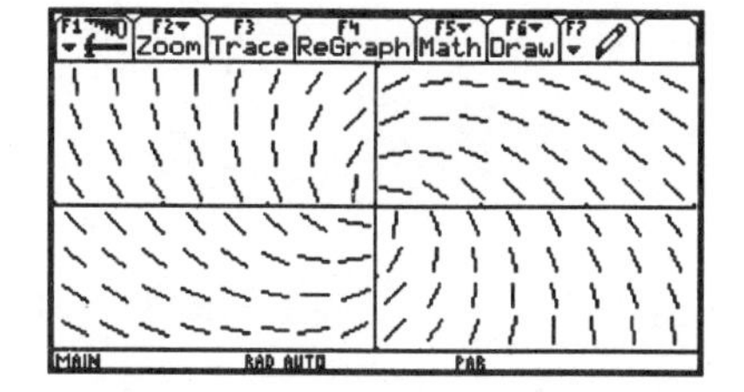

It is apparent from the direction field that graphs of solutions of the system spiral either in toward or out from the origin. Because of the differential equation satisfied by x, it is evident that $\frac{dx}{dt} < 0$ for all (x, y) in the first quadrant. Therefore, solution curves are oriented right-to-left in the first quadrant. Consequently, all solution curves spiral *in toward* the origin.

• EXAMPLE 3. *In a* $[-3, 3] \times [-2, 2]$ *window, plot the direction field for the system*

$$\frac{dx}{dt} = x + y, \quad \frac{dy}{dt} = x - y.$$

Determine from the direction field the behavior of solutions of the system.

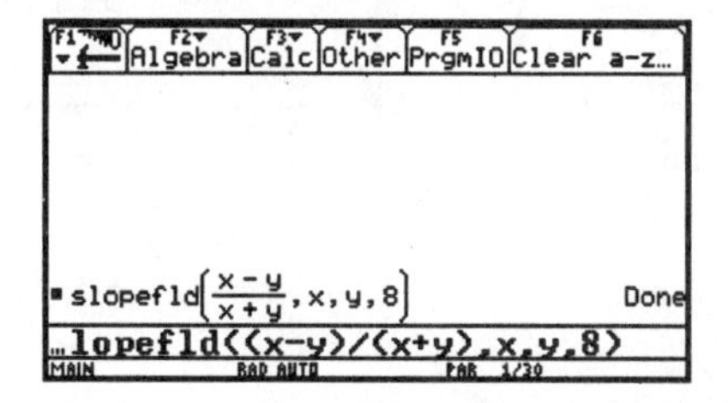

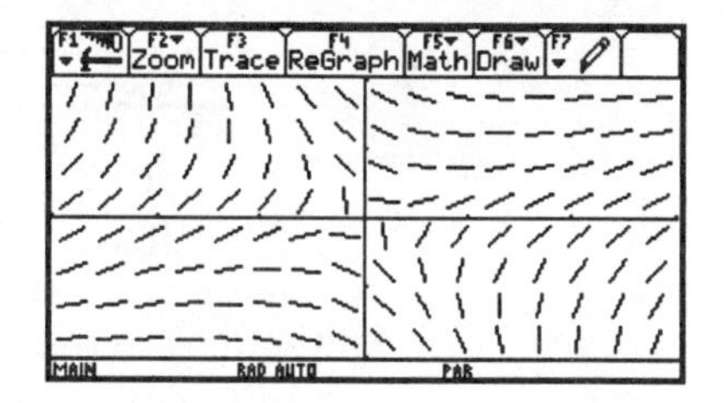

It is apparent from the direction field that graphs of solutions of the system have a hyperbolic character—approaching the origin but eventually veering away from it. Because of the differential equation satisfied by x, it is evident that $\frac{dx}{dt} > 0$ for all (x, y) in the first quadrant and $\frac{dx}{dt} < 0$ for all (x, y) in the third quadrant. Therefore, graphs of solutions are oriented left-to-right in the first quadrant and right-to-left in the third quadrant. Similarly, the differential equation satisfied by y indicates that graphs of solutions are oriented downward in the second quadrant and upward in the fourth quadrant.

Numerical approximation. The solution of a system

$$\frac{dx}{dt} = f(x, y), \quad x(0) = x_0,$$
$$\frac{dy}{dt} = g(x, y), \quad y(0) = y_0,$$

can be approximated by Euler's method, which in this case takes the form:

For $n = 1, 2, 3, \ldots$, *compute*

$$x_n = x_{n-1} + h\, f(x_{n-1}, y_{n-1}),$$
$$y_n = y_{n-1} + h\, g(x_{n-1}, y_{n-1}).$$

This is simple to implement on the **TI-89/92**. Let's use as an example the system

$$\frac{dx}{dt} = -x - y, \quad \frac{dy}{dt} = 2x - y,$$

from Example 2, with initial values $x(0) = -2$ and $y(0) = 2$.

First press **MODE** and set the **Graph** mode to **SEQUENCE**. Then in the Y= Editor, enter the formulas as shown. (You will need to go to the Home screen to issue **DelVar h,x0,y0** if those variables currently have values.) Also, press **F7** and set **Axes: CUSTOM**, **X Axis: u1**, and **Y Axis: u2**.

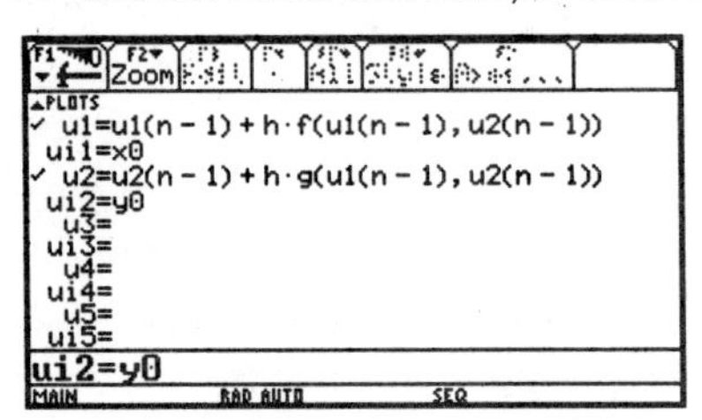

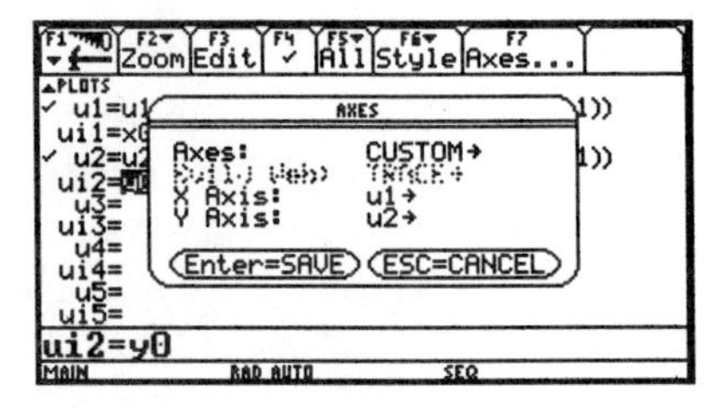

After setting up the formulas in the **Y=** Editor, go to the Home screen and define **f(x,y)**, **g(x,y)**, **x0**, **y0**, and **h**. We'll use **h**=.05, for starters. Finally we need to set window variables. Let's do 100 steps of the method (which will take us to $t = 5$, since $h = .05$) and plot the graph in a $[-3, 3] \times [-2, 2]$ window. Once these are set, press ◇**GRAPH**.

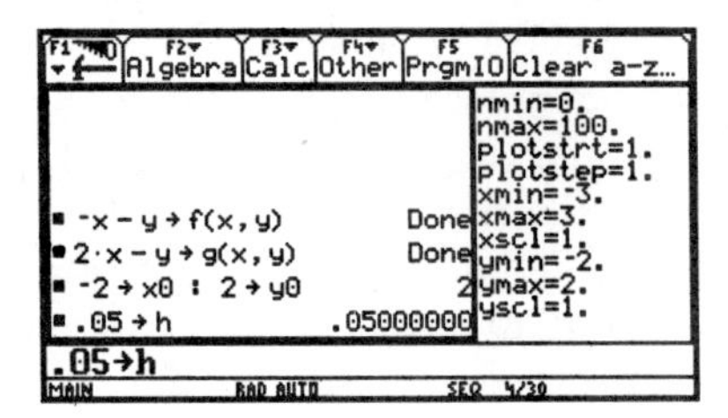

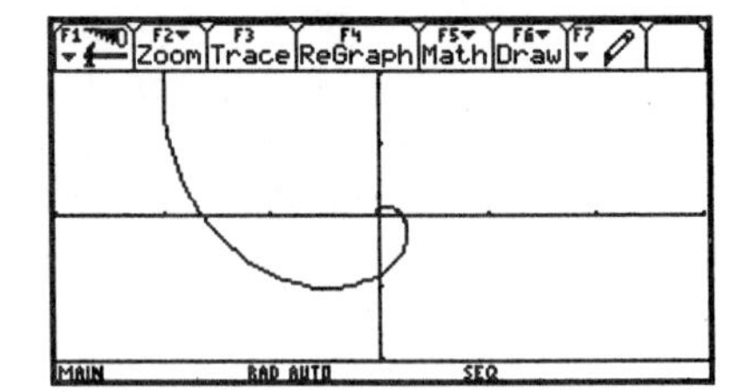

So we see a solution curve that is consistent with the direction field we saw in Example 2. In fact, we can use the **slopefld()** program to plot the direction field together with the approximate solution curve.

```
-x - y → f(x, y)                     Done
2·x - y → g(x, y)                    Done
-2 → x0 : 2 → y0                        2
.05 → h                         .05000000
slopefld(g(x, y)/f(x, y), x, y, 8)   Done
...lopefld(g(x,y)/f(x,y),x,y,8)
```

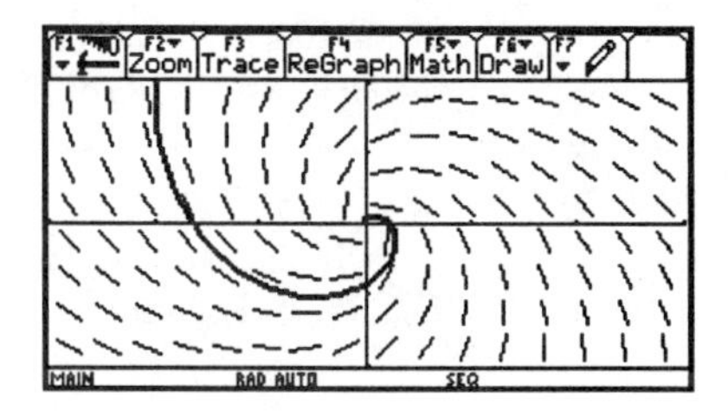

• EXAMPLE 4. *Plot the direction field for the system*

$$\frac{dx}{dt} = x + y, \quad \frac{dy}{dt} = x - y,$$

together with an approximate solution with $x(0) = -0.8$ *and* $y(0) = 2$, *in a* $[-3, 3] \times [-2, 2]$ *window. Use* $h = .05$ *and set* **nmax**$= 100$. *Then repeat for initial values* $x(0) = -0.85$ *and* $y(0) = 2$.

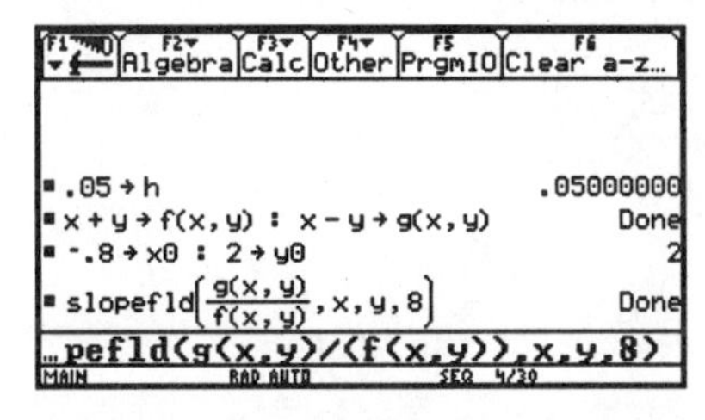

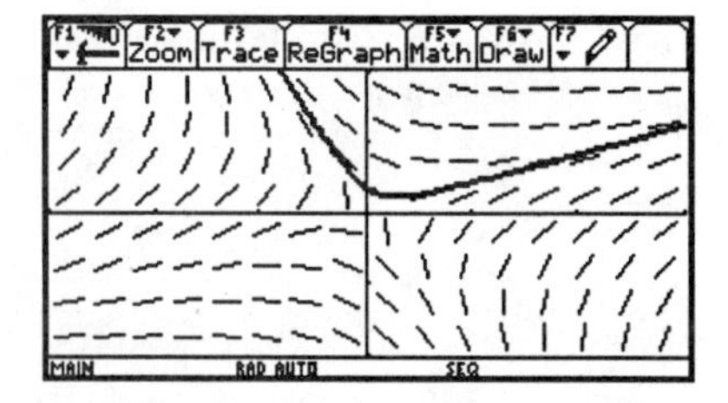

```
.05 → h                         .05000000
x + y → f(x, y) : x - y → g(x, y)    Done
-.8 → x0 : 2 → y0                       2
slopefld(g(x, y)/f(x, y), x, y, 8)   Done
-.85 → x0 : 2 → y0                      2
slopefld(g(x, y)/f(x, y), x, y, 8)   Done
...pefld(g(x,y)/(f(x,y)),x,y,8)
```

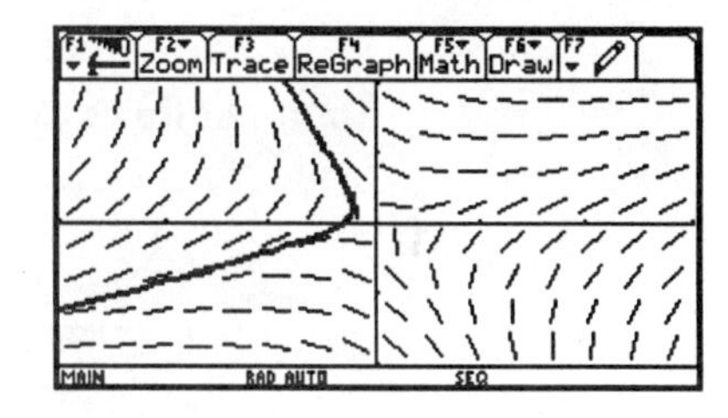

Improved Euler. Euler's method is a useful and simple method; yet unless we use a *very* small stepsize, it lacks the accuracy needed to produce approximate solutions that reflect the true behavior of many interesting systems. We need a method that is more accurate at moderately small stepsizes. Such a method is the so-called *Improved Euler's method* (a.k.a. *Heun's method*).

Consider a single differential equation

$$\frac{dy}{dt} = f(y).$$

If we integrate both sides of this equation from $t = t_{n-1}$ to $t = t_n = t_{n-1} + h$, we arrive at

$$y(t_n) - y(t_{n-1}) = \int_{t_{n-1}}^{t_n} f(y(t))dt.$$

Using a simple left-endpoint approximation to the integral (using only one subinterval) and isolating $y(t_n)$, we get

$$y(t_n) \approx y(t_{n-1}) + h\, f(y(t_{n-1})),$$

which leads to Euler's method. If we use a trapezoidal approximation instead, we get

$$y(t_n) \approx y(t_{n-1}) + .5h\Big(f\big(y(t_{n-1})\big) + f\big(y(t_n)\big)\Big).$$

The difficulty now is that the right side of this depends on $y(t_n)$, the very quantity we wish to approximate. So we use ordinary Euler's method to estimate it. This gives approximation

$$y(t_n) \approx y(t_{n-1}) + .5h\Big[f\big(y(t_{n-1})\big) + f\Big(y(t_{n-1}) + h\, f\big(y(t_{n-1})\big)\Big)\Big],$$

in which the right side depends only on $y(t_{n-1})$. This approximation leads to "Improved Euler:"

$$y_n = y_{n-1} + .5h\Big(f(y_{n-1}) + f\big(y_{n-1} + h\, f(y_{n-1})\big)\Big).$$

For a system

$$\frac{dx}{dt} = f(x, y), \quad x(0) = x_0,$$
$$\frac{dy}{dt} = g(x, y), \quad y(0) = y_0,$$

Improved Euler takes the form

$$x_n = x_{n-1} + .5h\Big(f_{n-1} + f\big(x_{n-1} + h\, f_{n-1},\, y_{n-1} + h\, g_{n-1}\big)\Big),$$
$$y_n = y_{n-1} + .5h\Big(g_{n-1} + g\big(x_{n-1} + h\, f_{n-1},\, y_{n-1} + h\, g_{n-1}\big)\Big),$$

where f_{n-1} and g_{n-1} are short for $f(x_{n-1}, y_{n-1})$ and $g(x_{n-1}, y_{n-1})$, respectively.

Now, let's implement Improved Euler on the **TI-89/92**. First we need to clear the variables **f**, **g**, and **h**. Then, to represent the terms added to x_{n-1} and y_{n-1} on the right side of each of the equations for x_n and y_n, we'll define functions **imeuf()** and **imeug()** in terms of **h**, **f(x,y)**, and **g(x,y)**:

h*(f(x,y) + f(x+h*f(x,y), y+h*g(x,y)))/2 →imeuf(x,y)

h*(g(x,y) + g(x+h*f(x,y), y+h*g(x,y)))/2 →imeug(x,y)

Once these are defined, it's an easy job to set up the sequences in the **Y=** Editor as seen in the screen on the right below. (And as long as **imeuf()** and **imeug()** aren't deleted, you'll never have to enter them again!)

```
■ DelVar f,g,h                                      Done
■ h·(f(x,y) + f(x + h·f(x,y), y + h·g(x,y)))/2 ▸
                                                    Done
■ h·(g(x,y) + g(x + h·f(x,y), y + h·g(x,y)))/2 ▸
                                                    Done
...y),y+h*g(x,y)))/2→imeug(x,y)
MAIN        RAD AUTO        SEQ 3/30
```

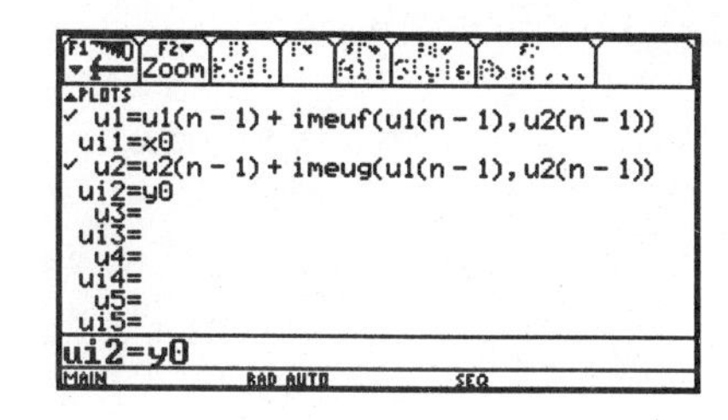

Let's try this out on the simple system

$$\frac{dx}{dt} = y, \quad \frac{dy}{dt} = -x,$$

with initial values $x(0) = 1$ and $y(0) = 0$. We only need to define the functions **f(x,y)** and **g(x,y)**, the initial values **x0** and **y0**, and the stepsize **h** before setting window variables and plotting the graph.

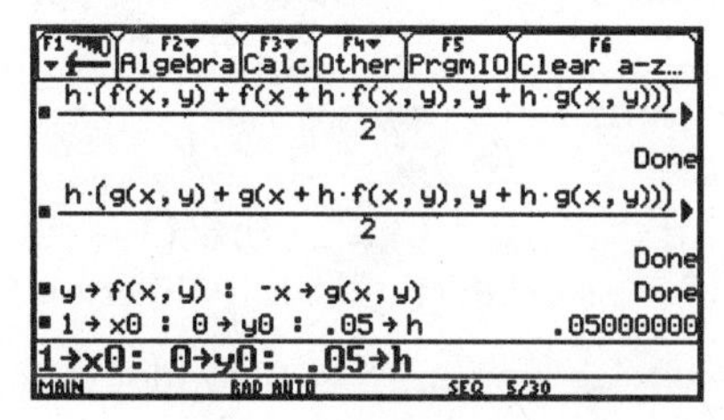

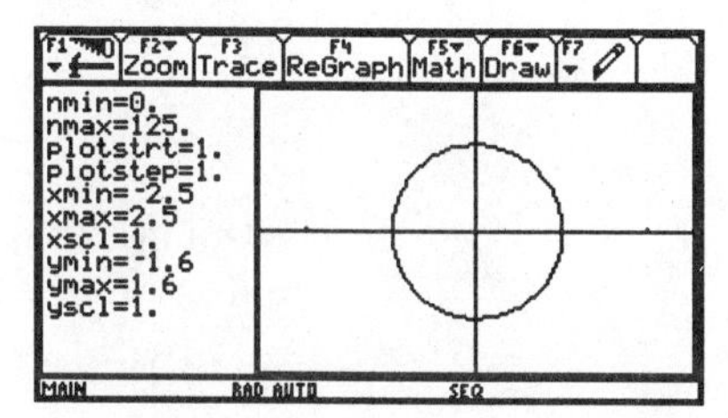

The solution of this initial value problem problem is $x = \sin t$, $y = \cos t$. So the exact solution curve would be a (clockwise) parameterization of the unit circle. It is a testament to the accuracy of Improved Euler that the approximate solution curve traced (approximately) back to its starting point at $(1, 0)$. (In fact, it would have done a reasonably good—and faster—job with $h = .1$ rather than .05.)

The accuracy of Improved Euler (or better) is especially needed when dealing with *nonlinear* problems such as in the following example.

• EXAMPLE 5. *Plot the direction field for the "predator-prey" system*

$$\frac{dx}{dt} = x(10 - y), \quad \frac{dy}{dt} = y(x - 10),$$

together with an approximate solution with $x(0) = y(0) = 5$, *in a* $[0, 25] \times [0, 25]$ *window. Use* $h = .01$ *and set* **nmax**$= 70$.

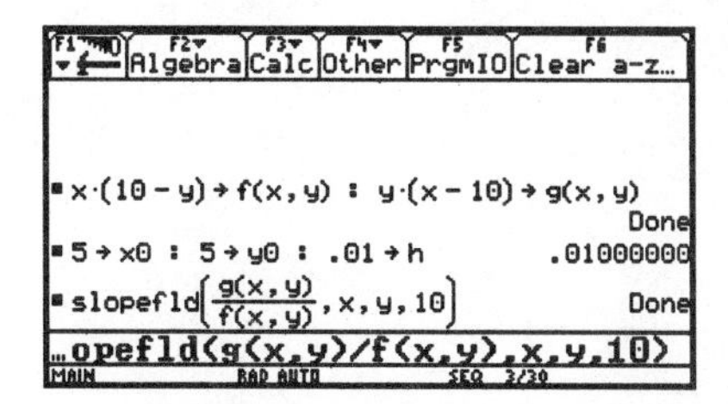

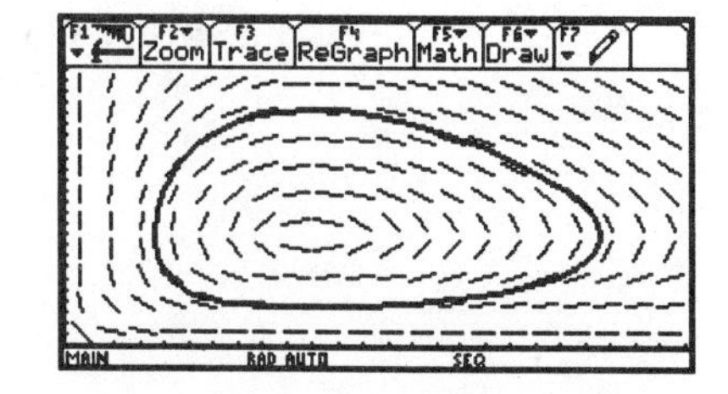

Exercises

1. Verify that $x = \sin 2t + \cos 2t$ and $y = \sin 2t - 3\cos 2t$ satisfy
$$\frac{dx}{dt} = x + 5y, \quad \frac{dy}{dt} = -x - y.$$
Plot this solution for $0 \le t \le \pi$, along with the direction field for the system.

2. Verify that $x = -e^t \sin 2t$ and $y = e^t \cos 2t$ satisfy
$$\frac{dx}{dt} = x - 2y, \quad \frac{dy}{dt} = 2x + y.$$
Plot this solution for $0 \le t \le \pi$, along with the direction field for the system.

3. Use Euler's method with stepsize $h = .05$ to plot an approximate solution of the system
$$\frac{dx}{dt} = y, \quad \frac{dy}{dt} = -x - y^3$$
with initial values $x(0) = 0$ and $y(0) = 2$. Use a $[-2, 2] \times [-2, 2]$ window and set **nmax**=250. Using **slopefld()**, plot the direction field also.

4. a) Use Euler's method with stepsize $h = .1$ to plot an approximate solution of the system
$$\frac{dx}{dt} = -y, \quad \frac{dy}{dt} = x$$
with initial values $x(0) = 1$ and $y(0) = 1$. Use a $[-2, 2] \times [-2, 2]$ window and set **nmax**=63.

   b) Repeat with $h = .05$ and **nmax**=126.

   c) Repeat with $h = .025$ and **nmax**=252.

   d) Repeat using Improved Euler with $h = .1$ and **nmax**=32.

   e) Verify that the exact solution of the initial value problem is
$$x = \cos t - \sin t, \quad y = \cos t + \sin t$$
and therefore that $x^2 + y^2 = 2$ for all points (x, y) on the solution curve.

5. Rework Example 5 using ordinary Euler's method. (But don't plot the direction field.) Try $h = .01$ with **nmax**=70 and then $h = .001$ with **nmax**=700. What do you observe?

8 Parametric and Polar Curves

Parametric curves, polar coordinates, and polar curves and areas are discussed in Chapter 11 of Stewart's *Calculus*. Here we will first explore the **TI-89/92**'s capability for plotting parametric curves and then look at some interesting polar curves.

8.1 Parametric curves

• EXAMPLE 1. *A cycloid is a curve traced out by a fixed point on a circle as it rolls along a straight line. Plot the cycloid given by*

$$x = t - \sin t, \quad y = 1 - \cos t.$$

Plotting such a parametric curve requires setting the Graph **MODE** to **PARAMETRIC**. The screen shown on the right below shows the rolling circle and the parameter t, which is the angle of rotation of the circle.

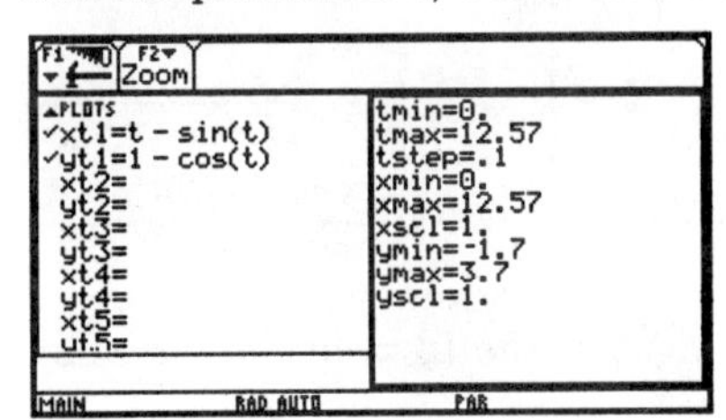

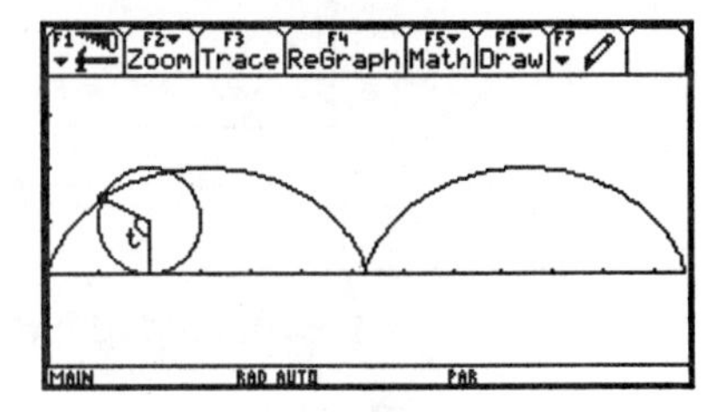

The circle was drawn using tools from the "pencil" menu (**F7**).

• EXAMPLE 2. *Lissajous figures are parametric curves of the form*

$$x = a\cos t, \quad y = b\sin nt,$$

where $n > 1$. *Take* $a = b = 1$ *and plot the Lissajous figure with* $n = 5$.

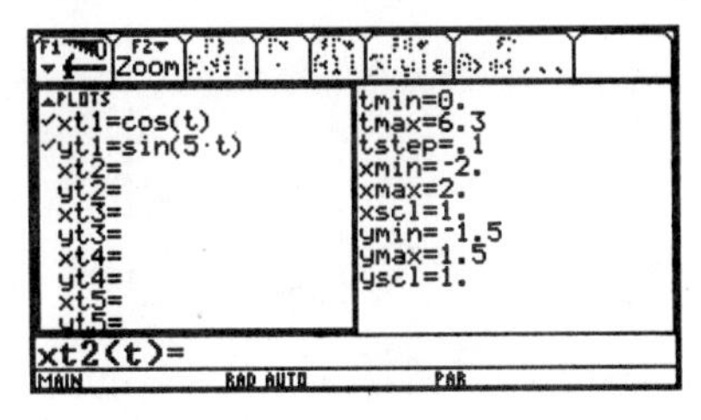

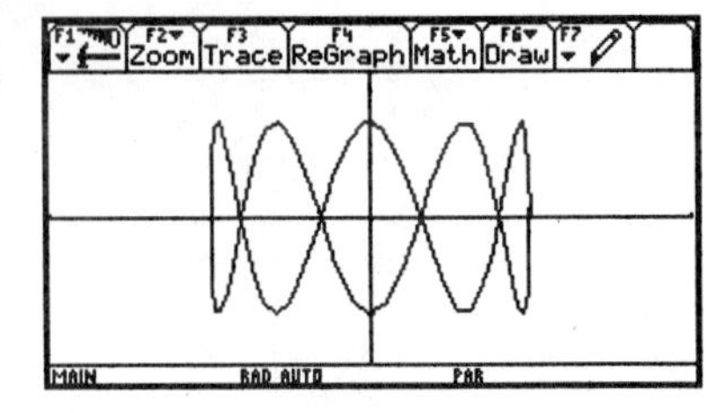

Arc length. The length of a parametric curve

$$x = f(t), \quad y = g(t), \quad a \le t \le b,$$

is given by

$$L = \int_a^b \sqrt{\left(\frac{dx}{dt}\right)^2 + \left(\frac{dy}{dt}\right)^2}\, dt,$$

provided that the curve is traced out exactly once as t increases from a to b. (Otherwise, L may still be interpreted as total distance travelled between time $t = a$ and time $t = b$ by a particle whose position at time t is given by $(f(t), g(t))$.)

• EXAMPLE 3. *Find the length of one arch of the cycloid in example 1.*

The required derivatives are easily seen to be

$$\frac{dx}{dt} = 1 - \cos t \quad \text{and} \quad \frac{dy}{dt} = \sin t,$$

so we simply enter the arc length formula containing these expressions, integrating from $t = 0$ to $t = 2\pi$.

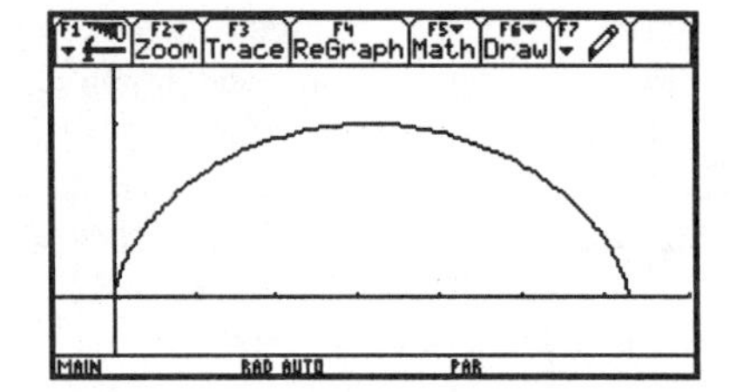

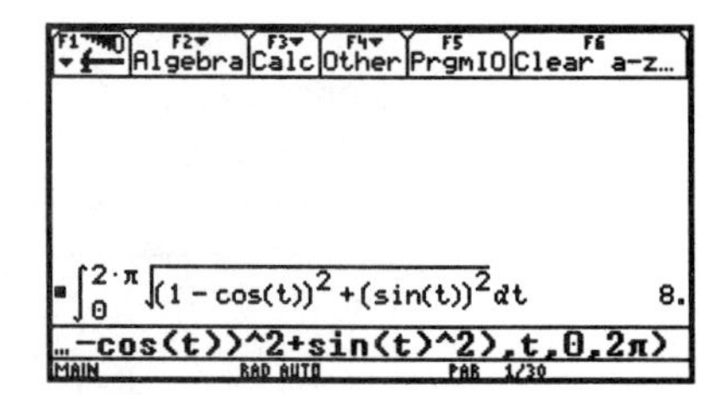

• EXAMPLE 4. *Find the length of the Lissajous figure in example 2.*

The required derivatives are easily seen to be

$$\frac{dx}{dt} = -\sin t \quad \text{and} \quad \frac{dy}{dt} = 5\cos 5t,$$

so we simply enter the arc length formula containing these expressions.

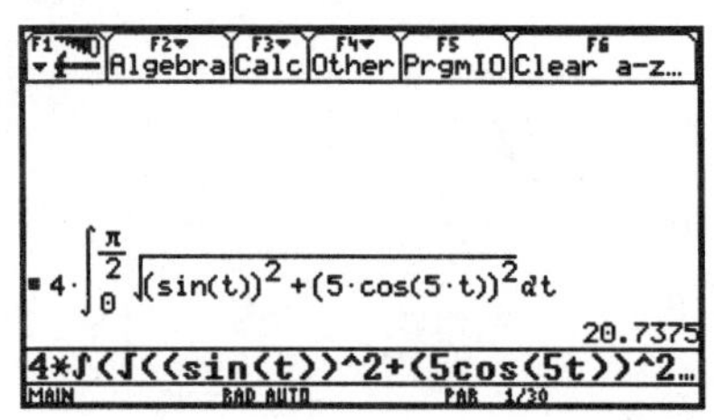

Note that in this calculation, we have taken advantage of symmetry—for the sake of efficiency—by integrating from 0 to $\pi/2$ and multiplying the result by 4. Generally speaking, the numerical approximation of definite integrals is less time-consuming over a shorter interval. Try for yourself doing the same calculation without using symmetry; i.e., by integrating from 0 to 2π.

Exercises

1. Plot and find the length of the curve

$$x = \cos t, \quad y = \sin t^2, \quad 0 \le t \le \sqrt{3\pi}.$$

2. Plot and find the length of the curve

$$x = 2\cos t, \quad y = \sin 2t.$$

3. A baseball player hits a long home-run in which the ball's path in flight is given by

$$x = 522\left(1 - e^{-t/3}\right), \quad y = 588\left(1 - e^{-t/3}\right) - 96t + 3.$$

   Find the time t at which the ball hits the ground, and find the distance that the ball travels in flight. Then compute the average speed of the ball during its flight.

4. The planet Fyodor has a circular orbit about its sun, and Fyodor's moon Theo has circular orbit about Fyodor in the same plane. Theo revolves about Fyodor ten times per Fyodoran "year," and the distance from Theo to Fyodor is one-seventh the distance from Fyodor to its sun. Determine appropriate parametric curves and simulate the motion of this system. Assume that the sun is located at the origin and use the **Animate** Path Style. (*Note*: Physics is totally ignored by this problem!)

8.2 Polar curves

A *polar curve* is the graph of an equation $r = f(\theta)$ where r and θ are standard polar coordinates defined by

$$x = r\cos\theta, \quad y = r\sin\theta.$$

• EXAMPLE 1. *Plot the graph of* $r = \ln(\theta + 1)$, $0 \le \theta \le 5\pi$.

To graph such a curve with the **TI-89/92**, we first set the Graph **MODE** to **POLAR**. Then we enter the function in the **Y=** Editor and set appropriate values for window variables.

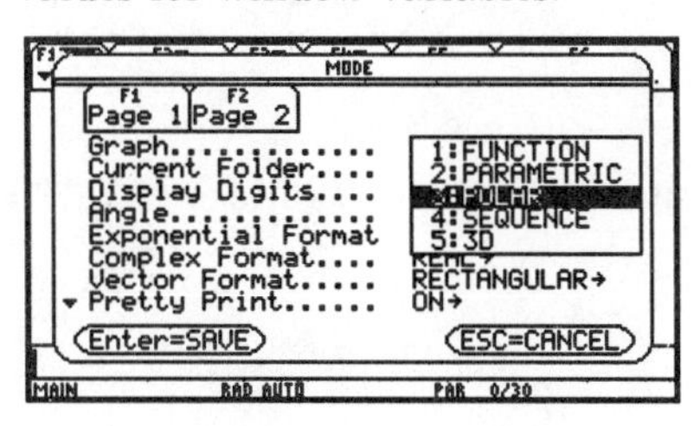

With that done, we're ready to press ◇**GRAPH** to see the plot.

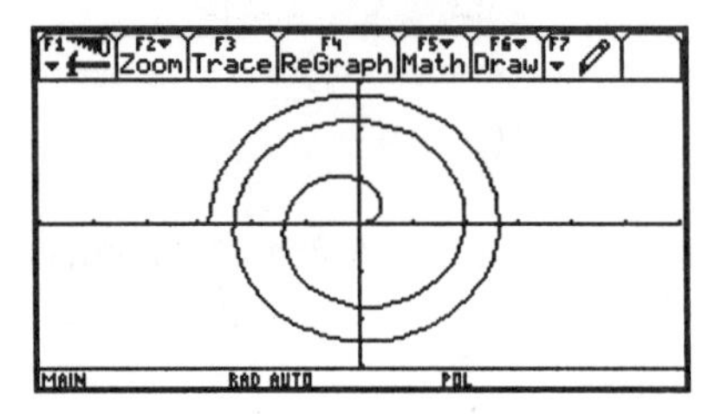

• EXAMPLE 2. *Plot the "five-leaved rose,"* $r = \sin 5\theta$.

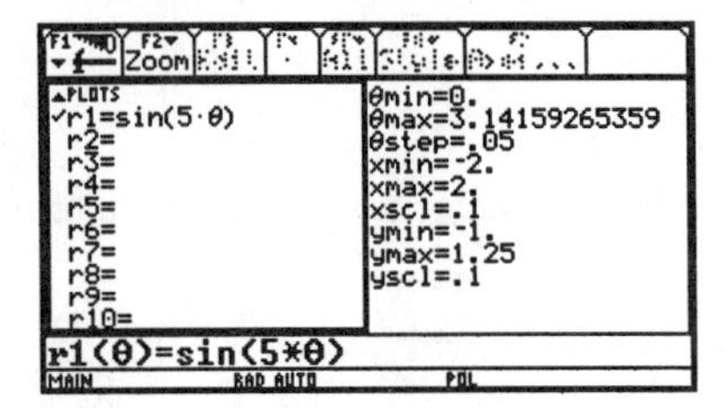

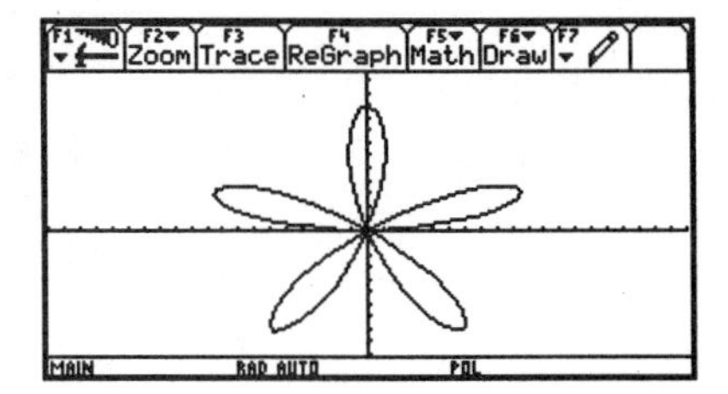

Note that the five-leaved rose was completely drawn with $0 \le \theta \le \pi$. Because $\sin 5(\theta + \pi) = -\sin 5\theta$, the curve would be trace out twice with $0 \le \theta \le 2\pi$.

• EXAMPLE 3. *Plot the graph of* $r = \sin(9\theta/4)$, $0 \le \theta \le 8\pi$.

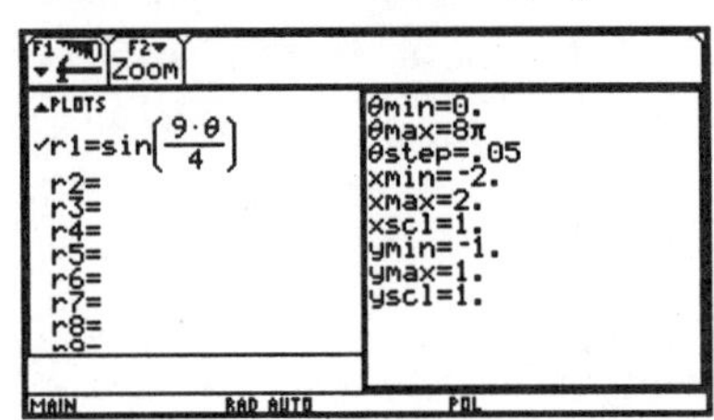

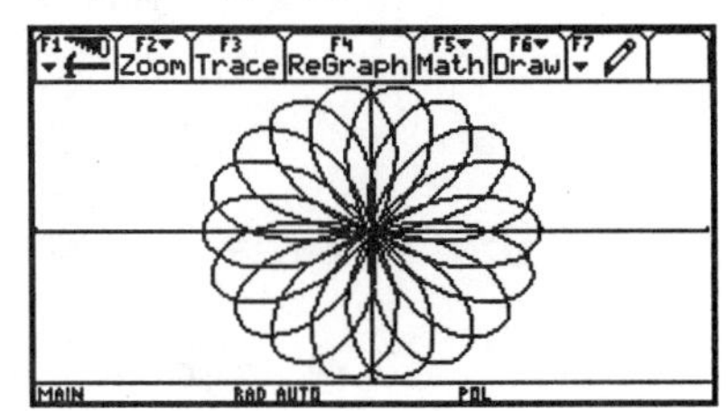

Areas. The area of a region bounded by the graph of $r = f(\theta)$ and the rays $\theta = a$ and $\theta = b$ is given by

$$\int_a^b \frac{1}{2} f(\theta)^2 \, d\theta,$$

assuming that f is continuous and nonnegative and that $0 \le b - a \le 2\pi$.

• EXAMPLE 4. *Find the area enclosed by one loop of the three-leaved rose*

$$r = \sin 3\theta.$$

Since one loop of the figure is drawn as θ rotates from $\theta = 0$ to $\theta = \pi/3$, the area of the enclosed region is

$$\int_0^{\pi/3} \frac{1}{2} \sin^2 3\theta \, d\theta = \frac{\pi}{12}.$$

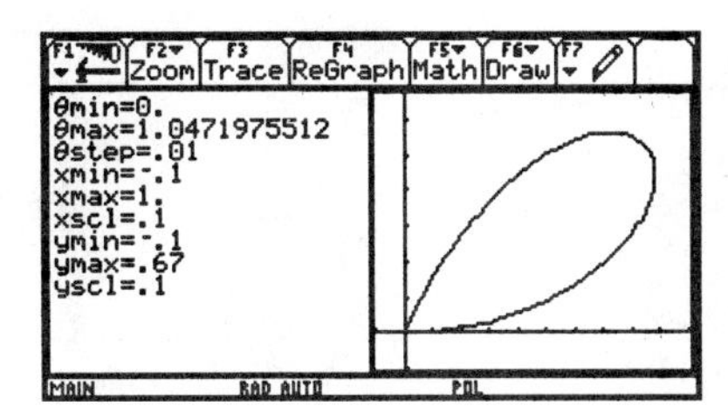

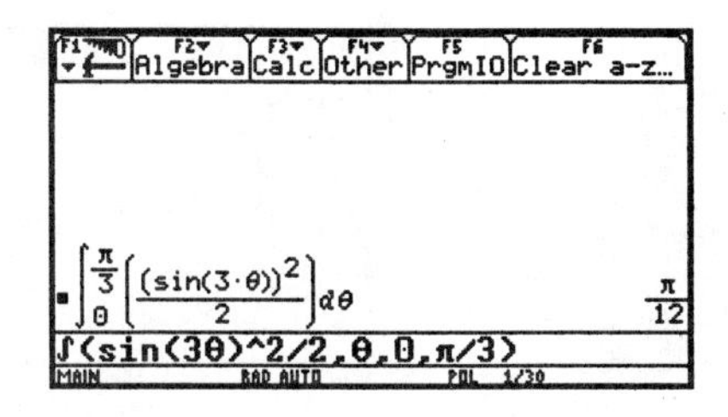

The area of a region bounded by two polar curves of $r = f(\theta)$ and $r = g(\theta)$ and the rays $\theta = a$ and $\theta = b$ is given by

$$\int_a^b \frac{1}{2}\left(f(\theta)^2 - g(\theta)^2\right) d\theta,$$

assuming that f and g are continuous and nonnegative, $0 \le b - a \le 2\pi$, and $f(\theta) \ge g(\theta)$ for $a \le \theta \le b$.

• EXAMPLE 5. *Find the area of the region that lies inside the circle* $r = 3\sin\theta$ *and outside the circle* $r = 2$.

The region is shown in the graph below, in which $0 \le \theta \le \pi$. (*Suggestion*: Set **Graph Order** to **SIMUL** (**F1-9**) to observe the curves as they intersect.) The curves intersect when $3\sin\theta = 2$. This gives two solutions in the interval $0 \le \theta \le \pi$, namely $\theta = \sin^{-1}(2/3)$ and $\theta = \pi - \sin^{-1}(2/3)$.

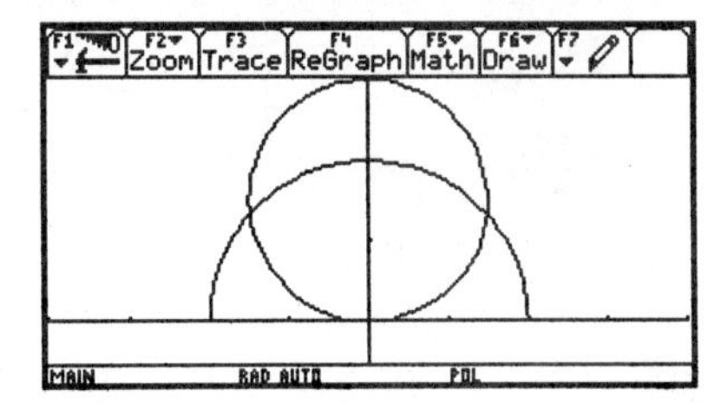

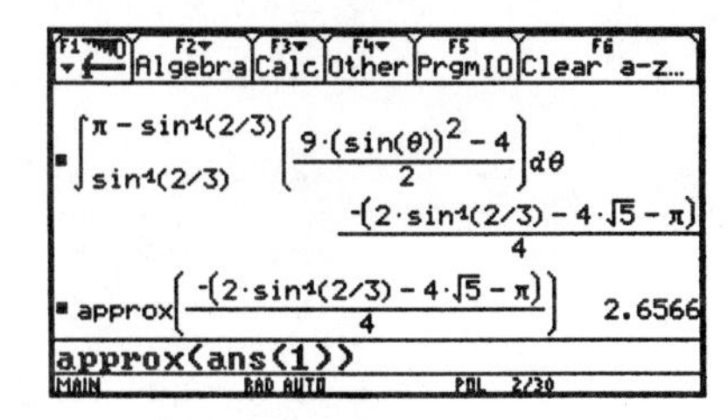

Arc Length. The length of a polar curve $r = f(\theta)$, where $a \le \theta \le b$, is

$$L = \int_a^b \sqrt{f(\theta)^2 + f'(\theta)^2}\, d\theta.$$

• EXAMPLE 6. *Find the length of the four-leaved rose* $r = \cos 2\theta$.

Using symmetry, we know that one-eighth of the length is traced out as θ rotates from $\theta = 0$ to $\theta = \pi/4$. So let's integrate over $0 \le \theta \le \pi/4$ and multiply the result by eight.

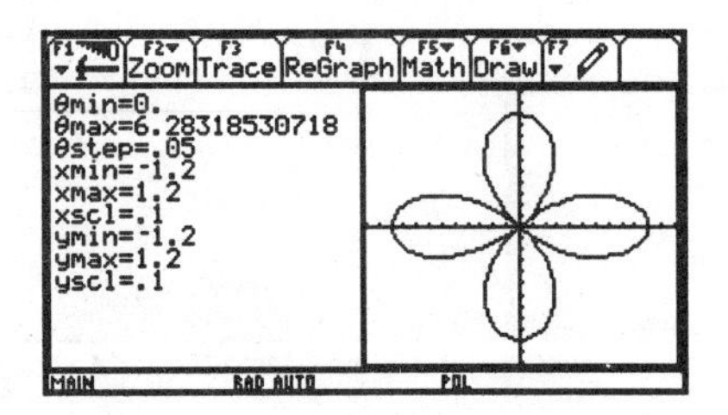

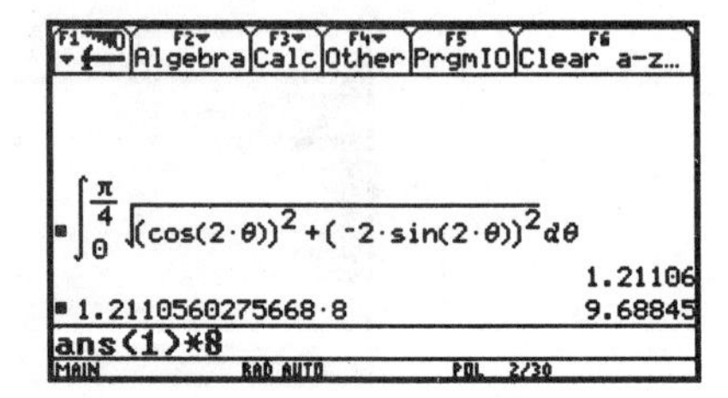

Our final example for this section involves an improper integral.

- EXAMPLE 7. *Find the length of the spiral* $r = e^{-\theta/5}$, $0 \leq \theta < \infty$.

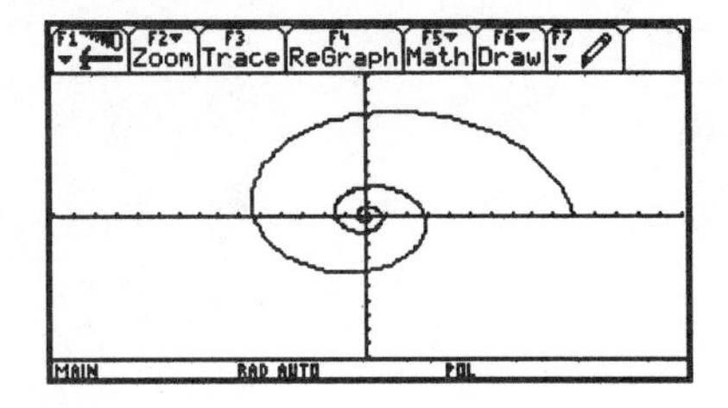

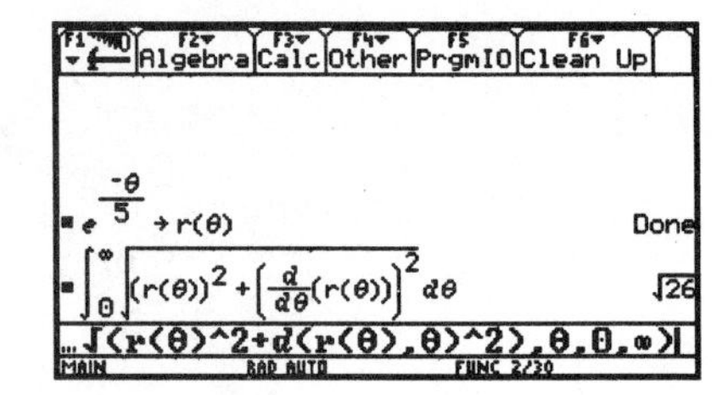

Exercises

1. Find the area of each region inside a loop formed by the graph of $r = \theta(2-\theta)(3-\theta)$.
2. Find the area inside one loop of the five-leaved rose $r = \sin 5\theta$.
3. Find the area inside the *cardioid* $r = 1 - \cos\theta$.
4. Find the area inside the inner loop of the *limaçon* $r = 1 + 2\cos\theta$.
5. Find the area of the region inside the circle $r = 3\cos\theta$ and outside the circle $r = 2\sin\theta$.
6. Plot and find the length of the polar curve $r = \sin\theta\,\cos 2\theta$.
7. Find the circumference of the *limaçon* $r = 3 + 2\cos\theta$.

9 Sequences and Series

The **TI-89/92** has graphical, algebraic, and numerical capabilities that lend themselves to the study of sequences and series. This chapter is an introduction to these capabilities and a supplement to Chapter 12 of Stewart's *Calculus*.

9.1 Sequences

This section is devoted to the ways in which the **TI-89/92** enables us to study sequences. First we'll look at the **TI-89/92**'s ability to generate sequences and find limits symbolically.

• EXAMPLE 1. Consider the sequence defined by

$$a_n = \frac{2n^2}{3n^2 - n + 1}, \quad n = 1, 2, 3, \ldots .$$

The **seq()** command, which is found under **List** in the **MATH** menu (**[2nd][5]**), can be used to display a number of terms of the sequence. Exact values are displayed if **EXACT** is selected as the **Exact/Approx MODE**.

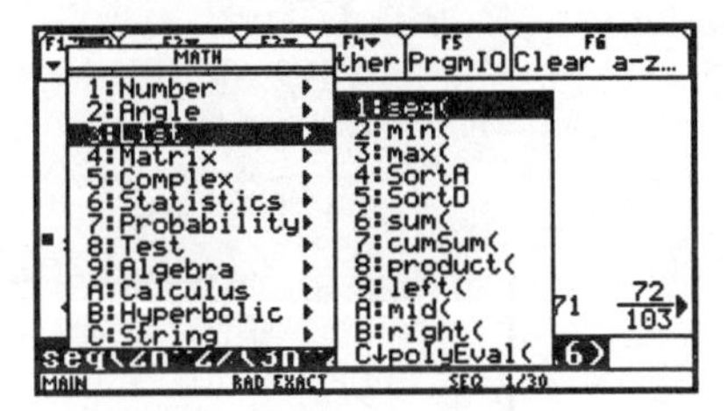

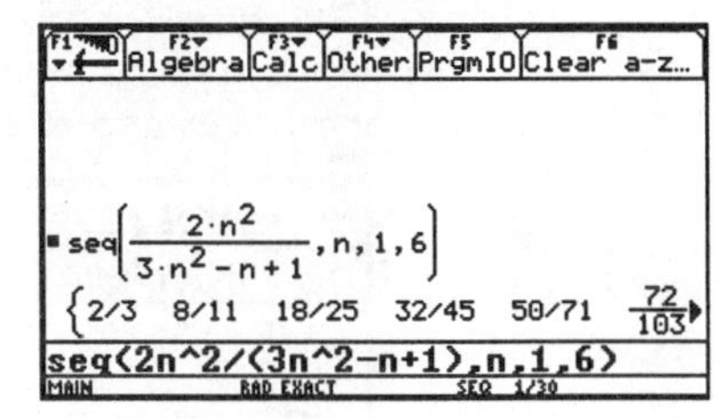

The **limit()** command, found in the **Calc** menu (**F3** from the Home screen), computes the limit of the sequence.

```
■ seq(2·n^2/(3·n^2-n+1), n, 1, 6)
{2/3  8/11  18/25  32/45  50/71  72/103
■ lim (n→∞) 2·n^2/(3·n^2-n+1)                2/3
limit(2n^2/(3n^2-n+1),n,∞)
MAIN   RAD EXACT   SEQ 2/30
```

This also works with sequences whose terms involve symbolic parameters, as seen in the following screens.

Plotting sequences. Sequences can be plotted easily on the **TI-89/92**.

• EXAMPLE 2. Suppose that we wish to study the sequence defined by

$$a_n = \frac{3n + 7(-1)^n}{4n + 5}, \quad n = 1, 2, 3, \ldots .$$

The first step in plotting the sequence is to press the **MODE** key and select **SEQUENCE** as the **Graph** mode.

Then press ◇**Y=** to bring up the **Y=** Editor. There, enter the formula for a_n as **u1**. Also from the **Y=** Editor, press **F7** and specify **n** as the variable for the **X Axis** and **u1** as the variable for the **Y Axis**.

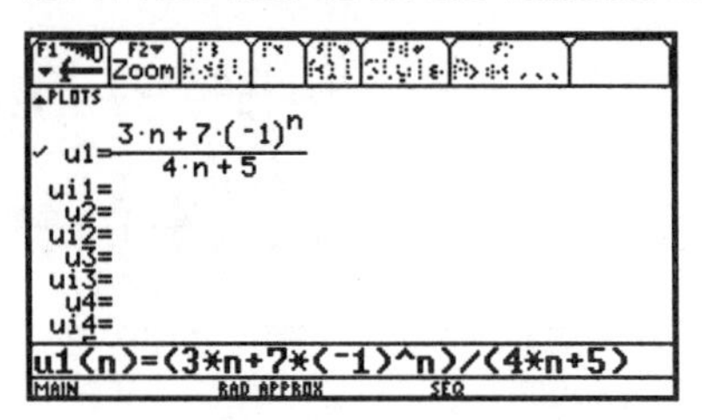

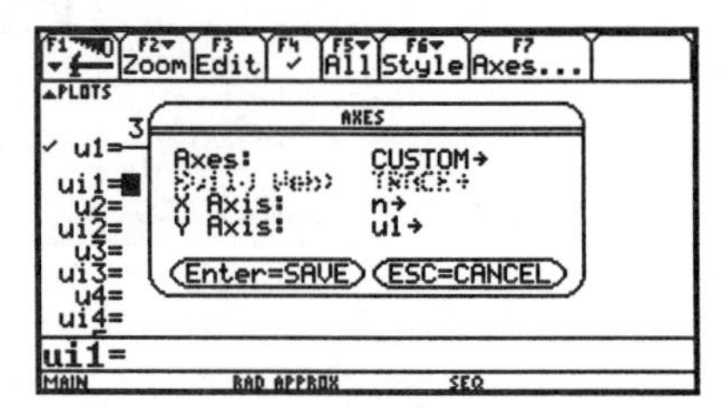

In order to get an idea what window bounds are appropriate, we can examine the terms in the sequence as a list. So press ◇**TblSet** and set each of the parameters **tblStart** and **Δtbl** equal to 1, and then press ◇**TABLE**.

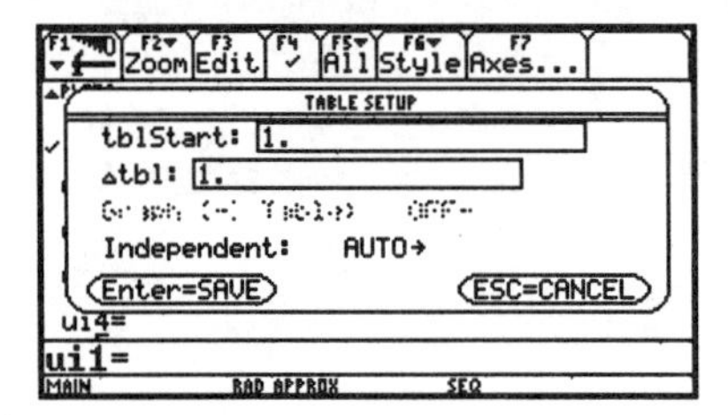

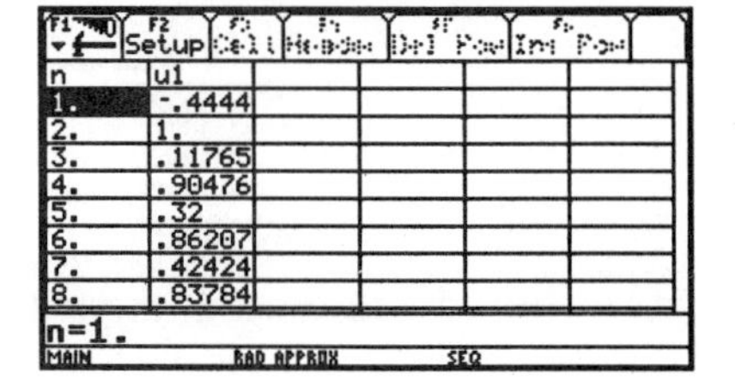

From the values in the table, it looks as if $-.5$ to 1.25 would be good choices for **ymin** and **ymax**. So press ◇**WINDOW** and enter window parameter values as shown below. Here we are plotting the first forty terms in the sequence. Then press ◇**GRAPH**. Notice that we have used **Trace** (**F3**) to display the value of the 36th term.

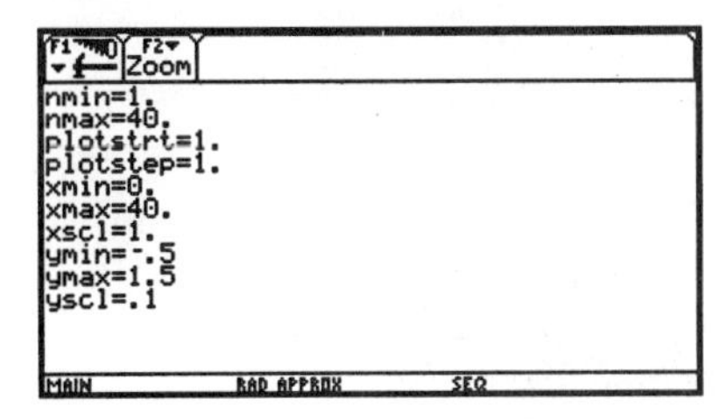

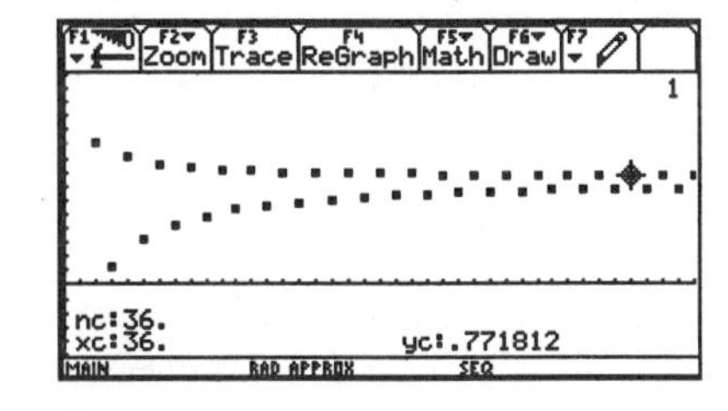

To better see the long-term behavior of the terms in the sequence, we can replot the sequence after setting **nmax** and **xmax** to 100.

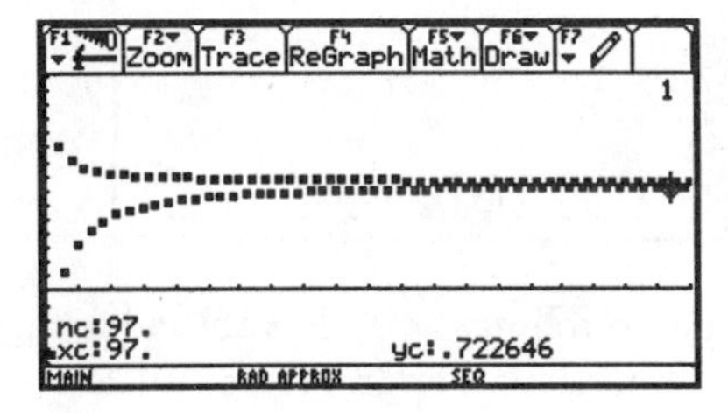

Finally, let's look at a plot of the first 200 terms. But first we'll change the plot **Style** to **Dot**.

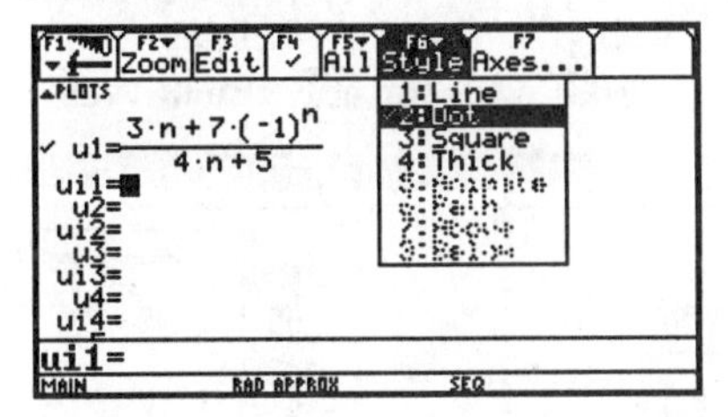

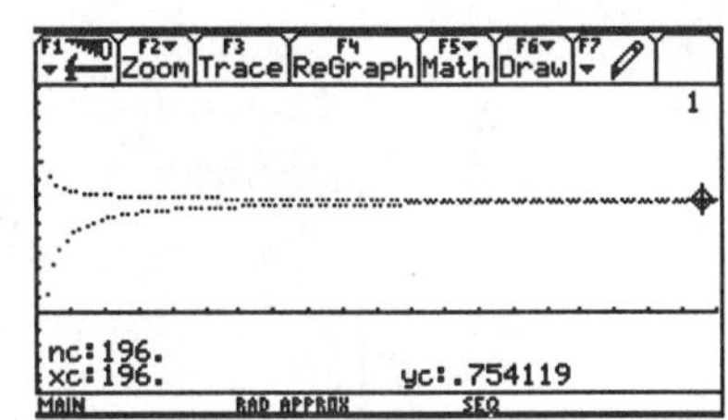

All of this provides numerical and graphical evidence that the limit of the sequence is $3/4$. To prove that this is indeed the case, first observe that

$$\frac{3n-7}{4n+5} \leq \frac{3n+7(-1)^n}{4n+5} \leq \frac{3n+7}{4n+5}$$

for all $n \geq 1$. Now the squeeze theorem tells us that

$$\lim_{n\to\infty} \frac{3n+7(-1)^n}{4n+5} = \frac{3}{4},$$

because

$$\lim_{n\to\infty} \frac{3n-7}{4n+5} = \lim_{n\to\infty} \frac{3n+7}{4n+5} = \frac{3}{4}$$

by application of l'Hôpital's Rule to the functions $f(x) = (3x - 7)/(4x + 5)$ and $g(x) = (3x + 7)/(4x + 5)$.

Recursive sequences. Often the nth term of a sequence depends not upon n, but upon previous terms in the sequence. Such a sequence is called a *recursive* sequence.

• EXAMPLE 3. The sequence $\{a_n\}_{n=1}^{\infty}$ defined by

$$a_1 = .5$$
$$a_n = 2\sin a_{n-1}, \; n = 2, 3, 4, \ldots$$

is a recursive sequence. In the **Y=** Editor, we enter a_n as **u1(n)** and then press ◇**TABLE** to view the first few terms.

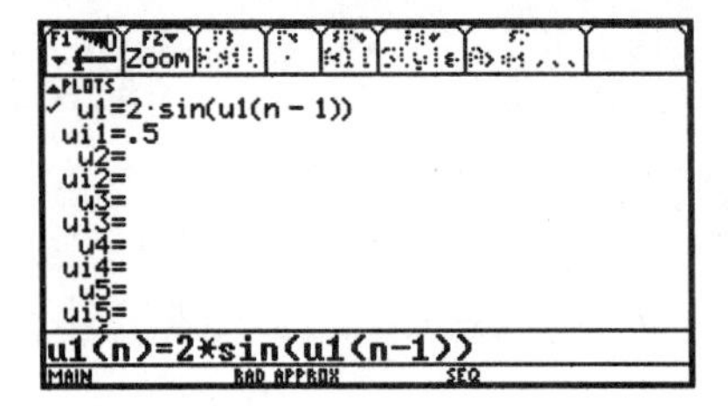

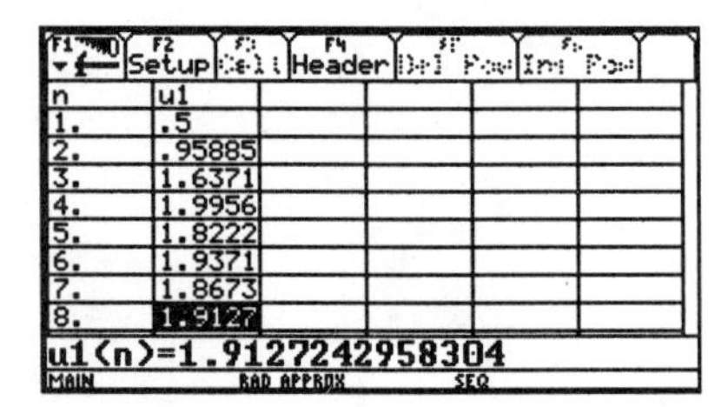

In fact, scrolling down the table a bit suggests that the sequence is converging to a limit that is approximately 1.8955. The generation of sequences such as this, in which $a_n = g(a_{n-1})$, with some given continuous function g and some given value of a_1, is often called *functional iteration*, or *fixed-point iteration*. There is a special graphical device, called a *web plot*, for visualizing such a sequence. To set up a web plot on the **TI-89/92**, press **F7** from the **Y=** Editor, and choose **Axes** = **WEB** and, for now, **Build Web** = **AUTO**.

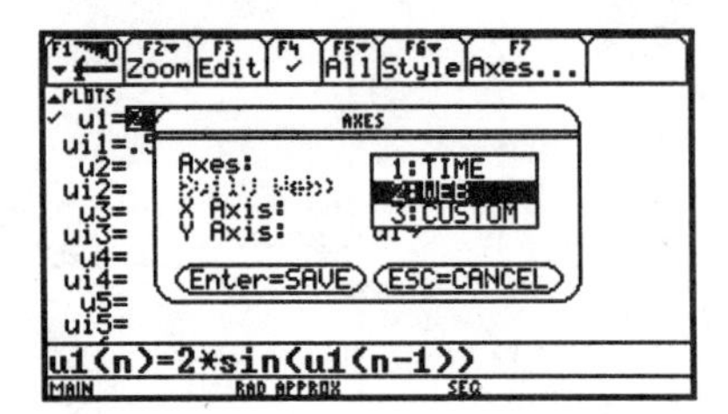

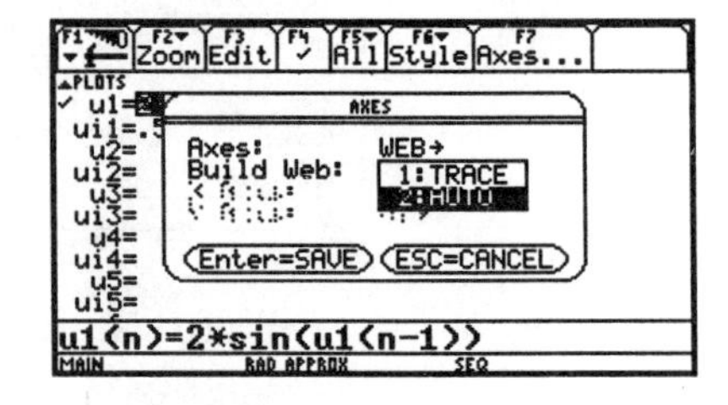

Now we set appropriate window variables and plot the graph.

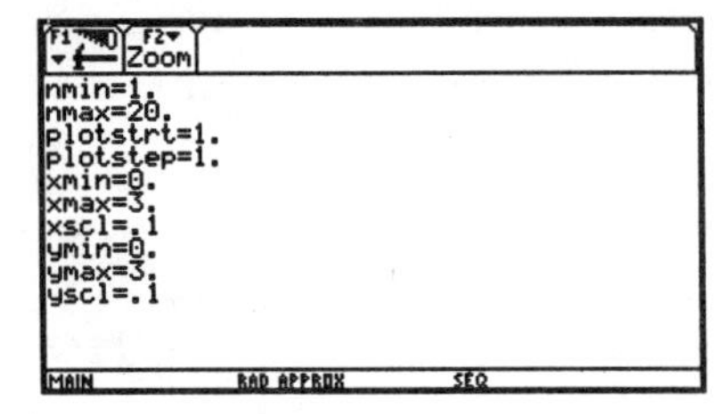

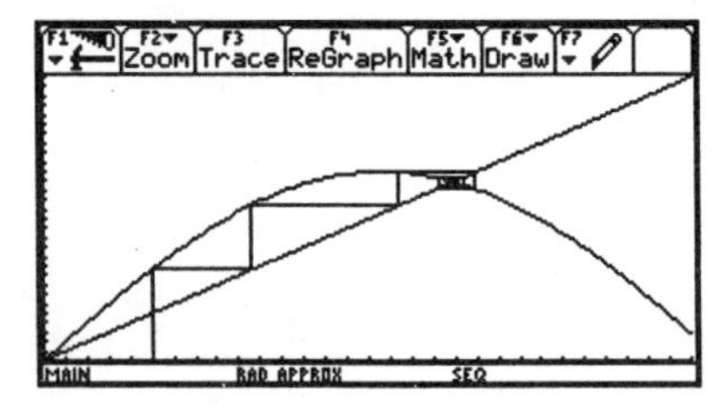

Note that the curve in the picture is the graph of $y = 2\sin x$ and the line is $y = x$. To get a better feeling for what this web plot is about, let's change to **Build Web** = **TRACE** in the **Axes** menu (**F7**) of the **Y=** Editor. Then pressing ◇**GRAPH** only shows the graphs of $y = 2\sin x$ and $y = x$.

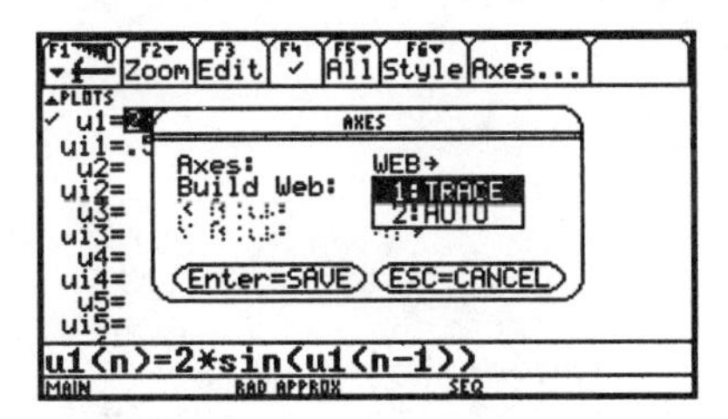

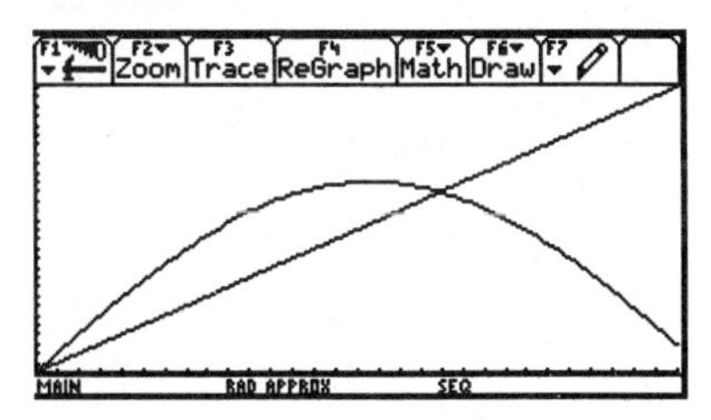

Now select **Trace** (**F3**) and press [⇒] once and then again.

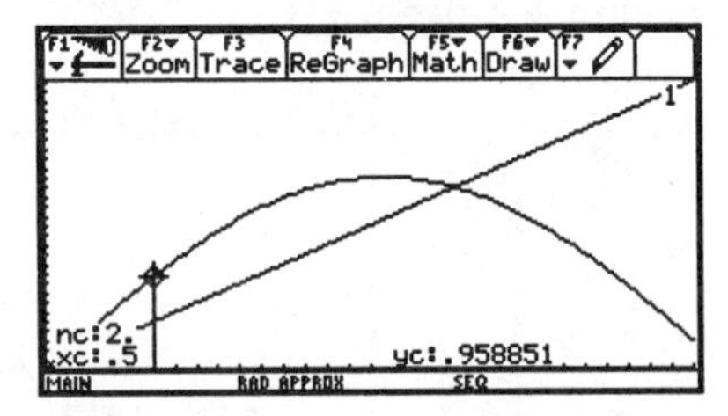

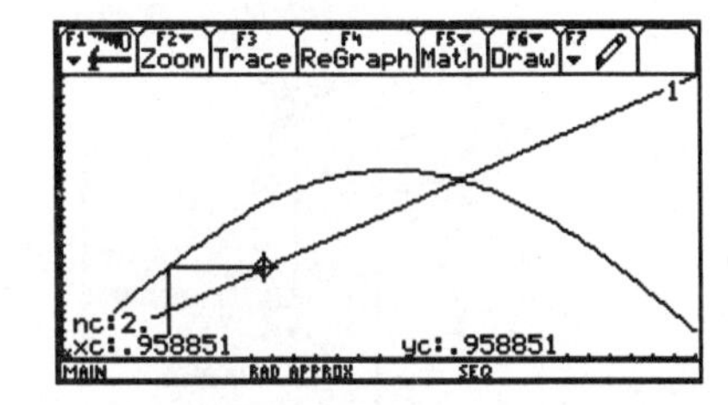

The first of these shows the computation of $a_2 = .958851$ from $a_1 = .5$, and the second shows the location of a_2 on the x-axis. Now press [⇒] twice more to find a_3 on both axes.

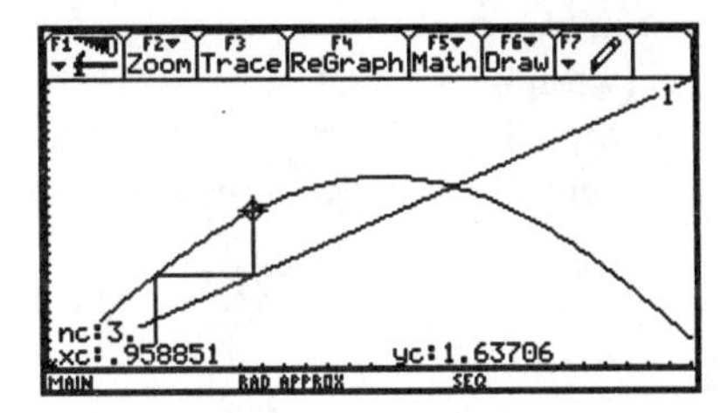

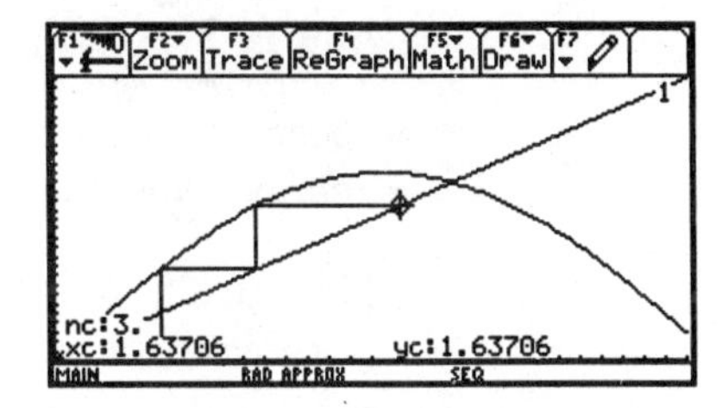

Continuing in this way, you will begin to see the "web" approach the point of intersection of the two graphs. It is also interesting to zoom in a bit to observe the behavior of the sequence once terms begin to get fairly close to the limit.

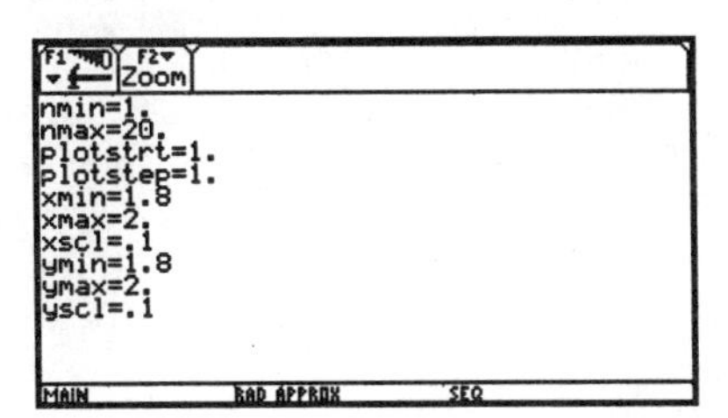

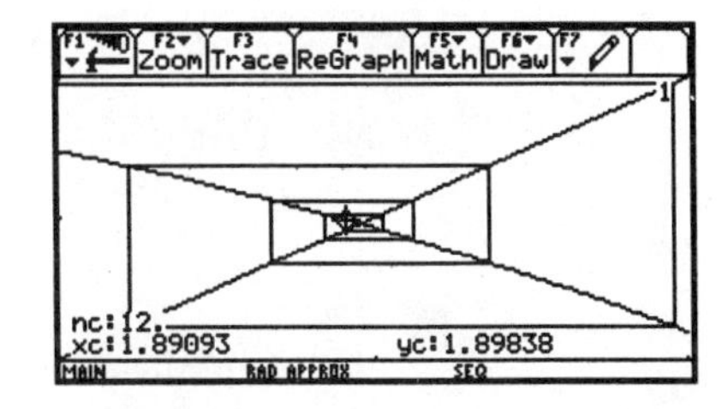

Our investigations so far indicate that

$$\lim_{n\to\infty} a_n \approx 1.8955.$$

Let x^* denote the exact value of this limit. Since

$$\lim_{n\to\infty} a_n = \lim_{n\to\infty} a_{n-1} = x^*$$

and $a_n = 2\sin a_{n-1}$ for all n, it must be true that

$$x^* = 2\sin x^*;$$

that is, the limit x^* is a solution of the equation $x = g(x)$, where $g(x) = 2\sin x$—that is, a *fixed point* of the function g. (Note that the points of our web plot converged to the point of intersection of the two graphs.) Let's check this by solving $x = 2\sin x$ with **nSolve()**.

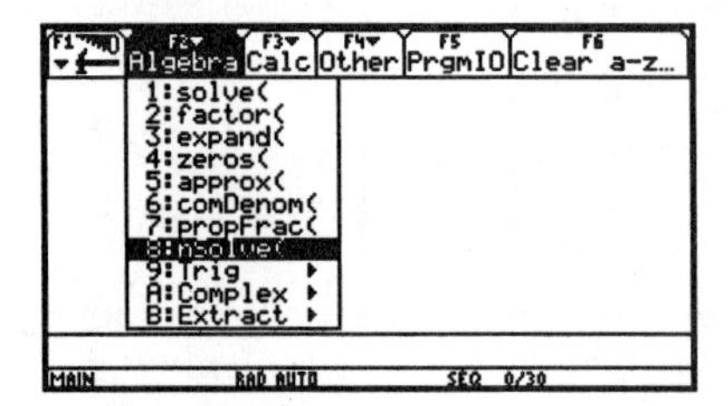

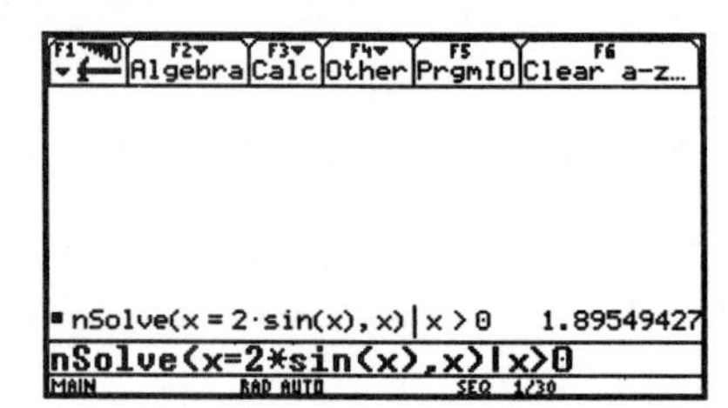

This is an example of a very important use of sequences. Exact solutions of many equations, such as $x = 2\sin x$, simply cannot be found. Thus we resort to some approximation procedure which amounts to generating a sequence that converges (we hope) to a solution. Provided that the sequence does converge, by computing enough terms in the sequence we can approximate the exact solution to any desired accuracy.

Newton's Method revisited. Recall that Newton's Method for approximating a solution of $f(x) = 0$ is described by

$$x_n = x_{n-1} - \frac{f(x_{n-1})}{f'(x_{n-1})}, \quad n = 1, 2, 3, \dots ,$$

where x_0 is some chosen initial approximation to the solution. This procedure generates a recursive sequence, since $x_n = g(x_{n-1})$ for $n \geq 1$, where

$$g(x) = x - f(x)/f'(x).$$

Note that if $x_n \to x^*$ as $n \to \infty$ and if $f'(x^*) \neq 0$, then

$$x^* = x^* - \frac{f(x^*)}{f'(x^*)},$$

which implies that $f(x^*) = 0$.

• EXAMPLE 4. Consider the problem of approximating the solution of

$$x^3 + x - 1 = 0.$$

Examination of the graph of the function $f(x) = x^3 + x + 1$ shows that there is exactly one solution and that this solution is between .5 and 1. So we'll take $x_0 = 1$ and use Newton's method, which here takes the form

$$x_n = x_{n-1} - \frac{x_{n-1}^3 + x_{n-1} - 1}{3x_{n-1}^2 + 1}, \quad n = 1, 2, 3, \ldots .$$

With **Graph MODE** = **FUNCTION**, enter $f(x)$ as y1 and $f'(x)$ as y2. Then switch to **Graph MODE** = **SEQUENCE**, and enter the Newton iteration formula as **u1**.

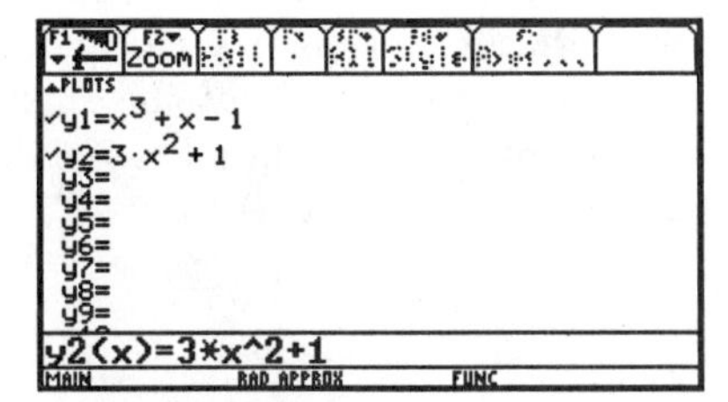

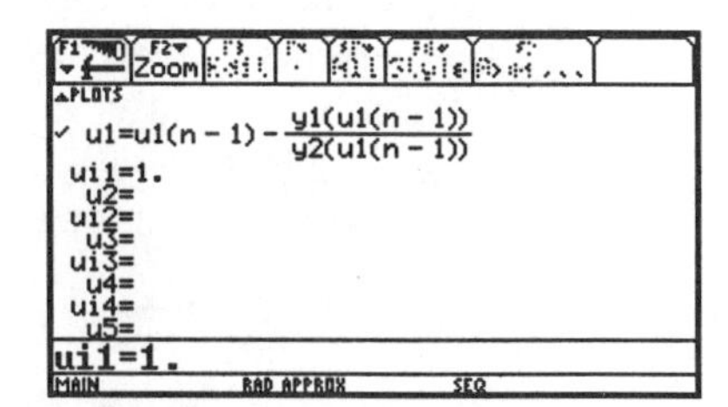

Press ◇ **TABLE** to see the very rapid convergence of the sequence to a limit $x^* \approx .68233$. Indeed, notice that we have at least five decimal places of accuracy after only four iterations. ◇ **GRAPH** shows the following picture on the left with window bounds of 0 to 2 on each axis and the picture on the right with window bounds of .5 to 1 on each axis.

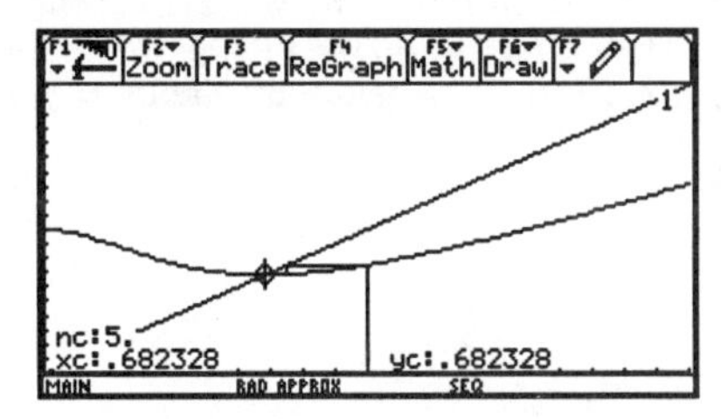

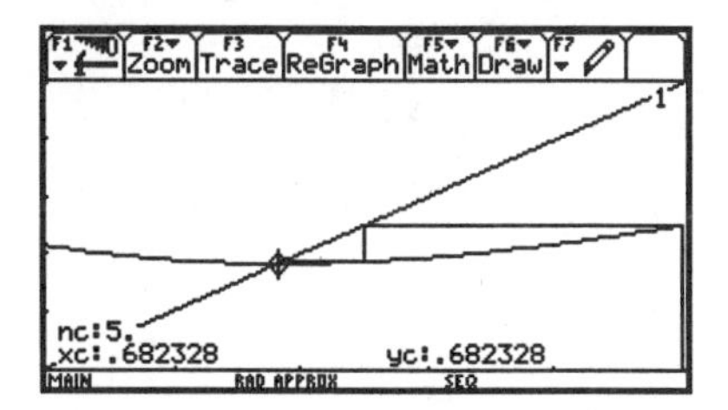

The curve in each of the pictures above is the graph of

$$g(x) = x - \frac{f(x)}{f'(x)}.$$

Notice that the extremely rapid convergence of the sequence is due to the fact that the graph of g has a horizontal tangent at the point of intersection. One of the exercises that follow will ask you to verify that this is a general property of Newton's Method.

Exercises

For Exercises 1–5, plot each sequence and try to determine whether the sequence has a limit as $n \to \infty$ and, if so, estimate the value of the limit or make a conjecture about its exact value. Then have your **TI-89/92** compute the limit with its **limit()** command.

1. $a_n = \dfrac{2+(-1)^n n^2}{n^2 - 3n + 4}$
2. $a_n = \dfrac{\ln(n)}{n^{1/3}}$
3. $a_n = \dfrac{n!}{200^n}$
4. $a_n = \dfrac{2n^3 - 6n^2 + 15}{n^4 + 18n^3 - 6}$
5. $a_n = (\cos^2 3n)^{1/n}$

In Exercises 6–10, make a web plot of each recursive sequence and try to determine whether the sequence has a limit as $n \to \infty$ and, if so, estimate the value of the limit or make a conjecture about its exact value. When possible, compute the exact value of the limit. In all cases, describe as well as you can the long-term behavior of the sequence.

6. $a_n = 1 + a_{n-1}/3, \quad a_1 = 0$
7. $a_n = 1 + a_{n-1}/3, \quad a_1 = 3$
8. $a_n = 2a_{n-1} - 3, \quad a_1 = 2.99$
9. $a_n = e^{-a_{n-1}}, \quad a_1 = 1$
10. $a_n = a_{n-1}(3 - a_{n-1}), \quad a_1 = 1.5$

In Exercises 11–15, use Newton's Method to approximate the smallest positive solution of the given equation to at least six decimal places.

11. $\sin^2 x - \cos(x^2) = 0$
12. $x - \cos x = 0$
13. $x = e^{-x}$
14. $\tan x = x$
15. $x^5 - x^4 = 1$

16. Let $g(x) = x - f(x)/f'(x)$, where f is a given twice-differentiable function.
    a) Show that $g'(x) = f(x)f''(x)/f'(x)^2$.
    b) Conclude that if $f(x^*) = 0$ and $f'(x^*) \neq 0$ (i.e., x^* is a simple root of f), then $g'(x^*) = 0$.
    c) Describe the effect of $g'(x^*) = 0$ on the web plot of the Newton's Method iteration.

9.2 Series

For any sequence $\{a_n\}_{n=1}^{\infty}$, there is an associated *sequence of partial sums* $\{s_n\}_{n=1}^{\infty}$ defined by

$$s_n = \sum_{k=1}^{n} a_k.$$

• EXAMPLE 1. If $a_n = 1/n^3$, then

$$\begin{aligned} s_1 &= 1, \\ s_2 &= 1 + \frac{1}{8} = \frac{9}{8}, \\ s_3 &= 1 + \frac{1}{8} + \frac{1}{27} = \frac{251}{216}, \\ s_4 &= 1 + \frac{1}{8} + \frac{1}{27} + \frac{1}{64} = \frac{2035}{1728}, \quad \cdots \end{aligned}$$

Notice that the terms of any such sequence of partial sums will satisfy

$$\begin{aligned} s_1 &= a_1, \\ s_n &= s_{n-1} + a_n, \quad n = 2, 3, 4, \ldots . \end{aligned}$$

This allows very efficient calculation of the partial sums.

• EXAMPLE 2. Consider again the sequence $a_n = 1/n^3$. To investigate the behavior of the partial sums, let's enter a_n as **u1** and s_n as **u2** in the **Y=** Editor and construct a table of values.

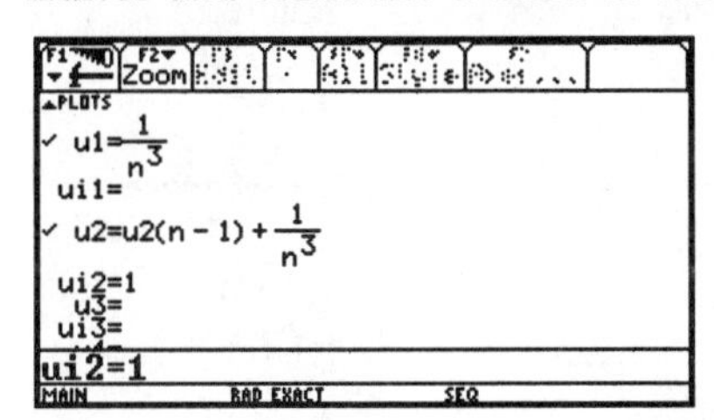

n	u1	u2
1.	1.	1.
2.	.125	1.125
3.	.037037	1.16204
4.	.015625	1.17766
5.	.008	1.18566
6.	.00463	1.19029
7.	.002915	1.19321
8.	.001953	1.19516

u2(n)=1.1951602435616

Values further down the table suggest that the partial sums are converging to a limit approximately equal to 1.2. To estimate this (supposed) limit more accurately, it is more efficient to plot the sequence of partial sums and use either **Trace** (**F3**) or the **Value** function from the **Math** menu (**F5-1**).

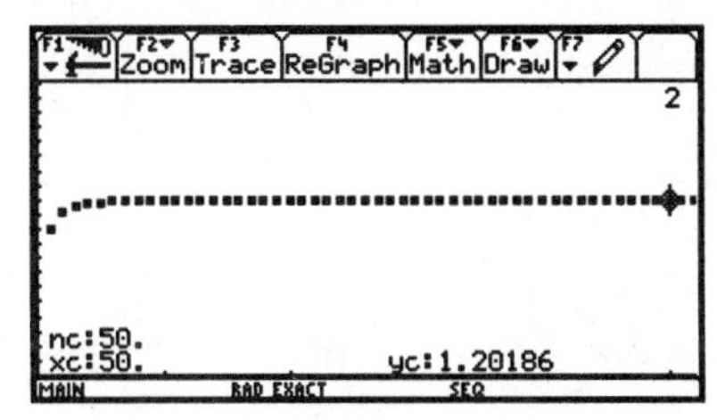

nc:200.
xc:200.
yc:1.20204

An alternative approach is to use the $\sum()$ operator from the **Calc** menu on the Home screen. (Be sure to switch **Exact/Approx MODE** to **APPROXIMATE**.)

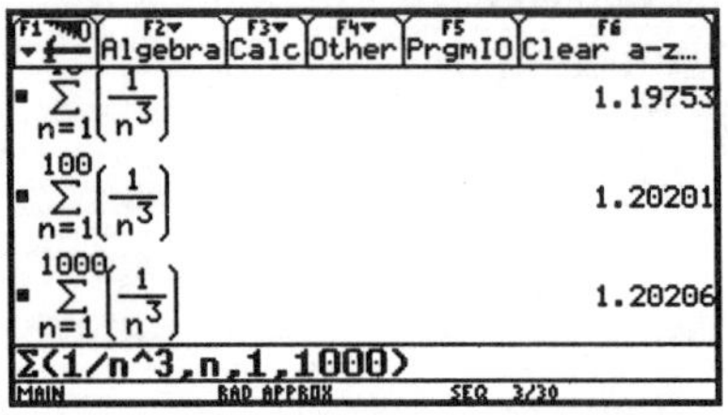

The numbers we are seeing suggest that this sequence of partial sums is a convergent sequence with a limit approximately equal to 1.202. However, this is an area where one must be very careful about such claims.

Given a sequence $\{a_n\}_{n=1}^{\infty}$, the limit of the associated partial sums is called an *(infinite) series* and is denoted by

$$\sum_{k=1}^{\infty} a_k = \lim_{n\to\infty} \sum_{k=1}^{n} a_k.$$

For example, our preceding investigation suggests that

$$\sum_{k=1}^{\infty} \frac{1}{k^3} \approx 1.202.$$

The **TI-89/92** can compute the exact value of certain series. This is done, again, with the $\sum()$ operator. As we see below, $\sum_{k=1}^{\infty} \frac{1}{k^3}$ is not a series that the **TI-89/92** can compute. However, $\sum_{k=1}^{\infty} \frac{1}{k^2}$ is. (Switch **Exact/Approx MODE** back to **EXACT**.)

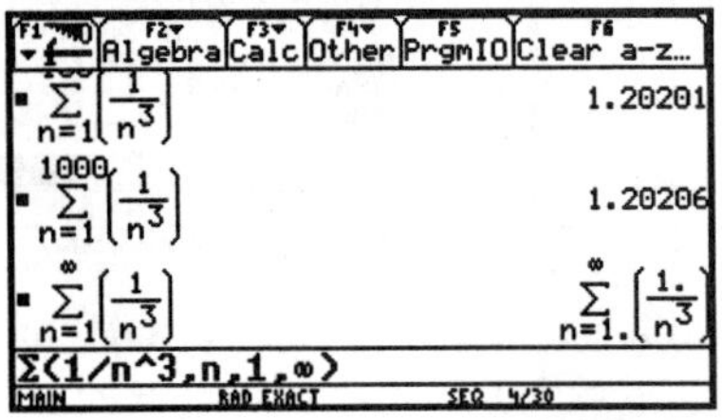

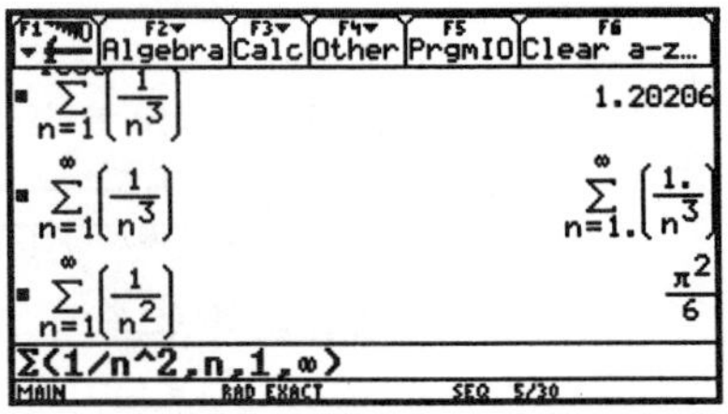

Among the series that the **TI-89/92** can compute exactly are *geometric series*.

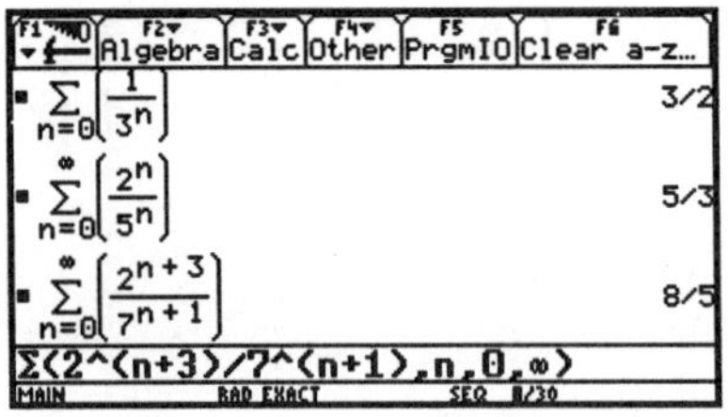

Convergence. It is often difficult to determine whether a sequence converges by numerical and graphical means. For example, it would be very difficult by graphical or numerical investigations to determine whether either the *harmonic series*

$$\sum_{n=1}^{\infty} \frac{1}{n},$$

or the p-series (with $p = 1.01$)

$$\sum_{n=1}^{\infty} \frac{1}{n^{1.01}},$$

is convergent. (The harmonic series is known to diverge, while p-series with $p > 1$ are known to converge.) For this reason, we need analytical tests to determine whether a series converges. Moreover, when dealing with a convergent series, it is very important to have an estimate of the error when estimating the series with a partial sum.

There are five primary tests for convergence. These are the integral test, the comparison test, the limit-comparison test, the ratio test, and the root test. Because it also provides a useful error estimate, we will illustrate only the ...

Integral test: *Let $a_n = f(n)$, where f is a positive, continuous, and decreasing function on the interval $[1, \infty)$. Then*

$$\sum_{k=1}^{\infty} a_k \text{ converges if and only if } \int_1^{\infty} f(x)dx \text{ converges.}$$

Moreover, we have the error estimate

$$\int_{n+1}^{\infty} f(x)dx \leq \sum_{k=n+1}^{\infty} a_k \leq \int_n^{\infty} f(x)dx$$

for the partial sum s_n.

• EXAMPLE 3. Consider the series

$$\sum_{k=1}^{\infty} \frac{1}{n \ln n}.$$

According to the integral test, this series converges if and only if the improper integral

$$\int_1^{\infty} \frac{dx}{x \ln x}$$

converges. The $\int()$ operator from the **Calc** menu lets us evaluate this integral easily. We see that an antiderivative of $(x \ln x)^{-1}$ is $\ln(\ln x)$, which clearly approaches ∞ as $x \to \infty$. Further verification is provided by the computation of the improper integral.

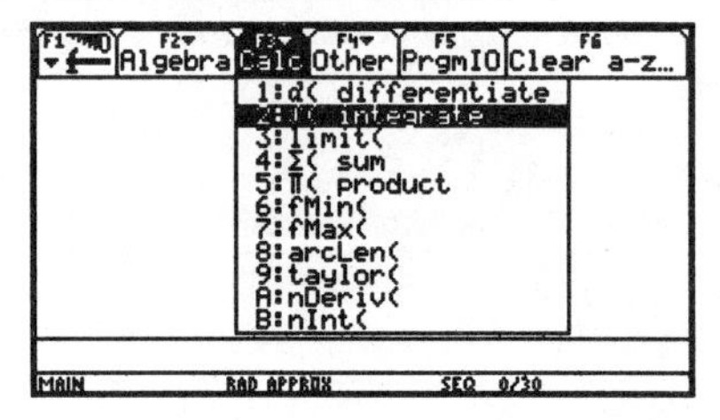

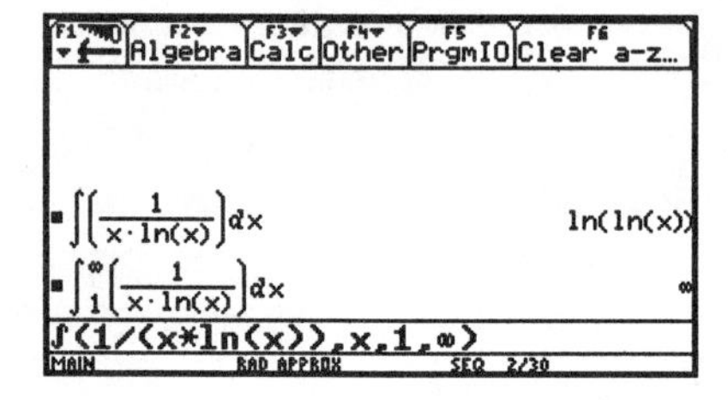

So we see that the series in question diverges.

• EXAMPLE 4. Consider the series

$$\sum_{k=1}^{\infty} \frac{n}{2^n}.$$

According to the integral test, the convergence or divergence of the series is determined by that of the improper integral

$$\int_1^{\infty} \frac{x}{2^x} dx.$$

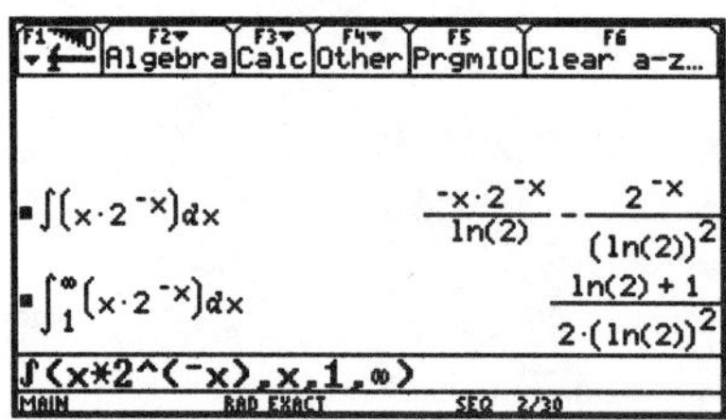

Thus we see that the series converges. Now suppose we wish to estimate the value of this series to within 0.001. How many terms must we sum in order to obtain that accuracy? According to the error estimate provided by the integral test, we need to find n such that

$$\int_n^{\infty} \frac{x}{2^x} dx \leq .001.$$

So first we compute $\int_n^{\infty} \frac{x}{2^x} dx$ in terms of n; then we set this equal to .001 and try to solve for n using **nSolve** from the **Algebra** menu. Note that **nSolve** is not successful until we restrict its search to $n > 1$, when it returns $n = 14.4883$.

Algebra Calc Other PrgmIO Clear a-z...
"no solution found"
nSolve(ans(1)=.001,n)
MAIN RAD EXACT FUNC 2/30

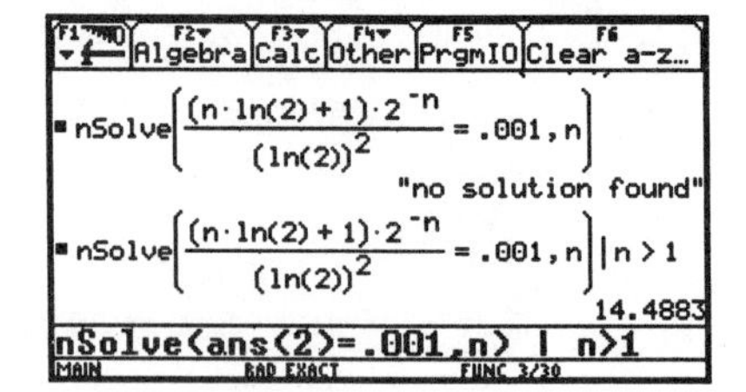

So $n = 15$ terms will provide the accuracy we desire. Computation of that partial sum suggests that the exact value of the series may be 2. In fact, the **TI-89/92** can compute the exact value of this sum, which is indeed equal to 2.

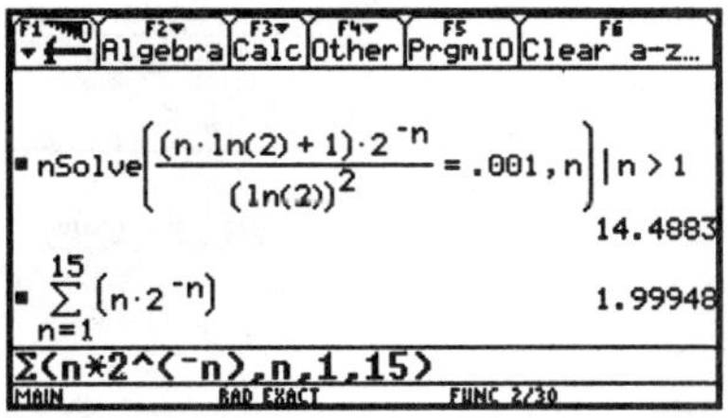

```
F1 F2 Algebra | F3 Calc | F4 Other | F5 PrgmIO | F6 Clear a-z...
■ nSolve((n·ln(2) + 1)·2^-n / (ln(2))^2 = .001, n) | n > 1      14.4883
■ Σ(n·2^-n), n=1..15                                            1.99948
Σ(n*2^(-n),n,1,15)
MAIN          RAD EXACT          FUNC 2/30
```

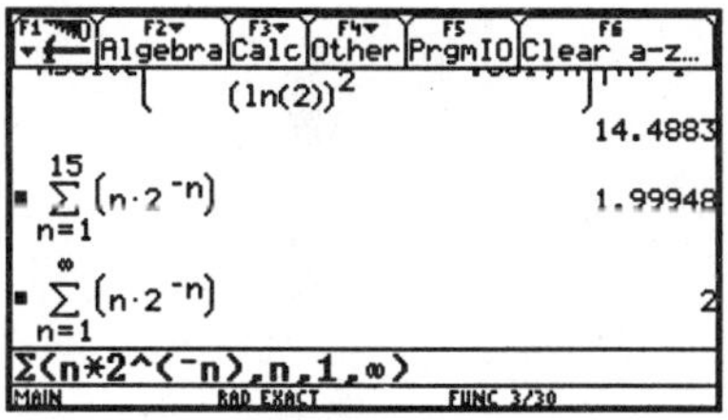

```
F1 F2 Algebra | F3 Calc | F4 Other | F5 PrgmIO | F6 Clear a-z...
(ln(2))^2                                                       14.4883
■ Σ(n·2^-n), n=1..15                                            1.99948
■ Σ(n·2^-n), n=1..∞                                             2
Σ(n*2^(-n),n,1,∞)
MAIN          RAD EXACT          FUNC 3/30
```

Power series. A power series, about the number a, is a function f defined, for some given sequence of coefficients $\{c_n\}_{n=0}^{\infty}$, by

$$f(x) = \sum_{n=0}^{\infty} c_n (x - a)^n$$

The domain $\mathcal{D}_f$ of such a function is the set of all x for which the series converges. There are three possibilities for the domain $\mathcal{D}_f$. Either:

(i) $\mathcal{D}_f = \{a\}$,
(ii) $\mathcal{D}_f = (-\infty, \infty)$, or
(iii) $\mathcal{D}_f$ is a bounded interval centered at a.

In cases (ii) and (iii), the domain is called the *interval of convergence*. In case (iii) it may be an open interval, a closed interval, or it may contain one of its two endpoints. Half the length of this interval is called the *radius of convergence* of the power series. In case (i), we say that the radius of convergence is zero; while in case (ii), we say that the radius of convergence is ∞.

• EXAMPLE 5. Consider the power series

$$f(x) = \sum_{n=1}^{\infty} \frac{x^n}{n} .$$

The ratio test shows that the series converges (absolutely) for $-1 < x < 1$. The series diverges when $x < -1$ or $x \geq 1$ and is a convergent alternating series when $x = -1$. By graphing several of the polynomials obtained by truncating the series, it is possible to visualize the convergence of the power series as a sequence of functions. Let's first graph the first through the fourth partial sums of the series; i.e., the first through the fourth degree approximations to the series, on the interval $[-2, 2]$.

y1=x
y2=x + x^2/2
y3=y2(x) + x^3/3
y4=y3(x) + x^4/4
y4(x)=y3(x)+x^4/4
MAIN RAD AUTO FUNC

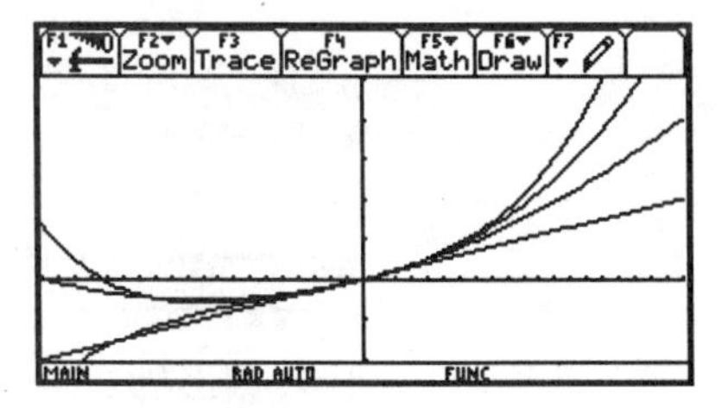

What should be noticed here is that the graphs are very close to each other near $x = 0$. Now let's graph the the fifth, tenth, and 20th degree approximations to the series on the interval $[-1.5, 1.5]$.

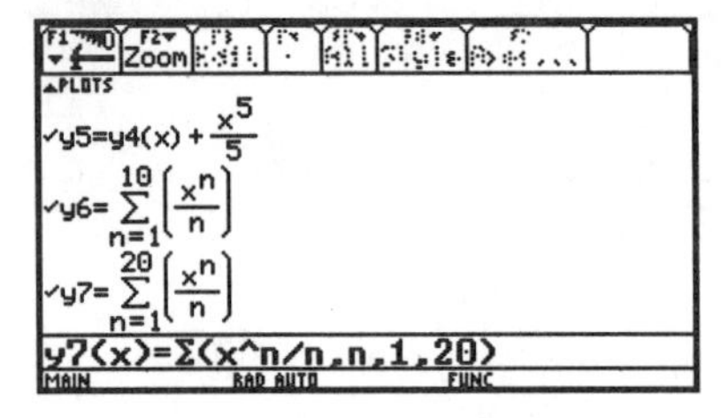

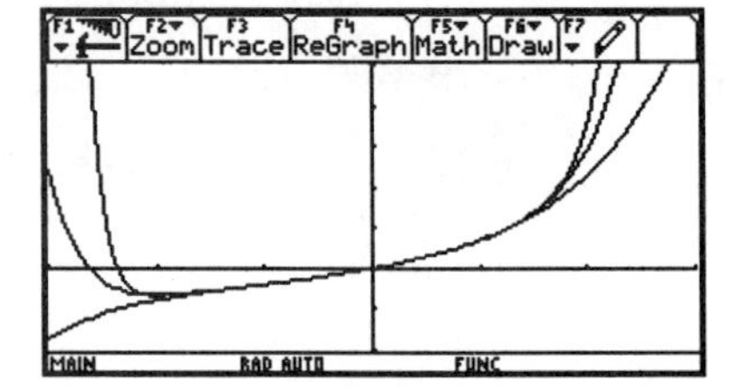

Finally, we'll graph the 200th partial sum on the interval $[-1.25, 1.25]$. (***Warning***: It takes the **TI-89/92** several minutes to do this.)

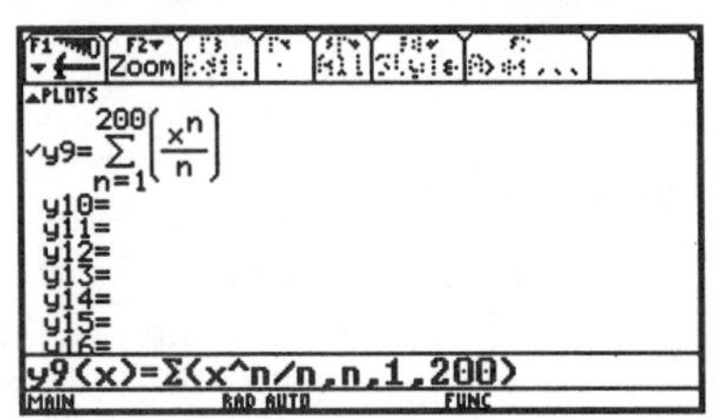

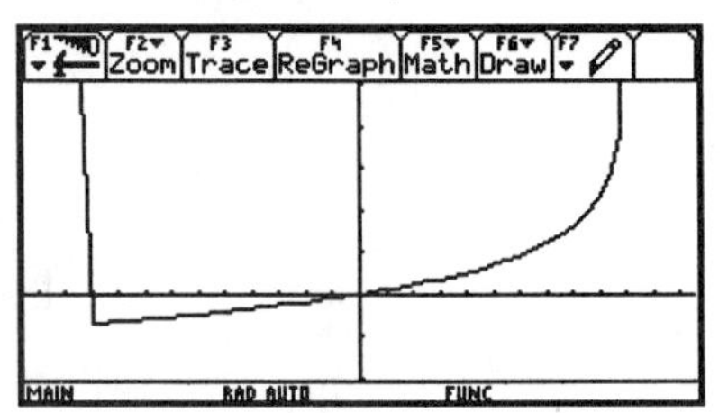

Notice that the derivative of function $f(x)$ in this example is

$$f'(x) = \sum_{n=1}^{\infty} x^{n-1}$$

$$= \sum_{n=0}^{\infty} x^n = \frac{1}{1-x} \quad \text{for } -1 < x < 1 \,.$$

Also, $f(0) = 0$. Therefore we can conclude that f has the "*closed form*"

$$f(x) = -\ln(1-x) \quad \text{for } -1 < x < 1 \,;$$

that is,

$$-\ln(1-x) = \sum_{n=1}^{\infty} \frac{x^n}{n} \quad \text{for } -1 < x < 1 \,.$$

Let's look at the graph of this function, together with, say, the seventh partial sum of the power series for comparison.

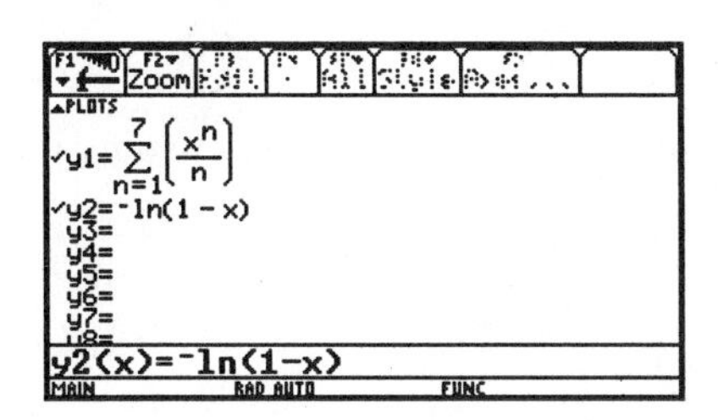

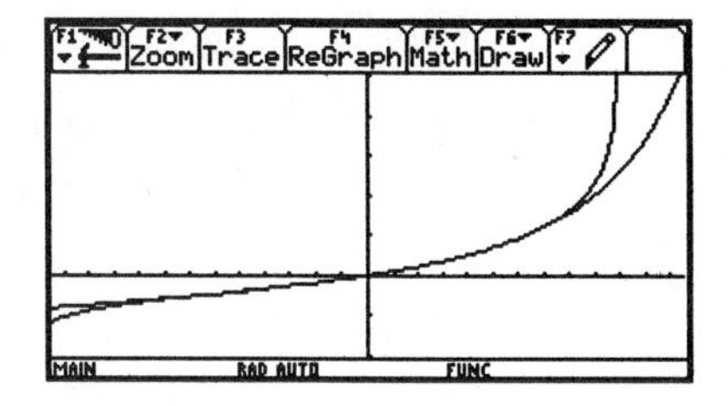

Note that the graph of $-\ln(1-x)$ is the one with the vertical asymptote at $x = 1$.

• EXAMPLE 6. Consider the power series

$$f(x) = \sum_{n=0}^{\infty} (-1)^n x^{2n}.$$

The ratio test reveals that the series converges (absolutely) for $|x| < 1$ and diverges for $|x| > 1$. It also clearly diverges if $x = \pm 1$. So let's first plot the second through the fifth partial sums on the interval $[-1.5, 1.5]$.

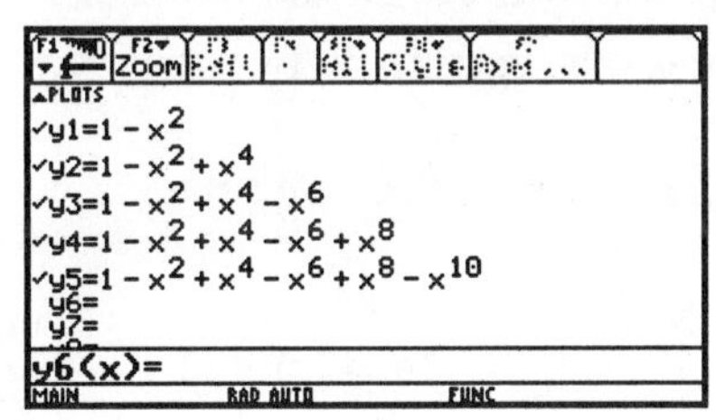

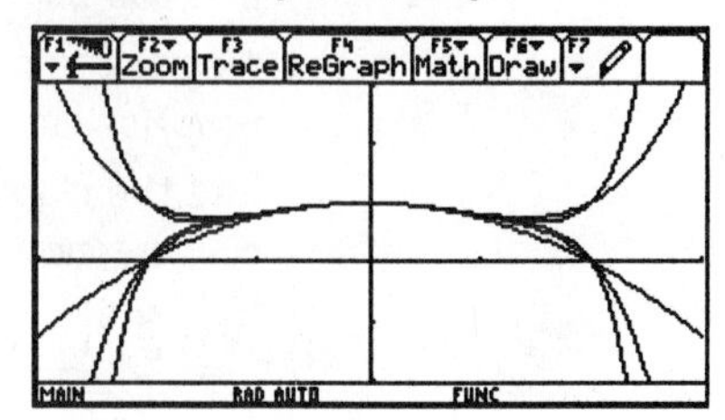

From what we know about the geometric series, it is not difficult to see the closed form

$$\sum_{n=0}^{\infty} (-1)^n x^{2n} = \frac{1}{1+x^2} \quad \text{for } -1 < x < 1.$$

So let's plot the eleventh partial sum together with $y = (1+x^2)^{-1}$, again on the interval $[-1.5, 1.5]$.

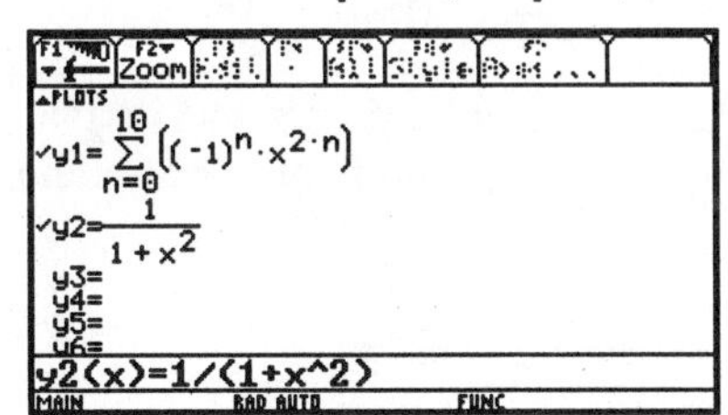

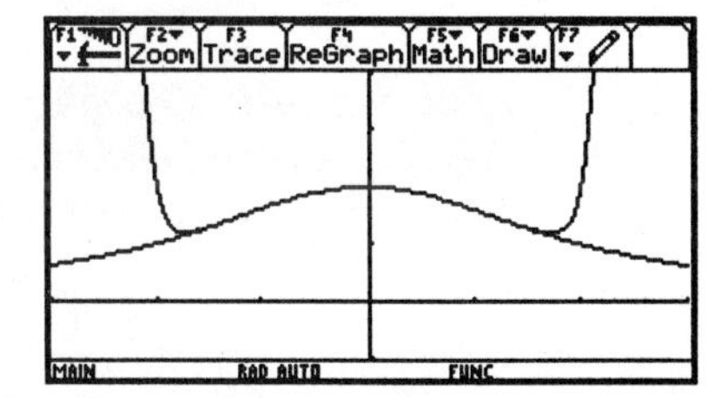

This graph does a very good job of illustrating the convergence of partial sums of a power series. In this case, the function represented by the power series is defined and bounded for all x, but the power series converges to it only on the power series's interval of convergence.

Exercises

For each series in 1–5, plot the sequence of partial sums and estimate the value of the series.

1. $\displaystyle\sum_{n=1}^{\infty} \frac{1}{n^2}$

2. $\displaystyle\sum_{n=1}^{\infty} \frac{n^2}{2^n}$

3. $\displaystyle\sum_{n=1}^{\infty} \frac{1}{n^n}$

4. $\displaystyle\sum_{n=0}^{\infty} \frac{1}{n!}$

5. $\displaystyle\sum_{n=0}^{\infty} \frac{(-1)^n}{(2n)!}\left(\frac{\pi}{3}\right)^{2n}$

For each series in 6–8, use the integral test error estimate to determine how many terms, when summed, will estimate the series to within 0.0001.

6. $\displaystyle\sum_{n=1}^{\infty} \frac{\ln n}{n^2}$

7. $\displaystyle\sum_{n=1}^{\infty} \frac{1}{n(n+1)}$

8. $\displaystyle\sum_{n=1}^{\infty} \frac{\ln(1+n^2)}{n^2}$

For each power series in 9 and 10,

a) find the interval of convergence with the ratio test;

b) plot the first five and the tenth partial sums of the series on an appropriate interval;

c) find the closed form of the series and plot the graph on its interval of convergence.

9. $\displaystyle f(x) = \sum_{n=1}^{\infty} \frac{(-1)^{(n+1)}x^{2n}}{n}$

10. $\displaystyle f(x) = \sum_{n=0}^{\infty} \frac{(-1)^n x^{2n+1}}{2n+1}$

9.3 Taylor polynomials

The nth degree Taylor polynomial about a point $x = a$ for a function f that is n times differentiable in an open interval containing a is given by

$$T_n(x) = \sum_{k=0}^{n} \frac{f^{(k)}(a)}{k!}(x-a)^k,$$

where $f^{(k)}$ denotes the kth derivative of f. If f has derivatives of all orders in an open interval containing $x = a$, then its Taylor polynomials are partial sums of the power series

$$f(x) = \sum_{k=0}^{\infty} c_k(x-a)^k$$

where the coefficients c_k are given by $c_k = f^{(k)}(a)/k!$. We call this series the *Taylor series* for f about $x = a$.

The **TI-89/92** has a built-in function for computing Taylor polynomials. It is the function **taylor()**, in the **Calc** menu.

• EXAMPLE 1. Consider the function

$$f(x) = \frac{x^2 - 1}{x^3 + 1}.$$

We'll find the third degree Taylor polynomial $T_3(x)$ about $x = 0$. In order to graph $T_3(x)$ along with $f(x)$, we assign the result of **taylor()** to the variable p3(x).

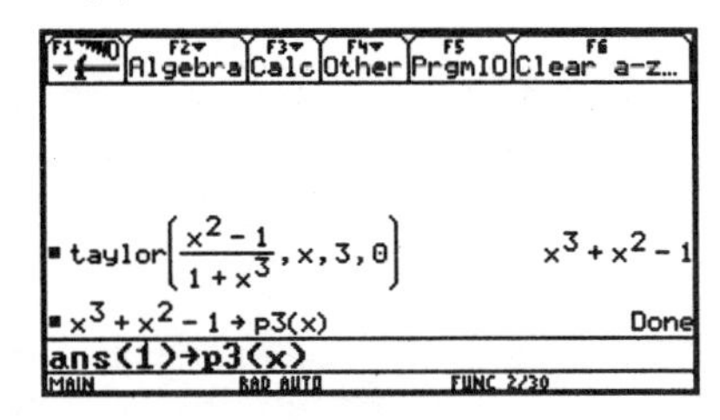

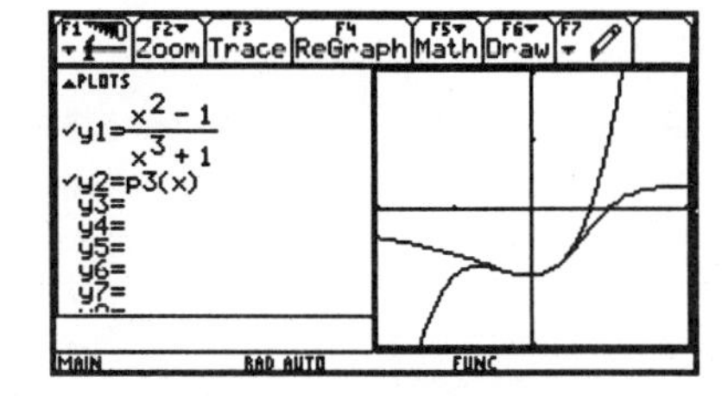

Finally, let's graph $f(x)$ along with $T_5(x)$ and $T_{10}(x)$. (Note that $T_{10}(x)$ is actually a polynomial of degree nine.)

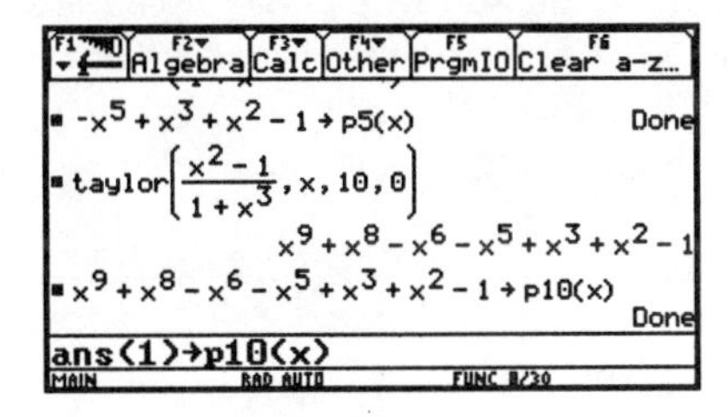

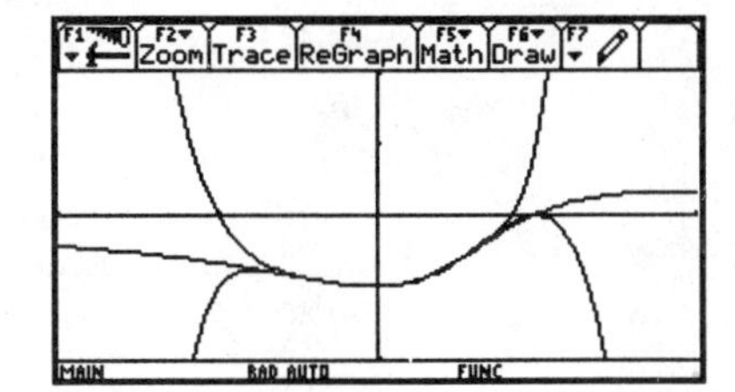

The remainder term. Taylor polynomials provide an effective way of approximating a function locally about a given point $x = a$. According to *Taylor's Theorem*, the remainder $R_n(x)$ in the approximation of $f(x)$ by $T_n(x)$ is given by the formula

$$f(x) - T_n(x) = R_n(x) = \frac{f^{(n+1)}(\xi_x)}{(n+1)!}(x - a)^{n+1},$$

where ξ_x is some number between x and a, provided that f is $(n+1)$-times differentiable on some interval containing a and x.

• EXAMPLE 2. Suppose that we approximate sin(.3) by computing the value of the cubic Taylor polynomial about $x = 0$ at .3:

$$\sin(.3) \approx .3 - \frac{(.3)^3}{6} = 0.2955.$$

Let's use the remainder term to estimate the error in the approximation. Since the fourth derivative of $\sin x$ is $\sin x$, and since $\sin x \geq 0$ for $0 \leq x \leq .3$, we have

$$|R_3(.3)| = \frac{\sin(\xi_{.3})}{4!}(.3 - 0)^4,$$

for some number $\xi_{.3}$ between 0 and .3. Using the very rough estimate $\sin(\xi_{.3}) \le 1$, we arrive at

$$|R_3(.3)| \le \frac{1}{4!}(.3)^4 = .0003375.$$

This estimate can be sharpened somewhat by using the fact that $\sin(\xi_{.3}) \le .3$. (Why is this true?) This gives the estimate

$$|R_3(.3)| \le \frac{.3}{4!}(.3)^4 = .00010125.$$

A still sharper error estimate can be found by realizing that the fourth degree Taylor polynomial for $\sin x$ about $x = 0$ is the same as the cubic Taylor polynomial about $x = 0$. Therefore we can estimate the error with

$$|R_4(.3)| = \frac{\cos(\xi_{.3})}{5!}(.3 - 0)^5 \le \frac{(.3)^5}{5!} = .00002025.$$

Finally, there is one more (simpler) method of estimating the error that is available to us. The full series for sin(.3) is an *alternating series.* Therefore, the error can be estimated by the absolute value of the next nonzero term, which here is $(.3)^5/5!$, the very same estimate that we obtained for $|R_4(.3)|$. The calculations on the left below indicate that this error estimate is quite sharp indeed. The plot on the right below shows the graphs of $\sin x$ and $x - x^3/6$ on the interval $0 \le x \le \pi$.

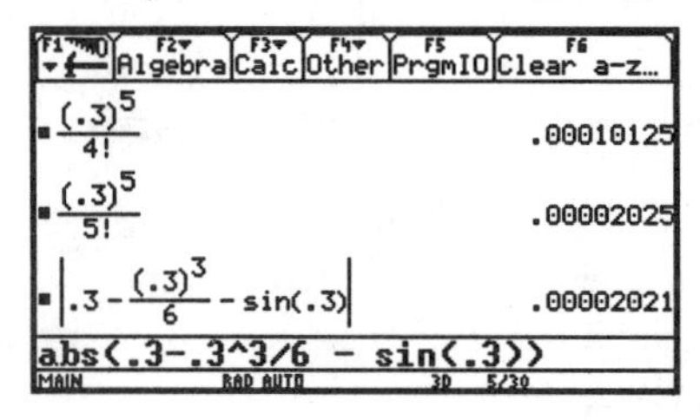

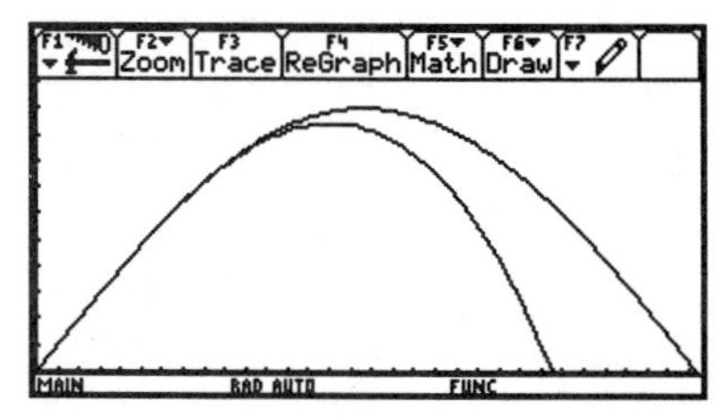

Because of the form of the remainder term, if we want to approximate $f(x)$ over an interval $[a - r, a + r]$, we have the error estimate

$$|R_n(x)| \le \frac{Mr^{n+1}}{(n+1)!} \quad \text{for } |x - a| \le r,$$

where

$$M = \max_{[a-r,\, a+r]} \left| f^{(n+1)}(x) \right|.$$

• EXAMPLE 3. Suppose we wish to find a polynomial approximation to $f(x) = \cos x$ uniformly on the interval $[-\pi, \pi]$ with an error of not more than 0.01. Since f and all of its derivatives have values between -1 and 1, we can take $M = 1$. So, for any x in $[-\pi, \pi]$, we have

$$|R_n(x)| \le \frac{\pi^{n+1}}{(n+1)!},$$

and therefore we want to find the least value of n such that

$$\frac{\pi^{n+1}}{(n+1)!} \leq 0.01.$$

By tabulating the sequence $\pi^{n+1}/(n+1)!$, we see that the desired value is $n = 10$.

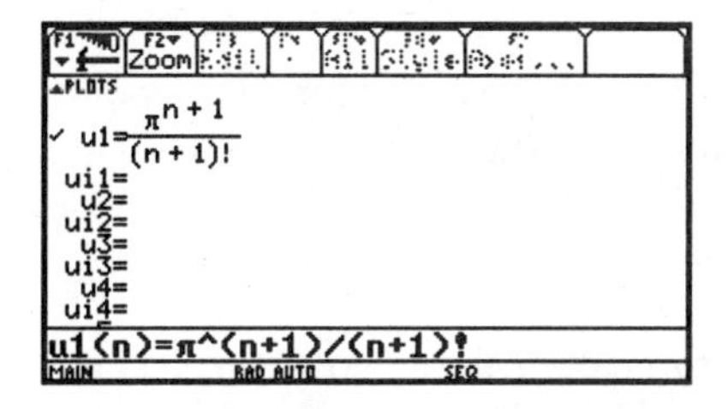

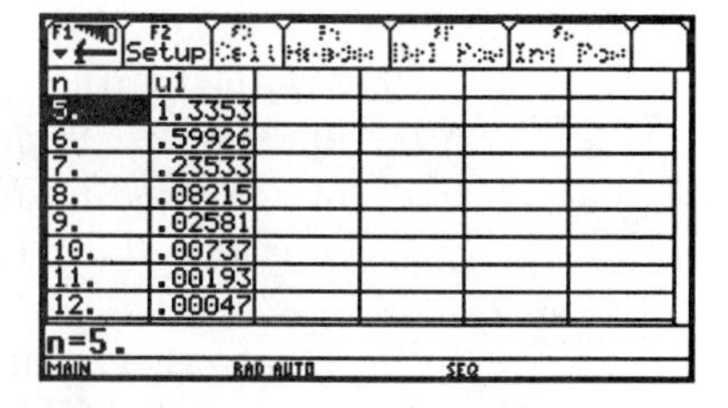

Now we plot $\cos x$ together with $p_{10}(x)$ on the interval $[-7, 7]$. Notice the closeness of the two graphs between $-\pi$ and π.

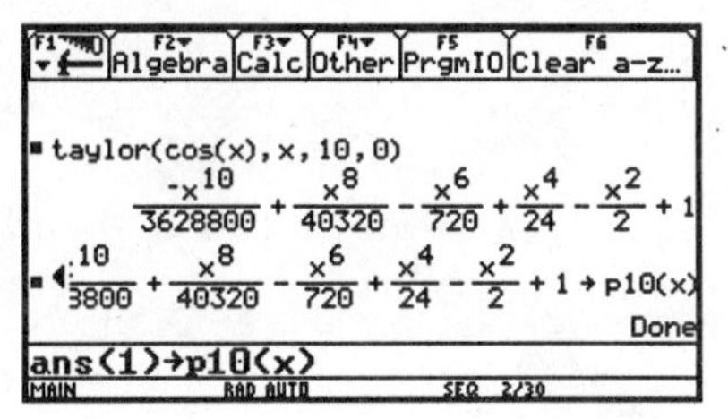

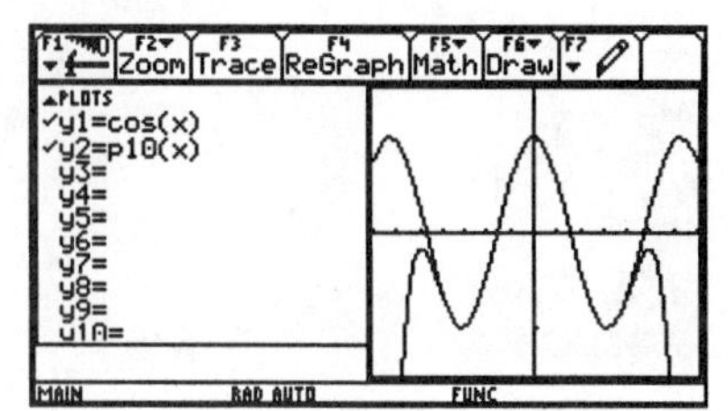

Exercises

1. Find the fifth and tenth degree Taylor polynomials for $f(x) = \cos x$ about $x = \frac{\pi}{2}$. Then plot them both along with $\cos x$ on the interval $[-4, 7]$.

2. Find the fifth and tenth degree Taylor polynomials for $f(x) = \cos(1 - e^x)$ about $x = 0$ and plot them along with $\cos(1 - e^x)$ on the interval $[-2, 3]$.

3. Find the fifth and tenth degree Taylor polynomials for $f(x) = (1 + x^2)^{-1}$ about $x = 0$ and plot them along with $(1 + x^2)^{-1}$ on the interval $[-2, 2]$.

4. Find a polynomial approximation to $f(x) = e^x$ on the interval $[-2, 2]$ with an error of not more than 0.01. Then plot both the polynomial approximation and e^x on the interval $[-4, 4]$.

5. Find a polynomial approximation to $f(x) = \ln x$ on the interval $[1/2, 3/2]$ with an error of not more than 0.001. Then plot both the polynomial approximation and $\ln x$ on the interval $[0.01, 4]$.

6. Find the third, fourth, and fifth degree Taylor polynomials for $f(x) = x^3 + x^2 + x + 1$ about $x = 0$ and also about $x = 1$. What do you observe?

7. Use the cubic Taylor polynomial for $e^{\sin x}$ to approximate $e^{\sin 0.25}$. Use the bound on the remainder term to estimate the error in the approximation.

8. Use the fourth degree Taylor polynomial for $\sqrt{1+x^4}$ about $x = 0$ to approximate the integral $\int_0^1 \sqrt{1+x^4}\,dx$. Estimate the error in the approximation.

9. Use the fourth degree Taylor polynomial for $\cos x$ about $x = 0$ to find an approximate formula for the positive solution of $\cos x = x^2 + k$, in terms of k, for $k < 1$. For what values of k does this give a reasonably accurate solution? *Hint*: *The quadratic formula is your friend.* ☺

10. What other important theorem does Taylor's Theorem become in the case $n = 0$? What does Taylor's Theorem tell us about the linear approximation to a twice differentiable function about $x = a$, i.e., in the case $n = 1$?

10 Projects

10.1 Two Limits

Objectives: The purpose of this project is to illustrate the notion of limit in two simple situations.

Background: You need only be familiar with the basic capabilities of the **TI-89/92** that are described in Chapters 1 and 2 of this manual.

I. *Suppose that a certain plant is one meter tall, and its height increases continuously at a rate of 100% per year. How tall will the plant be one year later?*

1. Suppose that instead of growing continuously the plant has monthly instantaneous growth spurts, in each of which its height increases by $\frac{100}{12}\%$. Compute the plant's height after twelve such monthly growth spurts.

2. Suppose that instead of growing continuously the plant has daily instantaneous growth spurts, in each of which its height increases by $\frac{100}{365}\%$. Compute the plant's height after 365 such daily growth spurts.

3. Suppose that instead of growing continuously the plant has n instantaneous growth spurts during the year, in each of which its height increases by $\frac{100}{n}\%$. Express the year-end height of the plant as a function $h(n)$.

4. Plot the graph of $h(n)$ in a $[1, 500] \times [2, 3]$ window. Use **Value** from the Graph screen's **Math** menu to find the values $h(12)$ and $h(365)$ that you computed in #1 and #2. Describe the behavior of the graph and estimate the number that $h(n)$ approaches as n gets larger and larger. Now give an (approximate) answer to the original question.

II. *Suppose that your friend asks for a loan of $2500 and offers to pay you back according to the following schedule. He will pay you $100 tomorrow, and on each day thereafter he will pay you 95% of the previous day's payment for the rest of your life (or until the daily payment amount is less than one penny.) Should you agree to the deal?*

1. The amount of the loan that has been paid back after n days is

$$A(n) = 100 + 100(.95) + 100(.95)^2 + 100(.95)^3 + \cdots + 100(.95)^{n-1}.$$

So in the **Y=** Editor, define

```
y1 =sum(seq(100*.95^k,k,0,x-1)
```

Then plot the graph of **y1** in a $[0, 30] \times [0, 2500]$ window, with **xres=4** to speed up the plot. How much of the loan has been paid back after 30 days?

2. The method used in #1 to compute $A(n)$ is very inefficient. Show that

$$A(n) - .95A(n) = 100 - 100(.95)^n = 100(1 - .95^n)$$

and consequently that

$$A(n) = \frac{100\,(1 - .95^n)}{.05} = 2000(1 - .95^n).$$

Redefine **y1** using this simpler formula. Then plot the graph in a $[0, 365] \times [0, 2500]$ window. What can you conclude about when the loan will be paid off?

3. After how many days will the daily payment become less than a penny? How much money you would get back if you did agree to the deal?

10.2 Computing π as an Area

Objectives: This project illustrates the notion of limit in the context of computing the area inside a circle by the "method of exhaustion."

Background: Basic trigonometry. You should also be reasonably familiar with the capabilities of the **TI-89/92** that are described in Chapters 1 and 2 of this manual.

In this project, we will explore a method for computing a that was known to the ancient Greeks. It is sometimes called the "method of exhaustion." (But with the **TI-89/92**'s help, hopefully we won't become exhausted. :-)) The method embodies the essence of calculus: *successively improved approximations.*

Let us suppose that we define π to be area inside a circle of radius 1. Then by approximating the area inside the circle, we approximate π. The way that we'll approximate this area is to compute the area of a regular polygon whose vertices lie on the circle.

1. Before we proceed with the computations, let's first create some pictures to illustrate what we're doing. Enter the following program, which draws a regular polygon with n sides and partitions the polygon into n identical triangles.

```
: polygon(n)
: Prgm: Local a,b
: ClrDraw:ClrGraph:FnOff:PlotsOff
: ZoomSqr
: For i, 1, n
```

```
:     2.π*(i-1)/n →a
:     2.π*i/n →b
:     Line cos(a),sin(a),cos(b),sin(b)
:     Line 0,0,cos(b),sin(b)
: EndFor
: EndPrgm
```

After entering the program, set window variables to see a $[-2,2]\times[-1,1]$ window. Execute the program from the Home screen with $n = 3$ by entering **polygon(3)**. Repeat with $n = 4, 5, \ldots, 10$ and then with $n = 16$, 24, and 36.

2. a) Now we need a formula for the area of each n-gon. Using basic trigonometry, and noting that each triangle can be divided up into two congruent right triangles, show that each of the n triangles within the n-gon has area

$$\sin\frac{180^\circ}{n}\cos\frac{180^\circ}{n},$$

and consequently the area within the n-gon is

$$A(n) = n\sin\frac{180^\circ}{n}\cos\frac{180^\circ}{n},$$

(*Note*: We are using degree measure for angles because radian measure requires us to know π.)

b) Graph the function $A(n)$ in a $[3,100]\times[0,4]$ window. Find the least number of sides the polygon must have in order that its area, rounded to three significant figures, is 3.14. (Make sure that your calculator is in degree mode.)

c) Find the least number of sides the polygon must have in order that its area, rounded to five significant figures, is 3.1416.

10.3 Lines of Sight

Objectives: The purpose of this project is to reinforce the concept of the tangent line and the slope of a curve. You will use the **TI-89/92**'s ability to draw tangent lines to a curve in order to approximate a "line of sight" and then use Calculus to solve the problem precisely.

Background: Before doing this project, you should be familiar with the basic definition of the tangent line to a curve and its slope. In particular, re-read the first parts of Sections 2.1 and 2.6 in Stewart's *Calculus*. You should also have become familiar with the capabilities of the **TI-89/92** that are described in Chapters 1–3 of this manual.

The problem: *Ant Man is standing 100 feet away from a tall building and 12 feet away from a greenhouse with semicircular cross-section of radius 12 feet, as shown in the figure. Ant Man's eyes are 5.6 feet above the ground. How high above the ground is the lowest point he can see on the side of the building, and how high above the ground is the highest point he can see on the roof of the greenhouse?*

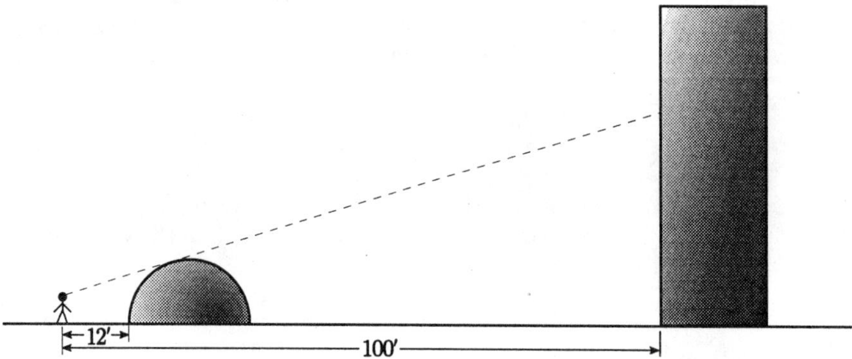

The following steps will lead to the solution of the problem.

1. In the **Y=** Editor, define **y1** $= \sqrt{144 - (x - 24)^2}$ and press **F1-9** to bring up the **GRAPH FORMATS** dialog box. There, set **Grid** to **ON**. Then set window variables to get a $[0, 105] \times [0, 50]$ window with **xscl** $= 5$, **yscl** $= 5$, and **xres** $= 1$. Press ⋄ **GRAPH** to create the plot. Now use **Shade** from the **Math** menu on the Graph screen (**F5-C**) to shade the semicircular region under the graph. (The lower and upper bounds on the region are $x = 12$ and $x = 36$.) Finally, press **F7** in the Graph screen, select **Vertical**, and draw the vertical line that is as close to $x = 100$ as possible.

2. Still on the Graph screen, select **Circle** from the "pencil" menu (**F7-4**) and draw a very small (semi)circle at the left edge of the screen at the point $E = (0, 5.6)$, where Ant Man's eyes are located. Then select **Tangent** from the **Math** menu (**F5-A**), and move the cursor to a point on the greenhouse roof where you think the tangent line might pass through E. Take note of the coordinates of that point, and press enter to see the resulting tangent line.

3. Repeat the second part of step 2 a few times until you have found a point on the greenhouse roof where the tangent line comes as close as possible to hitting E. Note the x- and y-coordinates of the point on the greenhouse roof each time. What is the equation of the best tangent line you could find this way?

4. Compute the y-coordinate when $x = 100$ on the graph of your final tangent line from step 3. Round to the nearest foot.

5. Let P denote the point on the greenhouse roof where Ant Man's line of sight is tangent to it. Let a denote the exact x-coordinate of P. Using the slope formula, find the slope of the line passing through E and P, in terms of a. On the Home screen, store this as a function **s(a)**.

6. Find the slope of the greenhouse roof (i.e., the derivative of **y1(x)**) at the point where $x = a$, and store this as **m(a)**.

7. Solve the equation **m(a)=s(a)** for a. Do this using **nSolve()** "with" the condition **a>12 and a<24**. Then compute the slope **m(a)** and write the equation of Ant Man's line of sight.

8. Compute and give the answers to the questions in the statement of the original problem.

10.4 Graphs and Derivatives

Objectives: The purpose of this lab is to reinforce the geometric interpretations of the first and second derivatives of a function f in terms of the graph of f.

Background: Before doing this project, you should be familiar with the definitions of the first and second derivatives of a function f and their relationship to geometric properties of the graph of f. In particular, re-read Section 4.3 in Stewart's *Calculus*. You should also have become familiar with the capabilities of the **TI-89/92** that are described in Chapters 1–3 of this manual.

Part I.

1. Graph the function $f(x) = x^4 - 6x^3 + 8x^2$ in a $[-2, 5] \times [-15, 20]$ window. Then, using **Trace** to find rough estimates for the endpoints (rounded to the nearest tenth), state the intervals on which:

   a) $f(x)$ is positive;
   b) $f(x)$ is negative;
   c) $f(x)$ is increasing;
   d) $f(x)$ is decreasing;
   e) the graph is concave up;
   f) the graph is concave down.

2. Add the graph of the derivative $f'(x) = 4x^3 - 18x^2 + 16x$ to the plot from #1. Then, using the same endpoint values as in #1, state the intervals on which

   a) $f'(x)$ is positive;
   b) $f'(x)$ is negative;
   c) $f'(x)$ is increasing;
   d) $f'(x)$ is decreasing.

3. Add the graph of the second derivative $f''(x) = 12x^2 - 36x + 16$ to the plot from #2. Then, using the same endpoint values as in #1, state the intervals on which

   a) $f''(x)$ is positive; b) $f''(x)$ is negative.

In 4–6, give interval endpoints rounded to the nearest thousandth.

4. Rework #1, using

   a,b) **Zero** from the **Math** menu (**F5-2**) to find endpoints of the intervals for parts a and b;

   c,d) **Minimum** or **Maximum** from the **Math** menu (**F5-3,4**) to find endpoints of the intervals for parts c and d;

   e,f) **Inflection** from the **Math** menu (**F5-8**) to find endpoints of the intervals for parts e and f.

5. Rework #2, using

   a,b) **Zero** from the **Math** menu (**F5-2**) to find endpoints of the intervals for parts a and b;

   c,d) **Minimum** or **Maximum** from the **Math** menu (**F5-3,4**) to find endpoints of the intervals for parts c and d;

6. Rework #3, using

   a,b) **Zero** from the **Math** menu (**F5-2**) to find endpoints of the intervals for parts a and b.

Part II.

Each of the following plots contains the graph of a function f together with graphs of f' and f''. Identify each graph as the graph of f, f', or f''. For each of the two plots, write a brief and concise paragraph explaining your choices.

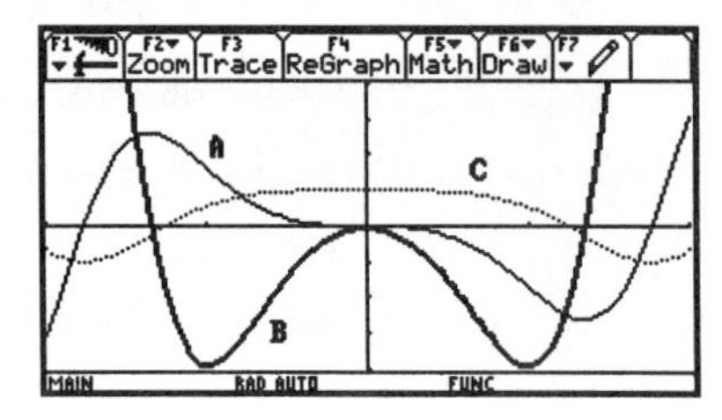

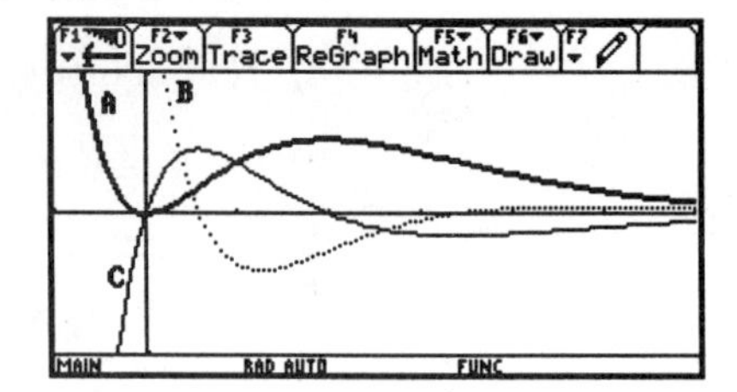

10.5 Designing an Oil Drum

Objectives: The purpose of this project is to guide the student through a series of realistic applied optimization problems.

Background: Before doing this project, you should be familiar with the concepts related to optimization problems. In particular, re-read Section 4.7 in Stewart's *Calculus*. You should also have read Chapters 1–4 of this manual.

General scenario: *You work for a company that manufactures large steel cylindrical drums that can be used to transport various petroleum products. Your assignment is to determine the dimensions (radius and height) of a drum that is to have a volume of 1 cubic meter while minimizing the cost the drum.*

The cost of the steel used in making the drum is $3 per square meter. The top and bottom of the drum are cut from squares, and all unused material from these squares is considered waste. The remainder of the drum is formed from a rectangular sheet of steel, assuming no waste.

Problem I. *Ignoring all costs other than material cost, find the dimensions of the drum that will minimize the cost.*

1. Using the fact that the drum's volume is to be 1 cubic meter, express the height h of the drum in terms of its radius r. Store the resulting expression as **ht(r)**.
2. Express the material cost first in terms of both r and h and then as a function of r alone. Store the resulting expression as **cost1(r)**.
3. Graph **cost1(r)** in a $[0, 2] \times [0, 70]$ window. Then use **Minimum** from the **Math** menu (**F5-3**) to find the value of r that minimizes cost. Round the result to the nearest hundredth.
4. What dimensions of the drum minimize the material cost?

Problem II. *The drum has seams around the perimeter of its top and bottom, as well as a vertical seam where edges of the rectangular sheet are joined to form the lateral surface. In addition to the material cost, the cost of welding the seams is $1 per meter. Find the dimensions of the drum that will minimize the cost of production.*

5. Add the seam-welding cost, in terms of r alone, to **cost1(r)** Store the resulting expression as **cost2(r)**.
6. Graph **cost2(r)** in a $[0, 2] \times [0, 70]$ window. Then use **Minimum** from the **Math** menu (**F5-3**) to find the value of r that minimizes cost. Round the result to the nearest hundredth.
7. What radius and height of the drum minimize the material cost plus the cost of welding the seams?

Problem III. *The cost of shipping each drum from your plant in Birmingham to an oil company in New Orleans depends upon both the surface area of the drum (which determines weight) and the sum of the drum's diameter and height (which affects how many drums can be transported on one vehicle). This cost is estimated to be $2 per square meter of surface area plus $1 per meter of diameter plus height. Find the dimensions of the drum that will minimize the total cost of production and transportation.*

8. Add the transportation cost, in terms of r alone, to **cost2(r)** Store the resulting expression as **cost3(r)**.
9. Graph **cost3(r)** in a $[0, 2] \times [0, 70]$ window. Then use **Minimum** from the **Math** menu (**F5-3**) to find the value of r that minimizes cost. Round the result to the nearest hundredth.
10. What radius and height of the drum minimize the total cost of production and transportation?

10.6 Newton's Method and a 1D Fractal

Objectives: The goal of this project is to illustrate some of the issues related to the convergence of Newton's Method and to reinforce the geometric interpretation of Newton's Method.

Background: Before doing this lab, you should be familiar with Newton's Method. In particular, re-read Section 4.9 in Stewart's *Calculus* and Section 4.4 in this manual. You should also have entered and used the program **newt()** from Section 4.7.

Materials: You'll need three different colored pencils, pens, or markers.

Overview: The simple cubic polynomial

$$f(x) = x^3 - x$$

has three distinct real zeros: 0 and ± 1. Our goal here is to get a visual picture of which initial guesses x_0 cause Newton's Method to converge to each of these three zeros of f.

1. In the **Y=** Editor enter the function given above as **y1**. Then graph the function, first in a $[-3, 3] \times [-10, 10]$ window and then in a $[-1.5, 1.5] \times [-2, 2]$ window.
2. For future reference, use **Minimum** and **Maximum** from the Graph screen **Math** menu to find the values of x at the local minimum and the local maximum that appear in the graph. Then use the derivative to find the exact values of these critical points.
3. The following function, **newtfn()**, returns the (approximate) number to which Newton's Method converges, given an initial guess x_0. In fact, we

want to plot its graph as a function of x_0. Enter **newtfn()** in your **myprogs** folder.

```
: newtfn(f, xvar, x0, tol)
: Func
: Local df, xnew, xi, err
: d(f,xvar) →df: x0 →xnew: 2*tol →err
: While err>tol
:    xnew →xi
:    xvar–f/df| xvar=xi →xnew
:    If xnew≠0 Then: abs((xi–xnew)/xnew) →err
:    Else: abs((xi–xnew)/tol →err
:    EndIf
: EndWhile
: Return xnew
: EndFunc
```

To be sure that **newtfn()** works properly, enter

myprogs\newtfn(x^3–x, x, x_0**, 10^(-5))**

for each of $x_0 = \pm 1.5, \pm .5, \pm .25$. Are the results sensible?

4. In the **Y=** Editor, define

   y1 = x^3–x ,

   y2 = myprogs\newtfn(t^3–t, t, x, 10^(-2)) .

   Then plot both graphs in a $[-1.5, 1.5] \times [-2, 2]$ window with **xres**=2. In light of the symmetry in the plot, replot in a $[0, 1.5] \times [-2, 2]$ window, this time with **xres**=1.

5. Based on the plot observed in #4, sketch the graph of $y = x^3 - x$ for $-1.5 \le x \le 1.5$, and then color-code the x-axis in some manner according to the behavior of Newton's Method on this problem. For example, you might let blue indicate convergence to -1, red convergence to 0, and green convergence to 1.

6. There is an interval of the form (a, ∞) that contains 1 and has the property that the Newton iterates converge to 1 whenever the initial guess x_0 is in that interval. The value of a is geometrically obvious. What is it? Notice that by symmetry the interval $(-\infty, -a)$ contains -1 and has the property that the Newton iterates converge to -1 whenever the initial guess x_0 is in that interval.

7. Let a have its value from #6. There is an interval of the form $(a - h,\, a)$ with the property that the Newton iterates converge to -1 whenever the initial guess x_0 is in that interval. If $a - h$ were used as x_0, what would be the resulting value of x_1? Use this to write and solve an equation for h. What then are the endpoints of the interval $(a - h,\, a)$? Notice that there is an analogous interval $(-a,\, -a + h)$ containing initial guesses for which the Newton iterates converge to 1.

8. There is an interval of the form $(-r, r)$ with the property that the Newton iterates converge to 0 whenever the initial guess x_0 is in that interval. If $-r$ were used as the initial guess x_0, what would be the resulting value of x_1? Use this to write and solve an equation for r. What then are the endpoints of the interval?
9. Graphically investigate the behavior of Newton's Method for initial guesses in the interval $(r, a - h)$. Once you're convinced you understand what's happening, redraw your color-coded x-axis and graph on a large sheet of paper. Also write a short paragraph describing what you have observed.

10.7 Optimal Location for a Water Treatment Plant

Background: Distance formula; the derivative; optimization problems. Review Section 4.7 in Stewart's *Calculus*.

Overview: *Three towns—Appalachee, Hull, and Eastville—agree to share the cost of constructing a new water treatment plant that will supply all three towns with water. The plant will be located on a nearby river. The main concern in deciding where to locate the plant is the total cost of installing the supply pipelines to all three towns and building an access road from the plant to a nearby interstate highway. The map below shows the center of each town, the river, and the interstate highway. The grid lines on the map are 1 mile apart.*

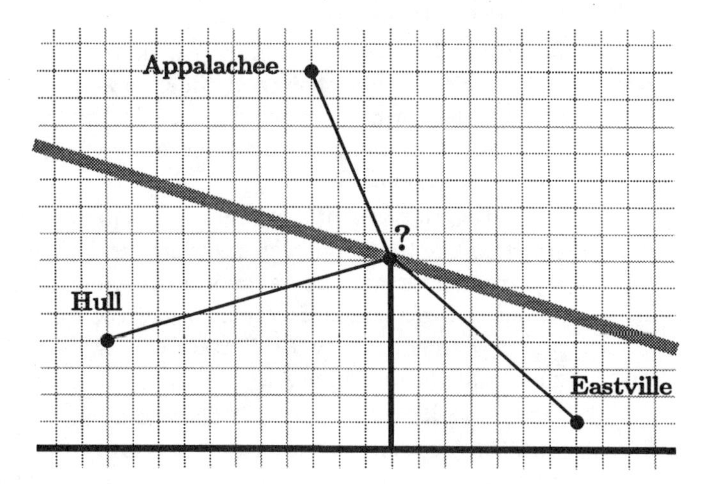

The cost of the pipeline to Eastville will be $15,000 per mile. The cost of installing pipelines to each of Appalachee and Hull will be $40,000 per mile, since they must go through heavily wooded areas. The cost of building the

road will be \$180,000 per mile. Assume that the supply pipelines extend to the centers of the three towns and ignore the width of the river.

Assignment: Find the point on the river where the new water treatment plant should be located in order to minimize the total cost of installing the supply pipelines and building the road.

10.8 The Vertical Path of a Rocket

Background: Antidifferentiation; rectilinear motion. Review Section 4.10 in Stewart's *Calculus* and Sections 4.1 and 5.1 in this manual.

The Situation: *A small rocket has a mass of 100 kg, including 30 kg of fuel. Its engine burns fuel at a constant rate of 1 kg/sec and produces a constant thrust of 980 Newtons until the fuel is exhausted. The engine is ignited at time $t = 0$, propelling the rocket upward on a vertical path.*

1. Express the mass m of the rocket, including fuel, as a function of t, for $t \geq 0$. After how many seconds does the rocket run out of fuel?
2. Let $y(t)$ denote the height in meters of the rocket at time t seconds. If we ignore air resistance, Newton's second law gives the equation

$$\frac{d}{dt}\left(m(t)\frac{dy}{dt}\right) = -9.8m(t) + 980,$$

   up until the instant when the rocket runs out of fuel. Find the height function $y(t)$ that is in effect up until the instant when the rocket runs out of fuel. What are the velocity and the height of the rocket at the instant when the rocket runs out of fuel?
3. After the rocket runs out of fuel, the problem becomes a simple one concerning the height of a projectile for which the only acceleration comes from gravity, and the initial height and velocity are known. Extend the height function $y(t)$ from #2 so that it is valid from time $t = 0$ until the rocket falls to the ground. Simulate the path with your **TI-89/92**.
4. Find the maximum height attained by the rocket.

10.9 Otto the Daredevil

Background: Antidifferentiation; rectilinear motion. Review Section 4.10 in Stewart's *Calculus* and Sections 4.1 and 5.1 in this manual.

The Situation: *Otto is an unusual daredevil. He jumps off of tall buildings with a small jet-pack strapped to his back. The jet-pack carries a small* *amount fuel—just enough to last for 10 seconds. The acceleration it provides is 14.4 m/sec^2. Moreover, the jet-pack can be switched on and then off only once. Thus, on each jump, Otto's goal is to program the jetpack to switch on and then off at precisely the right moments so that he lands with zero velocity.* (Throughout this project, ignore air resistance and assume that $g = 9.8$ m/sec^2.)

1. If Otto jumps off a 100 meter building, after how many seconds should he turn on his jet-pack? How many seconds after that does he land?
2. Repeat #1 for a 200 meter building.
3. How tall is the tallest building from which Otto should jump if he insists upon landing with zero velocity?
4. How tall is the tallest building from which Otto should jump if he doesn't mind landing with a velocity of -10 m/sec?

10.10 The Skimpy Donut

Background: Volume of a solid of revolution; area of a surface of revolution; optimization. Review Sections 4.7, 6.2, 6.3, and 9.2 in Stewart's *Calculus* and Sections 4.6, 6.2, and 6.3 in this manual.

Overview: *The* GETFAT DONUT COMPANY *makes donuts with a layer of chocolate icing. The company wants to cut costs by using as little chocolate icing as possible without changing the thickness of the icing or the weight (i.e., volume) of the donut. The problem, then, is to determine the dimensions of the donut that will minimize its surface area.*

Assume that the donut has the idealized shape of a torus—obtained by revolving a circle about its central axis—as shown below.

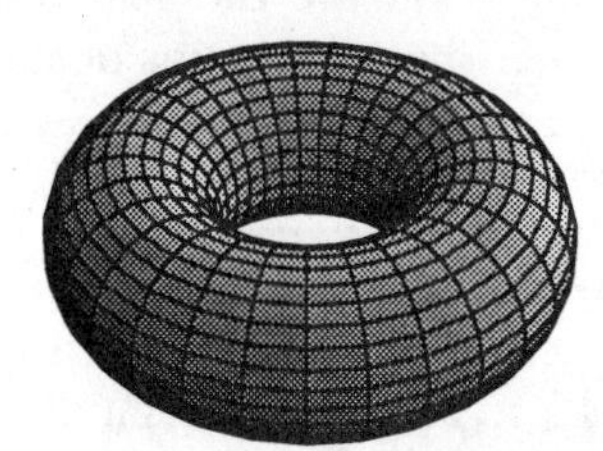

Such a torus can be obtained by revolving the circle $(x-a)^2 + y^2 = b^2$ about the y-axis, where b is the radius of the revolved circle, and a is the distance from the center axis to the center of the revolved circle. Previously, the donuts have been in the shape of such a torus with $a = 1$ inch and $b = 1/2$ inch.

1. Use the cylindrical shells technique to find the volume of the torus in terms of a and b. (Note that because of symmetry, it suffices to double the volume of the top half.) Now compute the volume V_0 that the donuts have had with dimensions $a = 1$ inch and $b = 1/2$ inch.

2. Find the surface area of the torus in terms of a and b. (Note that because of symmetry, it suffices to double the surface area of the top half.) As a check, the surface area should be $2\pi^2$ when $a = 1$ and $b = 1/2$.
3. With the volume fixed at its previous value V_0, determine the range of allowable values for each of a and b. (One important condition comes from the geometry of the torus.)
4. With the volume fixed at its previous value V_0, find the dimensions a and b that will minimize the donut's surface area.
5. Is there a maximum surface area for a fixed volume?

10.11 The Brightest Phase of Venus

Background: Trigonometry; optimization; area. Review Sections 4.7 and 6.1 in Stewart's *Calculus*.

Overview: *The brightness of Venus, as seen from Earth, is proportional to its visible area and inversely proportional to the square of the distance from Venus to Earth. In the figure below, note that as the angle ϕ increases from 0 to π, the visible area increases, thus increasing Venus's brightness. But the distance d from Earth to Venus also increases, which tends to decrease the brightness of Venus. Venus will appear brightest from Earth for some value of ϕ between 0 and π.*

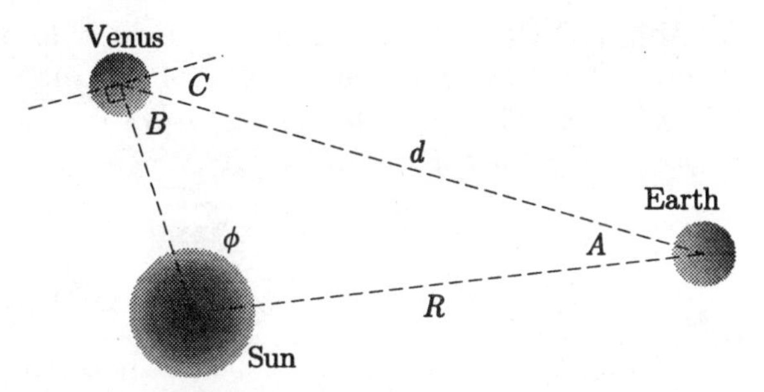

Assignment: Find the brightest phase of Venus—i.e., the angle ϕ for which Venus appears brightest from Earth—by following the steps outlined below. Carefully write up your solution, as usual, using complete sentences and graphs as needed. Use the **TI-89/92** wherever appropriate.

1. From basic geometry, show that $B = \pi - (\phi + A)$ and $C = (\phi + A) - \pi/2$.
2. Let a be the radius of Venus. Show that the visible portion of Venus lies between the curves $x = -\sqrt{a^2 - y^2}$ and $x = \sqrt{a^2 - y^2}\sin C$. Then show that the visible area of Venus is

$$\frac{\pi a^2}{2}(1 + \sin C).$$

3. Let β represent the brightness of Venus as viewed from Earth. As described in the overview above, β is proportional to the quantity

$$\frac{\text{visible area}}{d^2}.$$

Use this, together with the results of #1 and #2, to obtain a formula for β in terms of the angles ϕ and A and the distance d. Then use the Law of Cosines and the Law of Sines from Trigonometry to show that

$$d^2 = r^2 + R^2 - 2rR\cos\phi,$$
$$\sin A = \frac{r}{d}\sin\phi,$$

where r is the distance from the Sun to Venus, and R is the distance from the Sun to Earth.

4. Use the equations in #3 with the values $R = 93$, $r = 67$, and $a = 0.004$ (each in millions of miles) to express β in terms of the angle ϕ. Then find the value of ϕ between 0 and π that maximizes β.

10.12 Designing a Light Bulb

Background: Slope, continuity, and differentiability; optimization; the definite integral; solving systems of linear equations with **simult()**; using **when()** to define a piecewise-defined function.

Problem I. *The glass portion of a light bulb is to be in the form of a surface of revolution obtained by revolving the curve shown in the figure about the x-axis. The radius of the bulb's base is 1/2 inch, and the radius of its spherical portion is 1.25 inches.*

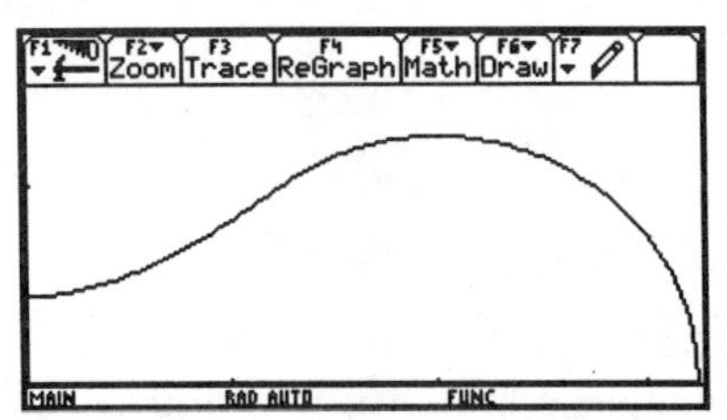

The curve is the graph of the piecewise-defined function:

$$f(x) = \begin{cases} ax^3 + bx^2 + cx + d, & \text{when } 0 \le x \le 1.25; \\ \sqrt{1.25^2 - (x-2)^2}, & \text{when } 1.25 \le x \le 3.25. \end{cases}$$

Find the coefficients a, b, c, and d. Then plot the graph of f in a $[0, 3.25] \times [0, 1.5]$ window.

Hint: At each of $x = 0$ and $x = 1.25$, there are two conditions given by 1) the point on the curve and 2) the slope of the curve.

Problem II. *This problem concerns the placement of the filament in the bulb. The filament will be located at a point P on the central axis of the bulb, r inches from the base, such that the distance from the point P to the entire inner surface of the bulb, as measured by*

$$\phi(r) = \int_0^{3.25} \sqrt{(r-x)^2 + f(x)^2}\, dx,$$

is minimized. Find r.

Hint: Being very patient, graph $\phi(r)$ on the interval $0 \le r \le 3.25$ after experimentally determining appropriate values of **ymax** and **ymin**. (Set **xres** to 10 to speed up the process; then go have lunch.) Two decimal places of accuracy for r is plenty.

Problem III. *Show that the distance from a point $P = (r, 0)$ to the semicircle $y = \sqrt{R^2 - x^2}$, as measured by*

$$\phi(r) = \int_{-R}^{R} \sqrt{(r-x)^2 + (R^2 - x^2)}\, dx,$$

is minimized by $r = 0$ (i.e., when P is the center of the circle).

10.13 Approximate Antidifferentiation with the Trapezoidal Rule

Background: Antidifferentiation; the Fundamental Theorem of Calculus; approximate integration; the Trapezoidal Rule. Review Sections 4.10, 5.3, and 8.7 in Stewart's *Calculus* and Sections 5.1 and 5.4 in this manual.

Overview: *The Fundamental Theorem of Calculus tells us that every continuous function f on an interval $[a, b]$ has an antiderivative $F(x)$, namely*

$$F(x) = \int_a^x f(t)dt.$$

In many cases, such as when $f(x) = \cos x,\ x^3,\ e^x,\ x\sin x^2$, and so on, $F(x)$ can be express in terms of other elementary functions. However, for many (in fact, most) functions, there is no simpler way of expressing an antiderivative. Our goal here is to use the trapezoidal rule to develop an efficient method of approximating $F(x)$.

Note that if we have a value for $F(x)$, then we can express $F(x + \Delta x)$ as

$$\begin{aligned} F(x+\Delta x) &= \int_a^x f(t)dt + \int_x^{x+\Delta x} f(t)dt \\ &= F(x) + \int_x^{x+\Delta x} f(t)dt. \end{aligned}$$

A two-point Trapezoidal Rule approximation to this last integral gives us

$$F(x+\Delta x) \approx F(x) + \frac{\Delta x}{2}\left(f(x) + f(x+\Delta x)\right).$$

Now, using this approximation, we can approximate values of F quite efficiently at a sequence of points $x_1,\ x_2,\ x_3, \ldots$, where $x_n = a + n\Delta x$, by using the recurrence formula

$$F_n = F_{n-1} + \frac{\Delta x}{2}\left(f(x_{n-1}) + f(x_n)\right),$$

where F_i represents an approximation to $F(x_i)$.

TI-92 Set-up: Follow the steps below to set up your **TI-89/92** to plot approximate antiderivatives.

1. Press **MODE** and set the **Graph** mode to **SEQUENCE**.
2. Go to the **Y=** Editor and set

   u1=u1(n–1)+dx, ui1=a, u2=u2(n–1)+step(u1(n–1)), and **ui2=0**.

   Highlight **u2** and select **Line** as the graph **Style** (**F6-1**). Finally, press **F7** and set **Axes: CUSTOM**, **X Axis: u1**, and **Y Axis: u2**. (Note that **u1** represents the variable x and **u2** represents $F(x)$.)
3. From the Home screen, enter

   Define step(x)=(f(x)+f(x+dx))*dx/2 .
4. Still in the Home screen, store a value in **dx** (typically .1 or .05), store a value in **a**, and define the function **f(x)**. Press ⋄**WINDOW** and set appropriate window variables. Then press ⋄**GRAPH** to plot the graph.

Assignment: First follow the set-up steps above. Test the set-up with $f(x) = \cos x$, $a = 0$, and **dx**$=.1$. You should see a good approximation to the graph of $\sin x$. Appropriate window variables for this test-case include **nmin**$=0$, **nmax**$=63$, **xmin**$=0$, **xmax**$=2\pi$, **ymin**$=-1$, **ymax**$=1$.

For each of the following, first plot the given antiderivative on the specified interval. Then switch the **Graph MODE** to **FUNCTION** and plot the integrand on the same interval. For each pair of graphs, describe in some detail the two expected relationships:

a) $\int_a^x f(t)dt$ gives the "net" area between the graph of f and the x-axis between a and x;

b) $f(x)$ is the derivative of $\int_a^x f(t)dt$.

1) $\int_0^x \sin(t^2)dt$, for $0 \le x \le 5$

2) $\int_1^x \sin(t^2)dt$, for $1 \le x \le 5$

3) $\int_0^x \frac{dt}{\sqrt{1+t^4}}$, for $0 \le x \le 10$

4) $\int_0^x e^{-t^2}dt$, for $0 \le x \le 3$

5) $\int_0^x e^{\sin t} dt$, for $0 \le x \le 4\pi$ 6) $\int_0^x \sin(t \cos t) dt$, for $0 \le x \le 2\pi$

7) $\int_0^x \sin(t + \cos t) dt$, for $0 \le x \le 4\pi$

Bonus: Modify the procedure described above so that it uses a three-point Simpson's Rule approximation rather than a two-point Trapezoidal approximation. Rework #1 and #6 above to test your modification.

10.14 Percentiles of the Normal Distribution

Background: The definite integral; approximate integration; improper integrals; Newton's Method. In Stewart's *Calculus*, read Section 9.5 and review Sections 8.7, 8.8, 5.3, and 4.9.

Overview: *The function*

$$g(x) = \frac{e^{-x^2/2}}{\sqrt{2\pi}}$$

is extremely important in Probability. Its graph is the standard normal (or bell) curve. If x represents an observable, random quantity (a random variable) that is "normally distributed" with mean value zero and standard deviation 1, then the area under the graph of g between $x = a$ and $x = b$ is the probability that a random observation of x will fall between a and b. If we define the notation $P(a \le x \le b)$ to represent this probability, then we can write

$$P(a \le x \le b) = \int_a^b g(x)dx.$$

Also,

$$P(x \le b) = \int_{-\infty}^b g(x)dx \quad \text{and} \quad P(a \le x) = \int_a^\infty g(x)dx.$$

Our goal here is to construct a short table of standard percentiles of the normal distribution—that is, a table of values of b corresponding to

$$P(x \le b) = .01,\ .05,\ .10,\ .25,\ .333,\ .50,\ .667,\ .75,\ .90,\ .95,\ .99.$$

Assignment: Carry out the following steps to determine the desired percentiles.

1. Plot g in a $[-3.5, 3.5] \times [0, .5]$ window with xres=1. Use the **∫f(x)dx** command in the **Math** menu to compute

$$\text{a)} \int_{-1}^1 g(x)dx, \quad \text{b)} \int_{-2}^2 g(x)dx, \quad \text{and c)} \int_{-3}^3 g(x)dx.$$

2. Go to the Home screen and compute $\int_{-10}^{10} g(x)dx$ with the $\int()$ operator to confirm that $\int_{-\infty}^{\infty} g(x)dx = 1$. Now, knowing that $g(x)$ is an even function, the value of $\int_{-\infty}^{0} g(x)dx$ should be 1/2. To confirm this expectation, compute $\int_{-10}^{0} g(x)dx$.

3. Let's try first to find the 75th percentile; that is, the value of b such that

$$\int_{-\infty}^{b} g(x)dx = .75.$$

Since $\int_{-\infty}^{0} g(x)dx = .5$, we can avoid the improper integral by instead finding b such that

$$\int_{0}^{b} g(x)dx = .25.$$

a) By trial and error, find a three decimal-place approximation to b.

b) Think of the current problem as that of finding a zero of the function $f(b) = \int_0^b g(x)dx - .25$. Using your approximation from part "a" as an initial guess, perform one step of Newton's Method to refine the approximation. Report the result rounded to five decimal places. (Hint: $f'(b) = g(b)$ by the Fundamental Theorem of Calculus.)

4. Repeat the process in #3 to find values of b corresponding to

$$\int_{-\infty}^{b} g(x)dx = .667, \ .90, \ .95, \ \text{and} \ .99.$$

5. Using symmetry and the percentiles already found, find the values of b for which

$$\int_{-\infty}^{b} g(x)dx = .01, \ .05, \ .10, \ .25, \ \text{and} \ .333.$$

10.15 Equilibria and Centers of Gravity

Background: Using the definite integral to compute arc length and centers of gravity; tangent and normal lines to a curve. Review Sections 9.1 and 9.3 in Stewart's *Calculus*.

Problem I: *Think of the x-axis as the surface of the floor, so that y is the vertical distance above the floor. A metal plate, with uniform thickness and mass density, occupies the region bounded by the parabola $y = x^2/4$ and the line $y = 4$ when balanced on its vertex. By symmetry, the center of gravity of the plate is at a point $(0, \bar{y})$ on the y-axis. Find $\bar{y}$.*

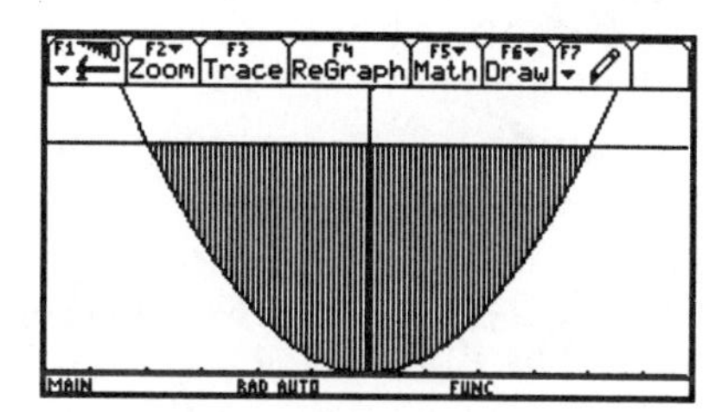

Problem II: *When the plate is balanced on its vertex at the origin, it is in an unstable position, because if the plate is disturbed ever so slightly, it will roll to one side and come to rest at a new equilibrium position. Suppose that the plate tips over to the right. Describe the new equilibrium position; in particular, find the new point where the plate touches the floor and the new coordinates of its center of gravity. Why is this position stable, while the original position was not?*

Problem III: *Suppose that the plate occupies the region bounded by the parabola $y = x^2/4$ and the line $y = 9/4$ when balanced on its vertex. Show that there is no other equilibrium position for this plate. Conclude that this position must be stable. (Why?)*

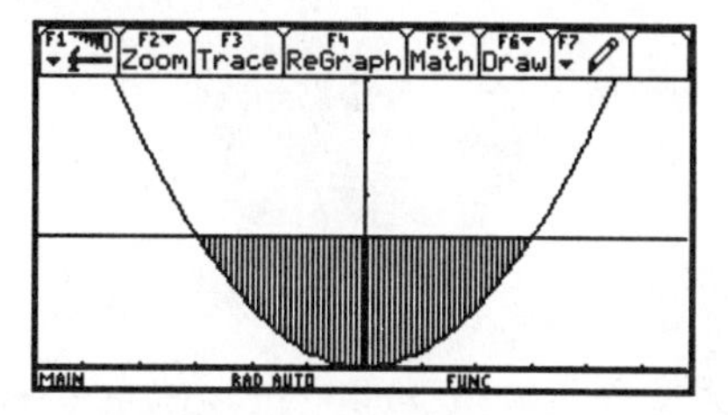

Problem IV: *Suppose that the plate occupies the region bounded by the parabola $y = x^2/4$ and the line $y = b$ $(b > 0)$ when balanced on its vertex. Find the greatest value of b for which being balanced on its vertex is the plate's only equilibrium position.*

Problem V: *Give simple geometric arguments for: a) why a semicircular plate has only one equilibrium position, and b) why every position of a circular plate is an equilibrium position.*

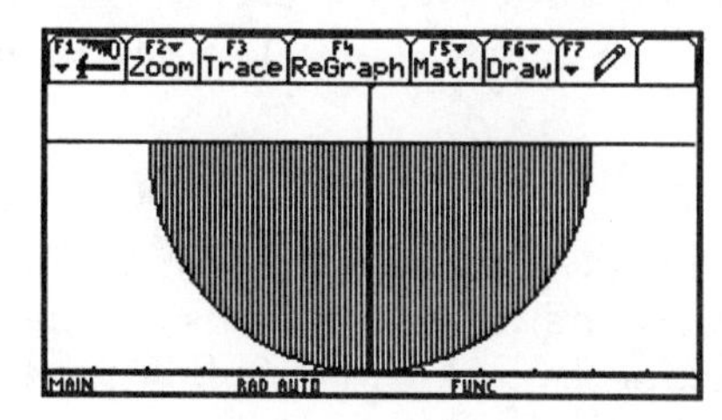

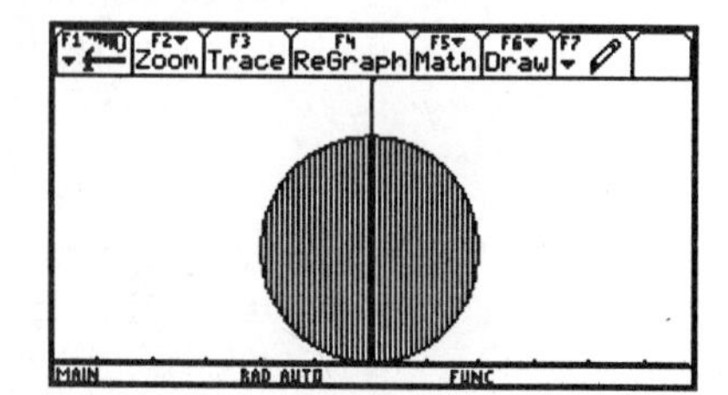

10.16 Draining Tanks

Background: Volume; separable differential equations. Review Sections 6.2 and 10.3 in Stewart's *Calculus* and Sections 6.2 and 7.4 in this manual.

Overview, part I: *Consider a tank whose horizontal cross-sectional area is described by $A(y)$, where y is the vertical distance to the bottom of the tank. For example, a tank in the shape of a right circular cylinder would have $A(y) = \pi r^2 =$ constant, and a tank in the shape of a cone (with vertex pointing down) would have $A(y) = \pi[\tan(\phi/2)y]^2$, where ϕ is the interior angle of the central vertical cross-section at the vertex.*

Toricelli's Law *states that the rate at which a fluid drains through a hole in the bottom of the tank is proportional to the square root of the depth of the fluid; that is,*

$$\frac{dV}{dt} = -k\sqrt{y}.$$

The constant k depends upon both the viscosity of the fluid and the size of the hole in the bottom of the tank. Since $V(y) = \int_0^y A(u)du$, this becomes

$$A(y)\frac{dy}{dt} = -k\sqrt{y},$$

of which the separated form is

$$\frac{A(y)}{\sqrt{y}}dy = -k\,dt.$$

1. A tank has the shape of a right circular cylinder (upright, i.e., flat side down) with radius $R = 1$ foot and height $H = 3$ feet. The tank is initially full of water and begins to drain through a hole in the bottom at time $t = 0$. Assuming that $k = 0.5$, find the depth $y(t)$ of water in the tank for $t \geq 0$. How much time does it take for the tank to empty?

2. A tank has the shape of a cone (with vertex pointing down), where the interior angle of the central vertical cross-section at the vertex is $\phi = \pi/3$. The tank is initially filled with oil to a depth of 3 feet and begins to drain through a hole in the bottom at time $t = 0$. Assuming that $k = .1$, find the depth $y(t)$ of oil in the tank for $t \geq 0$. How much time does it take for the tank to empty?

3. A tank with height 4 feet has the shape of the solid obtained by revolving the graph of $y = x^2$ about the y-axis. It is initially full of water and begins to drain through a hole at its vertex at time $t = 0$. Find, in terms of k, the depth $y(t)$ of water in the tank for $t \geq 0$. How much time (in terms of k also) does it take for the tank to empty?

Overview, part II: *If the initial depth of the fluid in the tank is y_0, then the time T that it takes for the tank to empty satisfies*

$$\int_{y_0}^{0} \frac{A(y)}{\sqrt{y}}\,dy = -k \int_0^T dt,$$

from which follows the formula

$$T = \frac{1}{k}\int_0^{y_0} \frac{A(y)}{\sqrt{y}}\,dy.$$

4. Consider a tank in the shape of a right circular cylinder with height H and radius R. Find the time required for the tank, initially full, to completely drain through a hole at the bottom, if:

   a) the tank is upright (i.e., flat side down);

   b) the tank is lying on its side.

5. For the tank in #4 suppose that $H = 2R$, so that the initial depth is the same for each of the two positions. In which of the two positions will the tank drain faster?

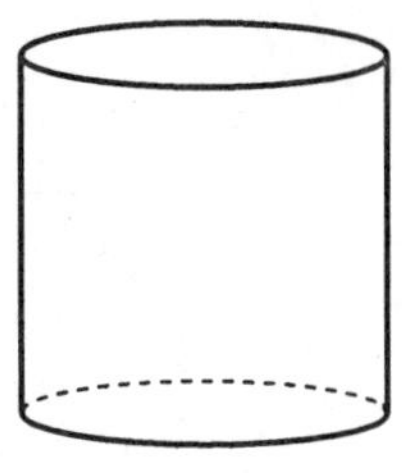
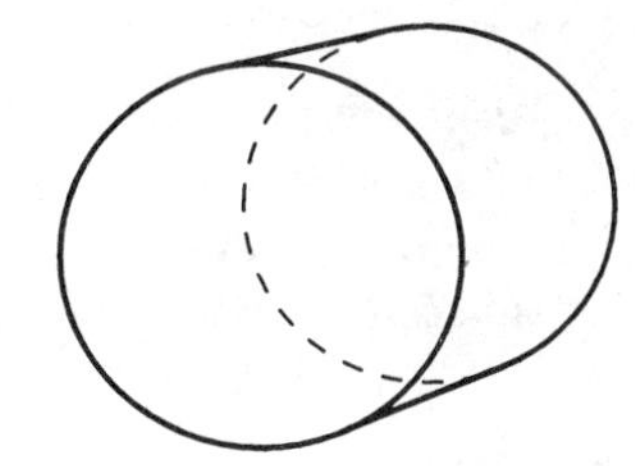

10.17 Parachuting

Background: First-order differential equations. Review Sections 10.1 and 10.3 in Stewart's *Calculus*.

Overview: *A sky-diver, weighing 70 kg, jumps from an airplane at an altitude of 700 meters and falls for T_1 seconds before pulling the rip cord of his parachute. A landing is said to be "gentle" if the velocity on impact is no more than the impact velocity of an object dropped from a height of 6 meters.*

The distance that the sky-diver falls during t seconds can be found from Newton's Second Law, $F = ma$. During the free-fall portion of the jump, we will assume that there is essentially no air resistance—so $F = -mg$, where $g \approx 9.8$ meters/sec^2 and m= 70 kg. After the parachute opens, a significant "drag" term due to the air resistance of the parachute affects the force F, *causing the force to become*

$$F = -mg - kv,$$

where v is the velocity and k is a positive drag coefficient. Assume that $k = 110$ kg/sec.

Assignment: Find the range of times T_1 at which the rip cord can be pulled that result in a gentle landing.

Hints: First find the range of impact velocities that result in a gentle landing. Next, find the velocity after T_1 seconds of free-fall. Use this velocity as the initial velocity for the initial value problem that governs the fall while the parachute is open. The resulting impact velocity can then be found as a function of T_1.

10.18 Spruce Budworms

Background: First-order differential equations. Review Sections 10.1–10.3 and 10.5 in Stewart's *Calculus* and Sections 7.1–7.3 in this manual. This project assumes that you have entered and tested the program **slopefld** from Section 7.2 of this manual.

Overview: *Spruce budworms are a serious problem in Canadian forests. Budworm "outbreaks" can occur in which balsam fir trees are quickly defoliated by hordes of ravenous budworms. A model of a budworm population leads to the differential equation*

$$\frac{dy}{dt} = y\,f(y) \quad \text{with} \quad f(y) = k\left(1 - \frac{y}{10}\right) - \frac{y}{1+y^2},$$

where y represents some fixed fraction of the number of budworms congregated in a certain area, and k is a positive parameter associated with birth and death rates. An "outbreak-threshold" value of y is a positive number y^* such that:

$$y(t) \text{ is decreasing for } t > 0 \text{ if } y(0) \approx y^* \text{ and } y(0) < y^*;$$
$$y(t) \text{ is increasing for } t > 0 \text{ if } y(0) \approx y^* \text{ and } y(0) > y^*.$$

The roots of f, which depend upon the value of k and correspond to positive equilibrium solutions, determine the outbreak threshold, should one exist.

1. Graph $y\,f(y)$ (i.e., **y1=** x*f(x)) in a $[0, 10] \times [-.75, .75]$ window for a few values of k between 0.1 and 1. What are the possibilities for the number of positive roots?

2. For each of $k = .3$, .45, and .6, plot the direction field in a $[0, 20] \times [0, 10]$ window and use Euler's Method to plot approximate solution curves with initial values $y(0) = 1$, 3, and 10. What happens to each of the solutions as $t \to \infty$?

3. For which of $k = .3, .45$, and $.6$ is an outbreak possible? What is the outbreak-threshold value of y in each case?
4. In order to find the values of k for which f has a "double root," solve simultaneously the equations $f(y) = 0$ and $f'(y) = 0$ for k and y. Then, using these values of k, repeat #2.
5. State precisely the values of k for which a budworm outbreak is possible.

10.19 The Flight of a Baseball I

Background: Parametric equations; velocity and acceleration; antidifferentiation; optimization. Review Sections 4.10 and 11.1 in Stewart's *Calculus* and Sections 2.2 and 8.1 in this manual.

Overview: *When a baseball player hits a fly-ball, many very complex physical phenomena affect its path—such as forces resulting from the spin of the ball, wind currents, and air resistance. However, we can learn a lot from considering a couple of idealized models of the path of the ball. The first of these ignores all forces other than gravitational force.*

Suppose that, at the instant when the ball is hit, it has an initial speed v_0, the tangent line to its path forms an angle ϕ with the ground, and its height above the ground is y_0. Let $x = x(t)$ and $y = y(t)$ be the parametric equations of the path, and assume that $x(0) = 0$ and $y(0) = y_0$. The only force on the ball is the downward vertical force of gravity. So we can write accelerations in each direction as

$$\frac{d^2x}{dt^2} = 0 \text{ and } \frac{d^2y}{dt^2} = -g.$$

The initial velocities in the two directions are $x'(0) = v_0 \cos\phi$ and $y'(0) = v_0 \sin\phi$. So integration of the accelerations results in

$$\frac{dx}{dt} = v_0 \cos\phi \text{ and } \frac{dy}{dt} = -gt + v_0 \sin\phi.$$

One more integration finds

$$x = v_0 t \cos\phi \text{ and } y = -\frac{g\,t^2}{2} + v_0 t \sin\phi + y_0.$$

For the remainder of this project, assume that length units are feet, so that $g \approx 32$ feet/sec^2.

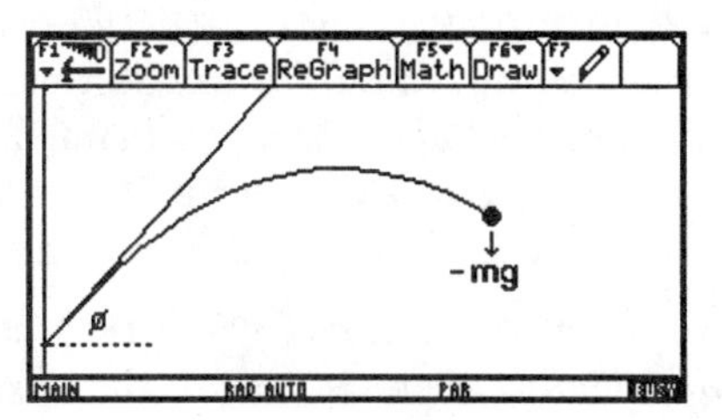

Assignment:

1. Taking $g = 32$ feet/sec^2, go to the the **Y=** Editor and enter the parametric equations as **xt1** and **yt1** in terms of v_0, ϕ, and y_0. Set the Graph **Style** to **Path**. Then on the Home screen define $\phi = \pi/6$, $v_0 = 100$ feet/sec, $y_0 = 3$ feet. Set **tmin** = **xmin** = **ymin** = 0 in the Window Editor. Start with **tmax= 2** and a $[0, 100] \times [0, 50]$ window, and adjust **tmax**, **xmax**, and **ymax** as necessary to plot the entire path and answer the following questions:

   a) Approximately how many seconds does the ball remain aloft?

   b) Approximately how far from home plate does the ball hit the ground?

2. What shape does the path appear to have? Solve the x-equation for t and substitute the result into the equation for y. Does this confirm your guess about the shape of the path?

3. Use the result of #2 to compute the distance x_{fin} from home plate to where the ball hits the ground—in terms of v_0, ϕ, and y_0. Treating v_0 and y_0 as constants, find the angle ϕ that maximizes x_{fin}. (As a check, you probably could have guessed the correct answer.)

4. Suppose that the home run fence is 20 feet high and 350 feet from home plate. Find the minimum initial velocity with which the ball must be hit in order to clear the fence. Use $y_0 = 3$ feet. Approximate the answer graphically (i.e., experimentally), and then find it by solving an equation.

5. Suppose that a player is not capable of hitting the ball with an initial velocity greater than 125 feet per second. Assuming that he does hit the ball with that initial velocity, what is the smallest angle ϕ that will carry the ball over a 10 foot fence that is 400 feet from home plate? Again, use $y_0 = 3$ feet. Approximate the answer graphically (i.e., experimentally), and then find it by solving an equation.

Related activities: The next project, **Flight of a Baseball II.** Also the Applied Projects *Which is Faster, Going Up or Coming Down?* and *Calculus and Baseball* in Chapter 7 of Stewart's *Calculus*.

10.20 The Flight of a Baseball II

Background: Parametric equations; velocity and acceleration; antidifferentiation; optimization. Review Sections 4.10 and 11.1 in Stewart's *Calculus* and Sections 2.2 and 8.1 in this manual. You should also have done the previous project, **The Flight of a Baseball I.** Optional background material: Sections 10.1–10.3 in Stewart's *Calculus*.

Overview: *A much more realistic picture of the path of a baseball comes from including air resistance in the model. A simple way of doing this is to* *assume that air resistance is a force that acts in the direction opposite that of the path and is proportional to the speed of the ball. The acceleration equations then become*

$$\frac{d^2x}{dt^2} = -k\frac{dx}{dt} \quad \text{and} \quad \frac{d^2y}{dt^2} = -g - k\frac{dy}{dt},$$

where k is a "drag coefficient" divided by the mass of the ball. Solution of these equations results in

$$x = \frac{v_0 \cos\phi}{k}\left(1 - e^{-kt}\right), \quad y = y_0 + \frac{1}{k}\left[\left(v_0 \sin\phi + 32/k\right)\left(1 - e^{-kt}\right) - 32t\right].$$

Assignment:

1. After entering **DelVar v0, ϕ, y0, k** from the Home screen, go to the the **Y=** Editor and enter these parametric equations as **xt2** and **yt2** in terms of v_0, ϕ, y_0, and k—keeping the functions **xt1** and **yt1** from the previous project, **The Flight of a Baseball I.** Set the Graph **Style** to **Path**. Then on the Home screen define $\phi = \pi/6$, $v_0 = 100$ feet/sec, $y_0 = 3$ feet, and $k = 0.005$ sec^{-1}. Plot both paths (with and without air resistance). Are the two paths significantly different? Repeat with $k = 0.01$, 0.05, 0.1, and 0.5.

2. In the **Y=** Editor, uncheck **xt1** and **yt1** so that only the path with air resistance accounted for is plotted. Let $v_0 = 100$ feet/sec, $y_0 = 3$ feet, and $k = 0.1$ sec^{-1}. Graphically (i.e., experimentally) find an estimate for the value of ϕ that maximizes the distance from home plate to where the ball hits the ground. Two-decimal-place accuracy is sufficient. Repeat with $k = 0.2$. Is the result different?

3. Suppose that the home run fence is 20 feet high and 350 feet from home plate. Find the minimum initial velocity with which the ball must be hit in order to clear the fence. Use $k = 0.1$ and $y_0 = 3$ feet. Approximate the answer graphically (i.e., experimentally). Be careful here—the angle ϕ must be taken into account. **Suggestion:** First fix ϕ at the value found in #2 and estimate the necessary initial velocity. Then increase and decrease ϕ by a small amount, say 0.02 (about 1°), and decrease your estimate of v_0 if necessary. Repeat until you have v_0 to two-decimal-place accuracy. Compare the result to your answer to problem 4 in **The Flight of a Baseball I.**

4. Suppose that a player is capable of hitting the ball with an initial velocity no greater than 125 feet per second. Assuming that he does hit the ball with that initial velocity, what is the smallest angle ϕ that will carry the ball over a 10 foot fence that is 400 feet from home plate? Again, use $y_0 = 3$ feet and $k = 0.1$. Approximate the answer graphically (i.e., experimentally). Compare the result to your answer to problem 5 in **The Flight of a Baseball I.**

5. Air density (which depends upon altitude above sea level) affects the value of k significantly. In particular, the value of k in Denver may be about 10% smaller than the value of k in Miami. Suppose that $k = 0.09$ in Denver and $k = 0.10$ in Miami. If a ball is hit with $y_0 = 3$, $v_0 = 135$, and $\phi = \pi/6$, how much farther from home plate would it land in Denver than in Miami? (Estimate to the nearest foot.)

6. **If you have studied Sections 7.2 and 7.3 in** Stewart's *Calculus*, derive the given parametric equations for the path from the acceleration equations.

Related activities: The Applied Projects *Which is Faster, Going Up or Coming Down?* and *Calculus and Baseball* in Chapter 7 of Stewart's *Calculus*.

10.21 Cannonball Wars

Objectives: The purpose of this project is to use the **TI-89/92** to investigate an interesting question concerning paths of projectiles.

Background: Before doing this project, re-read Section 11.1 in Stewart's *Calculus* and Sections 2.2 and 8.1 in this manual.

The Situation: *One cannon is perched atop a cliff, 50 meters above the ground below. Another cannon is on the ground, 100 meters from the base of the cliff. (See the figure below.) The inclination of each cannon differs from the line of sight between them by the same angle α (as shown in the figure). Each cannon fires a cannonball at exactly the same instant at exactly the same initial velocity v_0 m/sec.*

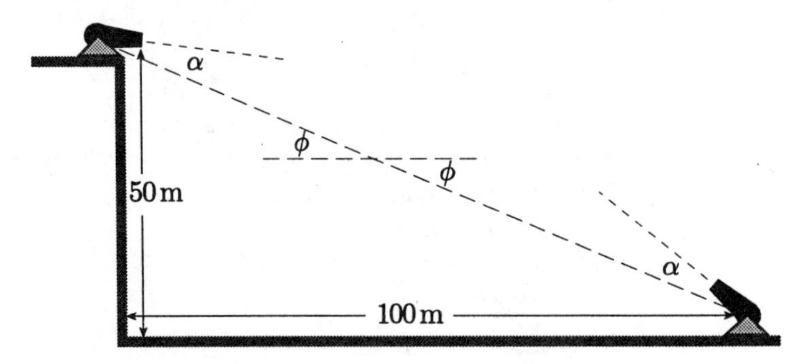

Let $\phi = \tan^{-1}(1/2)$. Ignoring air resistance, the paths of the cannonballs are described, respectively, by two pairs of parametric equations,

$$x_1 = v_0 t \cos(\phi - \alpha), \qquad y_1 = -4.9\,t^2 - v_0 t \sin(\phi - \alpha) + 50;$$

$$x_2 = 100 - v_0 t \cos(\phi + \alpha), \qquad y_2 = -4.9\,t^2 + v_0 t \sin(\phi + \alpha).$$

The Question: *Do the cannonballs collide?*

1. After setting **Graph MODE** to **PARAMETRIC**, go to the **Y=** Editor and enter the parametric equations in terms of v0 and α. Use the approximation 0.464 (radians) for ϕ. For each of **y1t** and **y2t** set the **Style** to **Path** (**F6-6**). Also press **F1** and set **Graph Order** to **SIMUL**.

2. Go to the Window Editor and set **tmin**=0, **tmax**=5, **tstep**=.1. Also set window variables to show a $[0, 150] \times [0, 100]$ window.

3. From the Home screen, enter **20** $\rightarrow$ **v0** : **.25** $\rightarrow$ α. Then press ◇**GRAPH** to view the paths. What happens? Do the cannonballs collide?

4. Keeping $v_0 = 20$, replot the paths with $\alpha = 0.5$, 0.75, and 1. Do the cannonballs collide each time? What can be said about when they don't?

5. Change the value of v_0 to 40, and plot the paths for $\alpha = -0.3$, 0.25, 0.75, and 1.25. Do the cannonballs collide each time? What can be said about when they don't?

In what follows, assume that $-\pi/2 < \alpha - \phi < \pi/2$ *so that the cannon on the cliff fires its cannonball "forward." Also assume that* $0 < \alpha + \phi < \pi$ *so that the cannon on the ground fires its cannonball into the air.*

6. Supposing temporarily that the paths of the cannonballs are never interrupted by any obstacle (such as the ground), show that there is a positive time t^* at which the two cannonballs are located at the same point; that is, a time at which $x_1 = x_2$ and $y_1 = y_2$. (You may need to look up some trigonometric identities for this—or perhaps **tExpand()** and/or **tCollect()** will help.) *Suggestion*: Try first doing the special case where $\alpha = 0$.

7. For what values of v_0 and α does the collision occur before the cannonballs hit the ground? Give your answer as a shaded region in the $v_0\alpha$-plane.

10.22 Taylor Polynomials and Differential Equations

Background: Taylor Series; basic differential equations. Review Sections 10.1, 12.10, and 12.12 in Stewart's *Calculus* and Sections 7.1 and 9.3 in this manual. This project assumes that you have entered and tested the program **derlist()** from Section 3.4 of this manual.

Overview: *Consider a first-order initial value problem*

$$\frac{dy}{dt} = f(t, y), \quad y(0) = y_0.$$

The initial condition obviously provides the value of the solution at $t = 0$. *Notice also that the differential equation provides the value of* $\frac{dy}{dt}$ *at* $t = 0$, *namely*

$$y'(0) = f(0, y_0).$$

Moreover, differentiation of each side of the differential equation with respect to t produces an expression for $y''(t)$, which can then be evaluated at $t = 0$, using known values of $y(0)$ and $y'(0)$. In principle, if this process is continued, we can determine the values at $t = 0$ of as many derivatives as we like and thereby obtain any Taylor polynomial of the solution about $t = 0$.

Problem I: Consider the initial value problem

$$\frac{dy}{dt} = t^2 - y^3, \quad y(0) = 1.$$

1. Create a list containing $y(t)$ and its first through fifth derivatives by entering

   augment({y(t)}, myprogs\derlist(t^2−y(t)^3,t,4))

2. Enter the following to eventually obtain a list of values at $t = 0$ of $y(t)$ and its first through fifth derivatives:

   ans(1) | *d*(*d*(*d*(*d*(y(t),t),t),t),t)=d4

   ans(1) | *d*(*d*(*d*(y(t),t),t),t)=d3

   ans(1) | *d*(*d*(y(t),t),t)=d2

   ans(1) | *d*(y(t),t)=d1 and y(t)=d0

   ans(1) | t=0 and d0=1

   ans(1) | d1=ans(1)[2]

   ans(1) | d2=ans(1)[3]

   ans(1) | d3=ans(1)[4]

   ans(1) | d4=ans(1)[5]

3. The last calculation should have resulted in a list of numbers. Each of those numbers divided by the appropriate factorial is a coefficient of the fifth degree Taylor polynomial. Enter the following to obtain the first through fifth Taylor polynomials.

   ans(1)/{1, 1, 2, 6, 24, 120} →coeffs

   ∑(coeffs[n+1]∗x^n, n, 0, 1) →p1(x)

   ∑(coeffs[n+1]∗x^n, n, 0, 2) →p2(x)

   ∑(coeffs[n+1]∗x^n, n, 0, 3) →p3(x)

   ∑(coeffs[n+1]∗x^n, n, 0, 4) →p4(x)

   ∑(coeffs[n+1]∗x^n, n, 0, 5) →p5(x)

4. Enter these Taylor polynomials as **y1**, . . . , **y5** in the **Y=** Editor and then plot all five graphs in a $[-1.5, 1.5] \times [-1, 2]$ window. Do this once with **Graph Order** set to **SEQ** and then again with **Graph Order** set to **SIMUL**.

Then to see the situation more clearly, uncheck (**F4**) all but **y4** and **y5**; then replot.

Problem II: Using the same process as in problem I, plot the fourth and fifth Taylor polynomials of the solution of

$$\frac{dy}{dt} = \cos(t\,y), \quad y(0) = 0.$$

Problem III: Using a similar process to that in problem I, plot the fourth and fifth Taylor polynomials of the solution of the second order problem

$$\frac{d^2y}{dt^2} = -t\,y, \quad y(0) = 1, \; y'(0) = 0.$$

10.23 Build Your Own Cosine

Objectives: The purpose of this project is to show how a Taylor polynomial can be used to construct a periodic function such as cosine.

Background: Before doing this project, review Sections 12.10 and 12.12 in Stewart's *Calculus* and Section 2.1 in this manual.

The Situation: *Suppose that you have a simple calculator whose only functions are the basic arithmetic functions: addition, subtraction, multiplication, and division. However the calculator does allow you to program your own functions, using the basic arithmetic functions as well as basic programming constructs such as if-statements and functions such as int and absolute value. Your assignment here is to create a cosine function for your calculator. The function must compute* $\cos x$ *to within* 5×10^7 *for any* x *(thus giving six decimal-place accuracy).*

Since the graph of $\cos x$ is periodic with period 2π, we first need to obtain an approximation on the interval $[0, 2\pi]$. Also, because of symmetries in the graph of $\cos x$ on $[0, 2\pi]$, we need only be concerned initially with obtaining an approximation on the interval $[0, \pi/2]$. So it is sensible to consider Taylor polynomials about $x = \pi/4$, the midpoint of that interval.

1. Show that the eighth-degree Taylor polynomial $T_8(x)$ for $\cos x$ about $\pi/4$ is the Taylor polynomial of least degree that will provide the accuracy we want on $[0, \pi/2]$.

2. Enter $\cos x$ as **y1** and $T_8(x)$ as **y2**. Then plot both graphs first in a $[0, \pi/2] \times [-.5, 1.5]$ window and then in a $[-2\pi, 2\pi] \times [-2, 2]$ window.

3. Now that we have a good approximation to $\cos x$ for $0 \le x \le \pi/2$, let's extend that approximation to the interval $0 \le x \le \pi$ by means of the identity $\cos x = -\cos(\pi - x)$. Enter the following to obtain a good approximation to $\cos x$ for $0 \le x \le \pi$.

y3 = when(x< π/2, y2(x), −y2(π−x))

Plot this function and $\cos x$ in a $[0, \pi] \times [-1.5, 1.5]$ window and then in a $[-2\pi, 3\pi] \times [-2, 2]$ window.

4. Now that we have a good approximation to $\cos x$ for $0 \leq x \leq \pi$, let's extend that approximation to the interval $0 \leq x \leq 2\pi$ by means of the identity $\cos x = \cos(2\pi - x)$. Enter the following to obtain a good approximation to $\cos x$ for $0 \leq x \leq 2\pi$.

   y4 = when(x< π, y3(x), y3(2π−x))

   Plot this function and $\cos x$ in a $[0, 2\pi] \times [-1.5, 1.5]$ window and then in a $[-2\pi, 4\pi] \times [-2, 2]$ window.

5. Now that we have a good approximation to $\cos x$ for $0 \leq x \leq 2\pi$, all we need to do is compose it with a function that will "shift" any x to its corresponding value in $[0, 2\pi]$. In the language of trigonometry, we simply need to find a coterminal angle between 0 and 2π for x radians. This is done by subtracting from x the greatest integer multiple of 2π that is less than or equal to x. This is accomplished by means of the **floor** function. Enter

   y5 = y4(x−2π*floor(x/(2π)))

   and plot the result along with $\cos x$ in a $[-12, 12] \times [-1.5, 1.5]$ window.

Index

◇ **F**

◇ **G**

◇ **H**

◇ **I**

◇ **L**

◇ **M**